CHICAGO PUBLIC LIBRARY

R00535 20936

RA
577.5
.I53   Indoor air and human
1985   health

DATE			

DISCARD

D1416490

© THE BAKER & TAYLOR CO.

INDOOR AIR
AND
HUMAN HEALTH

Proceedings of the Seventh Life Sciences Symposium
Knoxville, Tennessee
October 29 – 31, 1984

Sponsored by

Oak Ridge National Laboratory
U.S. Department of Energy
U.S. Environmental Protection Agency
Tennessee Valley Authority
Electric Power Research Institute

INDOOR AIR
AND
HUMAN HEALTH

Edited by

RICHARD B. GAMMAGE
STEPHEN V. KAYE
Health and Safety Research Division
Oak Ridge National Laboratory
Oak Ridge, Tennessee

Technical Editor
VIVIAN A. JACOBS
Oak Ridge National Laboratory
Oak Ridge, Tennessee

LEWIS PUBLISHERS, INC.
121 S. MAIN STREET, P.O. DRAWER 519, CHELSEA, MI 48118

Library of Congress Cataloging in Publication Data
Main entry under title:

Indoor air and human health.

　　　Proceedings of the Seventh Life Sciences Symposium, held in
Knoxville, Tenn., Oct. 29–31, 1984, sponsored by Oak Ridge National
Laboratory and others.
　　　Bibliography: p.
　　　Includes index.
　　　1. Ventilation — Hygienic aspects — Congresses.
2. Air — Pollution, Indoor — Toxicology — Congresses.
3. Air quality — Congresses.　I. Gammage, Richard B.
II. Kaye, Stephen V.　III. Jacobs, Vivian A.　IV. Life
Sciences Symposium (7th : 1984 : Knoxville, Tenn.)
V. Oak Ridge National Laboratory.　[DNLM: 1. Air
Pollutants — adverse effects — congresses.　2. Air
Pollution — adverse effects — congresses.
W3 LI441A 7th 1984i / WA 754 I404 1984]
RA575.5.I53　1985　　363.7'392　　　85-13134
ISBN 0-87371-006-1

2nd Printing
COPYRIGHT © 1985 by LEWIS PUBLISHERS, INC.
ALL RIGHTS RESERVED

Neither this book nor any part may be reproduced or transmitted in
any form or by any means, electronic or mechanical, including
photocopying, microfilming, and recording, or by any information
storage and retrieval system, without permission in writing from the
publisher.

LEWIS PUBLISHERS, INC.
121 South Main Street, Chelsea, Michigan　48118

PRINTED IN THE UNITED STATES OF AMERICA

PREFACE

This annual series of life sciences symposia was started in 1978. A different topic each year has focused on subjects of particular and current interest to the U.S. Department of Energy, the scientific community, and the public.

Following the first oil embargo in 1973, there began a concerted effort to use energy more efficiently, including finding better ways to heat and cool our places of work and residence. Subsequent changes in the ways we build, insulate, furnish, heat, cool, and ventilate our buildings have been accompanied by a rising number of complaints of discomfort or sickness by occupants. The subject of indoor air quality and its possible deterioration in the light of a national drive for more energy-efficient buildings has become a popular theme for a number of conferences held within the last few years.

The ability to demonstrate definitive association between the physical and chemical condition of indoor air and adverse health effects has, however, proved to be difficult and elusive in many instances. This incongruity, the importance of its resolution, and the fact that there had been no recent attempt to focus a symposium primarily on health implications, persuaded us that the topic of "Indoor Air and Human Health" would be appropriate for the Seventh ORNL Life Sciences Symposium.

The intent of the symposium was to bring together experts who would present data on indoor pollutant levels and health effects in humans and animals from five principal classes of pollutants:

- Radon
- Microorganisms
- Passive Cigarette Smoke
- Combustion Products
- Organics

The data have been presented in forms that can best permit evaluations of health implications. Alternately, the data help us identify gaps in knowledge that need to be filled before such evaluations can be made. The pollutant classes are examined from viewpoints such as measurement and source characterization, habitat studies, health effects, risk analysis, and future needs.

We gratefully acknowledge the efforts of other individuals who worked for the success of the symposium. They include C. R. Richmond and R. A. Griesemer, co-chairmen, and H. R. Witschi, of Oak Ridge National Laboratory; I. H. Billick, Gas Research Institute; M. Lippmann, New York University; W. M. Lowder, U.S. Department of Energy; P. W. Preuss, Consumer Product Safety Commission; and L. A. Wallace, U.S. Environmental Protection Agency. The generous financial support of the following sponsors is also recognized: U.S. Department of Energy, U.S. Environmental Protection Agency, Tennessee Valley Authority, and the Electric Power Research Institute.

We are especially grateful to Lois Szluha, Kim Gaddis, Kay Branam, and Beverly Selmer for providing secretarial and administrative support both before and after the symposium. Annie Brown has been instrumental in preparing much of the manuscript material for reproduction. For the management of the symposium and efficient arrangements with the conference hotel, we thank Bonnie S. Reesor of the ORNL Conference Office. We also thank Raleigh Powell for the editorial assistance rendered to us and to the authors in coordinating the preparation of preconference publications and the prompt publication of the proceedings.

R. B. Gammage

S. V. Kaye

RICHARD B. GAMMAGE is program manager for occupational health research in the Health and Safety Research Division at Oak Ridge National Laboratory. His honors, B.Sc. and Ph.D. in chemistry, were earned at Exeter University, England, which last year awarded him an earned D.Sc. for distinguished research. Dr. Gammage has a background in reactivity of solids with atmospheric constituents, including lunar dusts. His experience in health sciences began with research and development of solid-state dosimeters for ionizing radiation, and expanded to chemical pollutants associated with energy production and conservation. Recently studied were urea-formaldehyde foam insulation, emitted formaldehyde, and other volatile organic compounds in residences. He has published about 120 technical articles. Dr. Gammage serves on the American Industrial Hygiene Committees for Workplace Environmental Exposure Levels and Indoor Air Quality, and is a member of the American Chemical and Health Physics Societies.

STEPHEN V. KAYE has been Director of the Health and Safety Research Division at Oak Ridge National Laboratory since the division's organization in 1977. As director of the division, he has overall responsibility for a broad range of basic and applied research aimed at assessing the biophysical and health-related impacts of energy technologies. Previous to his present position, he was a member of the Laboratory's Environmental Sciences Division.

He has published extensively on environmental systems analysis modeling of pollutant transport through exposure pathways, radiological dosimetry, and environmental impact assessments. Much of his current effort is devoted to the management integration of unique biological and physical science programs to address problems involving the impact of energy technologies on human health.

Dr. Kaye received his Ph.D. in radiation biology from the University of Rochester Medical School. He has an M.S. in animal ecology from North Carolina State University and a B.A. in biology from Rutgers University. He is a member of several technical societies, including the Society for Risk Analysis, Health Physics Society, American Association for the Advancement of Science, and Sigma Xi.

CONTENTS

Introduction

1.

Indoor Air Quality—Electric Utility Concerns

Ralph M. Perhac

Environmental Assessment Department
Electric Power Research Institute
Palo Alto, California 94303

Although this introduction is entitled "Indoor Air Quality - Electric Utility Concerns," it really could have the title "Indoor Air Quality - Societal Concerns." This issue of indoor air quality is one in which, for all practical purposes, the interests of the utility industry are identical to those of the rest of society. Both industry and society have an interest in indoor air quality for two principal reasons. One relates to the potential for adverse health effects stemming from indoor air pollution. The other is our basic need to establish better indoor air quality standards regardless of the potential for health problems.

First, let us examine the question of health and indoor air quality. Let us also keep in mind that man spends perhaps 75% of his time indoors and, if that time is spent in an environment of poor air circulation, we should be concerned about the quality of air and the possible accumulation of pollutants, such as nitrogen oxides, carbon monoxide, and formaldehyde. That concern should grow as we, as a society, continually advocate more home insulation and tighter-sealed homes in an effort to reduce energy consumption.
The utility industry finds itself in an interesting and conflicting situation. On the one hand, it encourages increased use of insulation, but, on the other hand, it faces the problem of potential accumulation of pollutants indoors as a result of the reduced air exchange because of the tightening of homes. Similarly, those who advocate solar heating (especially through the use of passive systems) face a conflict. Passive solar heating certainly can conserve fuel, but passive systems require the use of a 12- to 18-in. thermal wall (almost certainly composed of earth material - soil, rock, and brick) in a home with an air exchange rate of perhaps 0.1 to 0.2 exchanges per hour. In such a tight home, we have to recognize the potential for the accumulation of radon (and daughter products), which is emitted from the thermal wall.

Nothing suggests that we necessarily face a serious health problem as a result of indoor air pollution, but we should recognize that a potential for concern does exist and that we should conduct the research needed to resolve the questions we face.

Our second reason for having an interest in indoor air quality is our basic need to have a more accurate means of establishing air quality standards. Even if we were to know as an absolute fact that we face no health problem at all as a result of indoor exposure, we still need data on indoor air quality. We have numerous air quality regulations or standards designed to protect human health, but these standards are primarily based on outdoor air quality, whereas man actually spends about three-fourths of his time in an indoor environment. Furthermore, our air quality standards are designed to protect the most sensitive segments of the population, that is, those with asthma or other pulmonary dysfunction. Yet these are the very people who typically spend the great majority of their time inside buildings. How effective then are our air quality standards if we spend a fourth or less of our time in the environment to which the standards apply?

Obviously, if we are to judge the efficacy of our present outdoor air quality standards, we need data on man's total exposure — exposure that involves both the indoor and outdoor environments. Such information is particularly important for epidemiological studies that attempt to relate human health to air pollution. So many of our present epidemiological studies are inadequate simply because the only data on air quality are those that come from outdoor monitors. Such data provide, at best, a poor estimate of man's total exposure to pollutants. Until we have good data on indoor air quality and use those data to develop total exposure models, our epidemiological efforts will continue to be fraught with serious inadequacies.

This symposium refers to indoor air and human health. Even if health concerns were not a consideration, a symposium of this sort would still be valid. We do need studies of indoor air quality whether or not an indoor health issue exists.

The Environmental Assessment Department of the Electric Power Research Institute is pleased to be able to participate in this symposium and welcomes the opportunity to be one of its sponsors.

2.

Evaluation of Changes in Indoor Air Quality Occurring Over the Past Several Decades

David T. Mage

U.S. Environmental Protection Agency,
Research Triangle Park, North Carolina   27711

Richard B. Gammage

Health and Safety Research Division,
Oak Ridge National Laboratory*, Oak Ridge, Tennessee   37831

INTRODUCTION

A healthy adult might wonder, "Why is there so much concern and publicity about indoor air quality (IAQ)? I have been unaware, in my lifetime, of marked changes in the indoor air that I breathe." Nevertheless, subtle changes in indoor air have indeed occurred inside residences and office buildings, affecting numerous unsuspecting adults with possibly unpleasant consequences.

In the popular press, increasing numbers of articles have appeared in the past two decades reporting a "sick building syndrome." Otherwise healthy workers in office buildings and residents of private homes describe symptoms of discomfort or illnesses that are usually of an undetermined nature and cause; within our present understanding these symptoms are probably associated with various unsatisfactory qualities of the indoor air.

On the other hand, many individuals whose health is subpar or who are chemically sensitive have suffered more obvious ill effects. In a few cases these ill effects can be critically severe.

This chapter attempts to address some of the unresolved questions that may be in the minds of the public. It also seeks to place in historical perspective the changing IAQ, the underlying

---

* Research sponsored by U.S. Department of Energy under contract
  DE-AC05-84OR21400 with the Martin Marietta Energy Systems, Inc.

causes, and the manner in which researchers are tackling the problem. Some forecasts of current and future trends in IAQ are also
made.

INDOOR VS OUTDOOR IAQ

The Donora, Pennsylvania, episode in 1948 and the "Great London
Smog" of 1952, ended for all time public complacency about the quality of the air that they breathed. Smoke and sulfur dioxide (SO₂)
measurements during the 1952 episode in London were made outdoors at
roof top levels (domestic coal burning and fossil fuel burning
industry were the source activities) [1]. Because it was assumed
that the indoor environment provided a shelter of sorts from the
outdoor pollution, the question was never asked, "What are the concentrations of smoke and SO₂ indoors, in the bedrooms of the aged
and infirm who suffered the fatal effects of the smog?" Since that
time, and with controls having been placed on many outdoor air pollutants, there has been a gradually developing awareness that chemical pollution indoors can be worse than it is outdoors.

Some of the earlier studies were already indicating that indoor
air pollution could be a health problem. In 1958, Phair et al.
recommended that IAQ should be considered in air health studies, and
they reported a study of indoor air pollution and pulmonary function
of respiratory cripples in Cincinnati [2]. In the winter of 1964,
Biersteker, et al. [3], measured smoke and SO₂ inside and outside 65
homes in Rotterdam. Their results, shown in Figure 1, indicate that

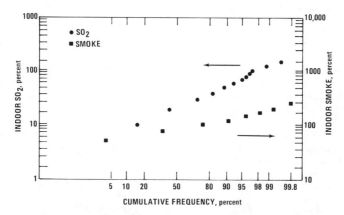

Figure 1.   Cumulative Frequency of Smoke and SO₂ Concentrations in
            Rotterdam Living Rooms as Percentages of the Concentrations
            Found Outdoors

the average indoor SO_2 was 20% of the outdoor concentration, but one home (with a faulty chimney) had a higher level of SO_2 indoors than outdoors. As measured by the reflectance of a smoke stain, the average indoor smoke was more than 80% of the outdoor value, and 20% of the homes had greater smoke values indoors than outdoors, as is also shown in Figure 1. The authors concluded, "The home of the modern inhabitant of Rotterdam, who spends on the average of only a few hours a day in the outdoor air, provides good protection against SO_2 but little protection against smoke. If 1 out of 100 houses has a failing chimney or a failing stove during smogs, one begins to wonder how far indoor concentrations of air pollutants play a role in causing premature death." In a later 1967 paper by Biersteker and de Graaf [4], they called indoor air pollution "a neglected variable in epidemiology." An earlier paper by Goldsmith et al. [5], discussed the uptake of carbon monoxide CO to form COHb in the blood of people going through a period of indoor and outdoor exposure to CO from automobile exhaust and cigarette smoke.

Although these early papers calling attention to the indoor environment as a possible important source of air pollution exposures stimulated research on IAQ, they had little effect on the conduct of epidemiological studies in the following decade. As late as 1972, Benson [6] commented that "indoor pollution levels appear to be controlled primarily by outdoor concentrations." In summary, in the 1950s and 1960s, reasonable protection from the outdoor environment was assumed for people who stayed indoors at home.

MODERN ERA FOR IAQ

It might surprise some readers that the acceptance of the spread of disease by the indoor airborne route is also a post-World-War II phenomenon. A statement was made in a 1944 American Public Health Association report that "conclusive evidence is not available at present that the airborne mode of transmission of infection is predominant for any particular disease [7]." In the 1950s and 1960s it became progressively clear that airborne spread of diseases by microorganisms can indeed occur and that the infectious particles are in the form of droplet nuclei. One landmark study in 1959 by Riley et al. [8] showed that vented air from a tuberculosis ward contained droplet nuclei capable of infecting guinea pigs. Riley's work can be regarded as starting the modern era for understanding of the indoor behavior of microorganisms in transmitting disease.

During the early 1970s, IAQ began to be investigated in a more directed manner. The first IAQ conference, "Improving Indoor Air Quality," was held at South Berwick, Maine, in 1972. The conferees indicated that the internal generation of pollutants from sources such as gas stoves was a more important problem than previously suspected [9]. One doctor reported that some of his highly sensitive patients were apparently responding adversely to unidentified compounds generated indoors.

Research results of an EPA (NAPCA) sponsored study that were reported in 1971 by Yocum et al. [10] related IAQ to outdoor pollutant concentrations and to the indoor generation of pollutants. Their results, shown in Figure 2, verified Biersteker's findings [3] that a leaky furnace could cause higher SO_2 in a home and that soiling particulates, measured by a smoke stain, were similar indoors and outdoors. In a follow-up report to EPA, Cote et al. [11] prophetically stated in 1974, "Indoor generation of air contaminants will assume greater importance as outdoor air quality improves." This report presented data on nitrogen dioxide (NO_2) to show that gas stoves produced stratified NO_2 concentrations in the kitchen and caused a wide variation of NO_2 throughout the home, as shown in Figure 3.

Thus, 1974 marked a beginning of an awareness of IAQ as a problem in its own right. The energy crisis of 1973 caused the IAQ situation to start changing dramatically, and a new era was begun.

MAJOR CONTRIBUTORS TO CHANGED IAQ

In the period following World War II, gradual and sometimes abrupt changes have taken place in people's life styles and in their homes and workplaces. The demand for improved housing for an increasing population with increased living standards soon outstripped the ability to meet it with traditional building materials such as natural woods. Increasing labor costs led builders to

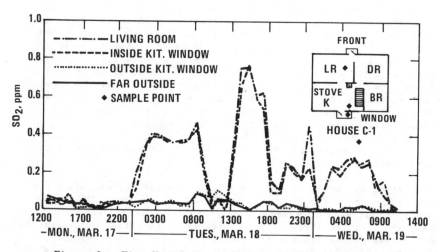

Figure 2. Time Variation of Sulfur Dioxide in a Home

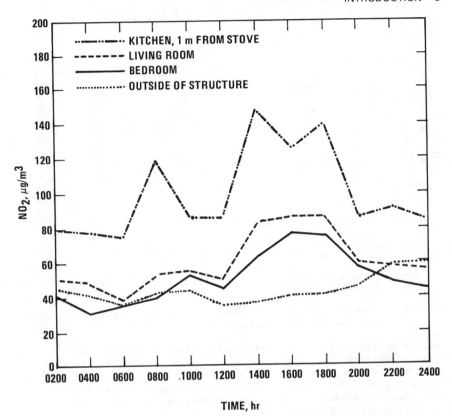

Figure 3.   Diurnal Indoor/Outdoor Pattern for NO_2 in a Home.
Spring-Summer, 1973 (Composite Day based on 6 Days of Data)

seek alternative materials that were cheaper and could be mass produced and processed.   Pressed-wood products and fiberboard were rou-
tinely incorporated into new home construction.   Household furnishings also changed in composition as plastics and artificial fibers
gained acceptance.   Household products such as cleansers, insecti-
cides, and personal health care and grooming aid products introduced
a host of new synthetic chemicals into the home environment with
unknown effects on IAQ and the health of the occupants.

In 1984, it was hypothesized that the introduction of chlorinated organic compounds and other chemicals such as phthalate ester
plasticizers into our homes, workplaces, and transport vehicles may
be associated with an observed decline in mean sperm density [12].
Figure 4 shows that chlorine production in the United States
increased rapidly after 1950 while mean sperm density began an

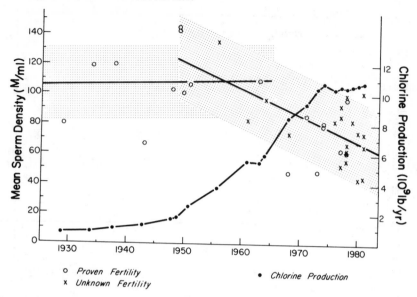

Figure 4.   Increasing Chlorine Production in the U.S. and the Apparent
Reduction in Human Sperm Density

apparent decline [13]. Many chlorinated chemical products such as
polyvinyl chloride are used indoors. Chlorine production probably
parallels the increasing quantity of synthetic organics that find
their way into indoor environments. New technology constantly
introduces new products such as carbonless copypaper and copy
machines that emit chemical vapors. One needs to be alert to these
kinds of insidious potentials that may arise from the battery of
synthetic trace organic vapors to which we are all exposed.

The worldwide energy crisis occurred in 1973-74. Suddenly, the
cost of oil increased from $3 to $30/bbl, and overnight people began
to worry about the cost of heating their homes and offices. To save
energy, building owners and operators began to tighten structures to
reduce the infiltration of outside air to reduce the energy expendi-
ture needed to heat or cool it to the desired temperature. Utilities began promoting their own residential energy conservation programs, with subsidies given to participating homeowners. The more
energy-efficient and air-tight house was perceived as a more desirable and saleable commodity.

Buildings became better insulated, and these barriers to the
exchange of heat and air between the building and the outside
environment led to what we now call the tight building syndrome
(TBS), a more appropriate name than the sick building syndrome.
This decrease in the number of air exchanges per hour (ACH) led to
an increase in concentration of the building contaminants being
emitted indoors. In addition, the tightness of the buildings and
the vapor barriers installed kept moisture within the structures;

this led to increases in the growth of microorganisms. Prior to
1975, a few isolated reports of TBS were published from the northern
European countries where colder weather had already led to tightened
buildings. However, after 1974, the sudden increase in tightened
and insulated buildings led to a rapidly expanding number of complaints from building occupants. Researchers began to recognize a
developing pattern of an endemic condition that would require a concerted effort before a solution could be found.

Table 1 lists the results of 118 investigations of TBS by the
U.S. National Institute for Occupational Safety and Health [14].

Table 1.   Presumed Source of the Problem in 118 Investigations

	n^a	%
Poor ventilation, thermal comfort, or humidity	63	53
Concern about office machines, copiers, VDT's	23	20
Contaminants from outside the building	12	10
Contaminants from structural materials	4	3
Miscellaneous or undetermined	16	14

^{a}Number of investigations associated with a problem category

Thermal comfort of building occupants is a factor in a large number
of the complaints (53%), and in some cases it may be a dominant
source of the problem.

The importance of thermal comfort for understanding the indoor
environment, the history of thermal comfort research, and the
development of a comfort equation to predict the comfortability of
people for a given physical activity, body clothing, air temperature, air humidity, air velocity and mean radiant temperature, are
reviewed by Fanger [15]. A 7-point subjective scale ranging from
cold to hot has been defined (i.e, -3 cold, < -2 cool, < -1 slightly
cool, < 0 neutral, < 1 slightly warm, < 2 warm, < 3 hot). If the
personal mean vote (PMV) is 0, 95% of the population will be comfortable (Figure 5). The American Society of Heating, Refrigerating, and Air-Conditioning Engineers (ASHRAE) comfort standards allow
a range for PMV of $-0.5 <$ PMV < 0.5, which will satisfy 90% of the
population. It is important to recognize that individual physiological differences will not permit the choice of any one thermal condition in a building that satisfies everyone, and this should be
stated in the standards for ventilation and building operation.

Stolwijk [16], in a recent survey of TBS episodes, reported
that complaints usually referred to "stuffy" buildings and "stale"
air. Wyon [17], in a 1984 review of building performance, discussed
how the lower room-air temperature settings from energy saving measures caused people to complain of drafts when air circulation is

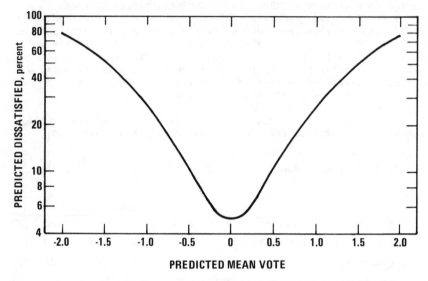

Figure 5.   Predicted Percentage Dissatisfied (PPD) as a Function of Predicted Mean Vote (PMV)

increased to alleviate stuffy or stale air complaints. Because thermal comfort conditions interact with air quality (ventilation vs draft), Wyon [17] recommended that the thermal comfort conditions should always be reported in studies of related IAQ complaints. This concept was taken one step further in a review of "sick" buildings in 1984 by Andersen [18] who recommended that future studies of TBS should only be done where the thermal comfort variables are inside normal limits (i.e., $-0.5 < PMV < 0.5$). This is because some of the symptoms of thermal discomfort may be similar to symptoms of pollutant effects, or the thermal discomfort might make the subjects' subjective evaluation of their symptoms more severe. Thus, any detailed epidemiologic study of a population in a state of mean thermal discomfort with PMV $\geq$ 0.5 may present such a severe confounding problem that the pollutant effects, if any, will be uninterpretable. This is an important consideration for any future indoor air study in homes and offices, and every effort should be made to equalize the thermal comfort states of exposed subjects and unexposed controls.

The historical pattern of ventilation standards is shown in Figure 6. As long ago as 1824, Tredgold [19] proposed that stuffiness could be prevented by an exchange of 4 cubic feet per minute (cfm) per person between indoor and outdoor air, which would reduce the CO_2 concentration in enclosed spaces. There was also the early recognition that for a comfortable and healthy indoor atmosphere, the CO_2 concentration should not be above 0.1%. This value, known

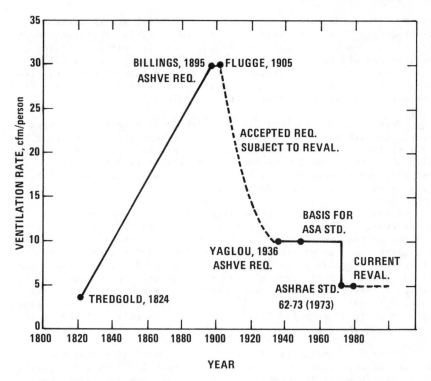

Figure 6.   Historical Development of ASHRAE Standard 62-73

as the Pettenkofer number, appeared more than 100 years ago. In
1887, Carnelly, Haldane, and Anderson [20] suggested that for
healthy air in dwellings and schools, a concentration of 20
organisms/L should not be exceeded. In 1893, Billings [21] recommended that 30 cfm per person was an acceptable minimum air exchange
rate, and the American Society of Heating and Ventilation Engineers
adopted it as a ventilation standard, which was incorporated in the
building codes of 22 states by 1925. In 1936, research by Yaglou
[22], on the threshold detection level of human body odors, led to a
recommendation of a 10 cfm per person air exchange rate, which was
adopted by the American Standards Association in 1946.

Prior to the energy crisis of 1973-74, buildings were designed
to meet an odor-based standard of ventilation. It is indeed ironic
that many of the earliest complaints about TBS based on odors were
disregarded on the grounds that odor is not a health problem. In
1973, the need for energy conservation led ASHRAE to reduce their
minimum recommended ventilation rate to 5 cfm per person, although
higher values were recommended for thermal comfort and odor elimination. In 1975, ASHRAE amended these standards to specify that the

minimum standards be used in ventilation design, and 45 states incorporated them into their building codes. This purposeful lowering of minimum ventilation requirements, coupled with new building designs and old building retrofits to reduce passive air infiltration, led to a gross reduction of the air supplied to building occupants. Further problems can arise if the buildings are not operated and maintained at their design specifications.

Meanwhile, on the home front, rising energy costs encouraged people to tighten up their residences. The caulking and weather stripping applied to cracks and seams in the home, and the installation of storm windows, were now economically justified. The U.S. Tax Code was altered to actively encourage these measures by making part of their costs tax deductible.

A comparison of average infiltration rates among newer and older homes is shown in Figure 7 [23]. The median rate varies from 0.9 ACH in the old homes to 0.5 ACH in the newer homes. Of course, any given home has widely variable ventilation depending on whether mechanical ventilation is in use, the wind speed and direction, and the temperature difference between the home and the outside air. Figure 8 [23] shows that a single home can have an order of magnitude variation in ACH (0.1 to 1.1), which can be coarsely translated into an order of magnitude variation in concentration of some airborne contaminants.

People's changing habits can have indirect and direct consequences on the indoor air that they breathe. There is a perception, difficult to quantify completely, that people spend more of their time indoors than they used to. This is certainly the case if the amount of time spent watching television is any guide [24]. The time that the TV set is left running each day has increased from 4–5 h in 1950 to 7 h in 1983 (Figure 9). Of course, the more time spent indoors, the greater the total exposure to indoor pollutants.

CHANGES IN MAJOR POLLUTANT CATEGORIES AND THEIR PERCEIVED HEALTH IMPLICATIONS

Radon

About 30 years ago scientists began expressing concern about radon gas being a health hazard inside dwellings [25]; already ^{222}Rn and its daughters had been recognized as the probable causative agent of the excess lung cancer found in uranium miners.

Initial interest focused on the use of natural building materials with high radium contents, such as the alum-shale concretes used in Sweden between 1929 and 1975. Other building materials have incorporated within them radium-containing industrial by-products, such as gypsum from the phosphate fertilizer industry [26]. Potable water derived from deep underground strata can in some areas contain

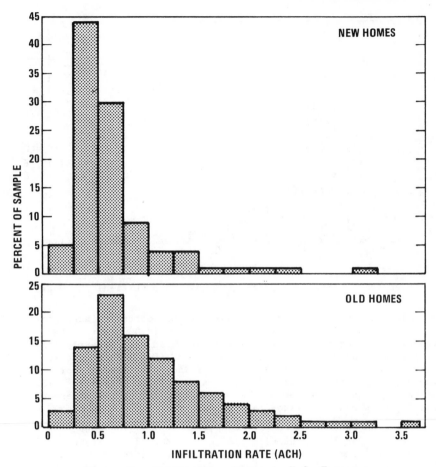

Figure 7.   Infiltration Rates in U.S. Houses

excessively high radon content [27].   Modern evidence suggests, however, that in general it is the radon in soil air that is the major
source of indoor radon, and that pressure-driven flow is the dominant transport process by which radon enters buildings [27].

   Radon is a pollutant whose indoor concentration varies with
approximate inverse proportionality to the rate of air exchange
[23].   The move to tightening of homes after the energy crisis of
1973-74 has led to the elevation of indoor levels of radon.   In
1978, Budnitz et al. [28] conservatively estimated that good insulation would approximately double radon exposures because of reduction
in ventilation.   This problem was also recognized by Cliff [29] in
1978; he anticipated that the dose from radon daughters would
increase with continuing efforts at energy conservation.

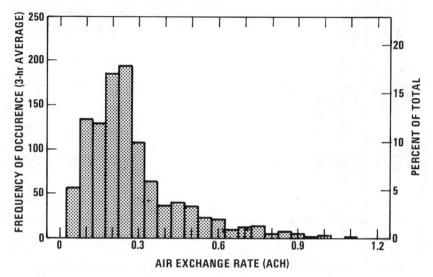

Figure 8.   Variations in Air Exchange in a Chicago, IL. House

    In 1980, Cohen [30] made assessments of increased lung cancers caused by reduced ventilation, and estimated 10,000 extra lung deaths in the United States.   It was estimated more recently [31] that indoor concentrations of ^{222}Rn of 1 pCi/L (37 Bq/m^3) would likely produce 1,000 to 20,000 cases of lung cancer each year among the U.S. population of 225 million.   A very recent assessment [32] of the Swedish population determined that 10–20% of the observed lung cancers could be radon induced.

    This potentially very serious health impact from indoor radon is still of largely hypothetical significance.   There is as yet little hard evidence that excess lung cancers are induced by exposures to indoor radon.   A recent study in Sweden [33] suggested that lung cancers occur more often in individuals who were smokers and were simultaneously exposed to higher levels of radon.   But the epidemiological studies conducted thus far have been too few in number, too small in their size, and too difficult to interpret for any firm conclusions to be drawn.

    Nevertheless, some safeguarding actions have been taken, particularly in Sweden.   Based on the potential dangers of radon, the established norms for maximum levels of radon daughters in Sweden since 1980 have been 400, 200, and 70 Bq/m^3 for existing, renovated, and newly built houses, respectively [34].   Favorable loans are extended for remedial actions in homes where the indoor levels are above 400 Bq/m^3.   It has been estimated that up to 4% of all Swedish homes have higher radon–daughter levels than 400 Bq/m^3 [35].   The issue of indoor levels of radon is of considerable importance to government agencies of many nations.

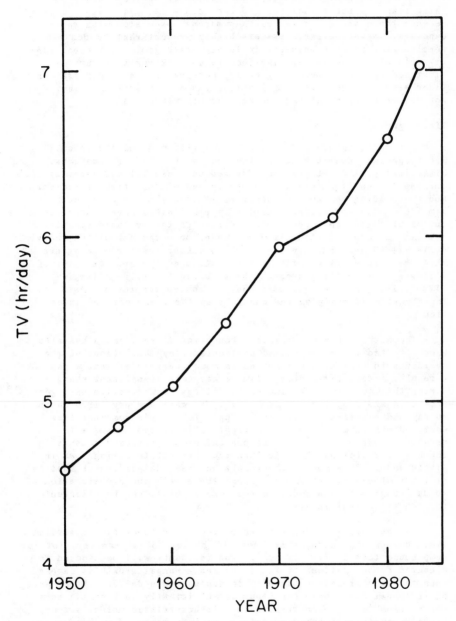

Figure 9. Increase in Time that U.S. TV Sets are Run Each Day

Among the general public, radon exposure has not raised anything like the furor that formaldehyde did a few years ago. This is because there are no acute health symptoms associated with radon exposure. Nevertheless, it seems highly prudent that "radon safe" construction be used, especially in high-risk geological areas [35]. It would also seem prudent that before an existing structure is tightened up, those promoting energy conservation should first consider measuring the existing level of radon to ensure that they are not exacerbating an already unsatisfactory situation.

Microorganisms

Microbial agents and allergens comprise most of the indoor microorganisms that can affect the health of building occupants. Historically, the interest was focused on bacterial and viral agents in hospital environments and on the spread of infectious diseases. Modern building technology, however, widened the scope of the problem [36]. The newer, large office-type buildings have controlled indoor climates. Air recirculating through faulty humidification systems and contaminated air ducts have, on occasion, placed individuals within such buildings at risk to infectious airborne agents that are spread efficiently by ventilation equipment. Examples of illnesses are <u>Legionella pneumophila</u>, humidity fever (allergic alveolitis), and aspergillosis [36]. Problem resolution rests largely with identifying and controlling the reservoirs of infection.

The nature of microorganisms found inside residences has also been affected in recent years, particularly by the actions of the residents to operate their houses in more energy-efficient ways. As a result, indoor residential climate has seen significant changes during this century. The indoor conditions in wintertime have gone from cold and damp (pre-World War II), to warm and dry (1970 cheap energy and higher standards of living), to today's somewhat less warm and more humid conditions (tighter, insulated houses with reduced exchange of air). A common energy conservation measure in the United States has been to turn the thermostat setting down in winter and up in summer. The result has been damper houses that are at times also warm enough to support the growth and proliferation of molds, fungi, and house-dust mites. Each entity can be allergenic for sensitive individuals.

In general, the fungal spore content of outdoor air is several times higher than indoors in "normal" houses [37]. However, for the damp house that is visibly moldy, the residents will be exposed to elevated concentrations of fungal spores. Fungal growth is favored when the local relative humidity in a house exceeds 75%. Under such moist conditions a mold can flourish sufficiently to become a serious problem [38]. With respect to moisture-related molds, a completely new type of building defect has been reported in Sweden [39]. An offensive mold odor and associated medical complaints were reported to be focused in those houses built in the mid-1970s. The suggested causes were combined reduction in ventilation together with increased dampness under the house and the use of new construction materials (e.g., self-leveling cement) which appear hospitable for mold growth.

Insects (mites and cockroaches) and pets (cats, dogs, and birds) produce antigens that cause respiratory allergies. Central air conditioning systems in modern homes are very efficient at distributing these antigens throughout the living zones [40]. It is very important to identify and control offending indoor allergens. Continued exposure in cases of induced asthma often leads to airway hyperirritability and persisting airway obstruction. In cases of hypersensitivity pneumonitis, uncontrolled exposure will lead to pulmonary fibrosis and even permanent disability.

Passive Cigarette Smoke

Cigarette smoking [41] increased sharply during the 1930s and 1940s with a peak in annual per capita consumption reached in 1963. Since the mid 1960s there has been a rather small, slow decrease in cigarette smoking (see Figure 10). The percentage of adult smokers in the U.S. population decreased from 41.7 to 32.6% in the period 1965 to 1980, with a corresponding drop in per capita consumption of cigarettes of about 10%. It is, however, a moot point whether overall exposures to passive cigarette smoke have actually declined;

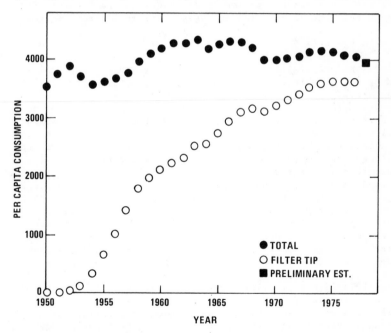

Figure 10. Annual Cigarette and Filter-Tip Cigarette Consumption Per Person Aged 18 Years and Over. 1950-1978

the effects of the reduced numbers of cigarettes smoked are coun-
tered by the trend towards lower rates of air exchange inside more
energy-efficient buildings.

Cigarette smoke has been established as a causative agent that
leads to lung cancer and heart disease in active smokers who inhale
the smoke into their lungs without dilution. The health effects of
passive smoking, however, have only recently been investigated, and
the seriousness of the health effects to nonsmokers is still a topic
of vigorous discussion [42, 43]. There seems to be gathering momen-
tum, however, to view the health effects of passive cigarette smoke
as more rather than less harmful. Weber [44], for example, claims
that smokers in the workplace can contribute from 30 to 70% of the
indoor carbon monoxide, nitrous oxide, and particulates and can be
annoying to 25 to 40% of the employees. Although there is an
increasing movement to isolate smokers from nonsmokers in workplaces
and public buildings, current ventilation operation standards may
not be sufficient to prevent the annoyance of cigarette smoke to
nonsmokers. Weber [44] proposed that limiting the increase of
indoor carbon monoxide due to smoking by 1.5 to 2.0 ppm could
prevent cigarette smoke reaching an unacceptable level. Recent
developments in analyses for nicotine and its metabolite cotinine
[45] indicate a sufficient passive uptake of tobacco smoke constituents in nonsmokers to possibly explain the associations of passive
smoking with lung cancer, pulmonary dysfunction, and the retarded
growth of children as reported in the scientific literature [41].

Combustion Products

The improper ventilation of home cooking and heating units,
caused by design defects and mechanical or structural failures of
the gas venting system, has been a problem that was recognized for
many years prior to 1973-74 and the energy crunch. Unvented cooking
fires have been used for centuries in certain parts of the world
where low standards of public health and medical care lead to low
life expectancies. Consequently, the health effects of acute exposures to combustion products may have been indistinguishable from
the endemic diseases and conditions in the population [46]. On the
other hand in a country such as Korea, several sudden deaths each
year are caused by carbon monoxide intrusion into homes through
cracks in heating passages under the floors that are warmed by
smoldering blocks of pressed charcoal [47]. These problems are by
no means restricted to developing nations. In a Dutch home, for
example, Biersteker et al. [3] found sulfur dioxide and smoke escaping from a defective vent, which was unknown to the occupants.

The 1973-74 energy crisis marked a significant change as ventilation of homes was decreased and auxilliary unvented combustion
devices, such as kerosene heaters, were used to minimize the usage
of expensive electricity and oil for heating. Patterns of home
heating changed in some residences, from heating the whole home centrally, to heating individual rooms with the portable unvented

heaters. A concern has developed that the increased concentrations
of NO_2, CO, H_2O, and CO_2 may lead to health problems for building
occupants. A typical problem is the use of unvented gas stoves to
boil water to supplement heat and humidification of small living
spaces in inner cities [48].

Volatile Organic Compounds

As a class, the volatile organic compounds (VOC) found indoors
have probably changed more dramatically in the past few decades than
any other pollutant category considered in this paper. We shall,
therefore, give expanded attention to this pollutant class.

Nonformaldehyde

The dramatic change in VOC indoors is due primarily to a post-
World War II proliferation in the types and quantities of synthetic
organic compounds that are used in building, furnishing, and household and personal consumer products. Many of these items emit trace
or higher amounts of a variety of VOC. Yet we know, in general, the
least about these VOC in terms of their indoor concentrations, emission characteristics, and health effects as compared to what we know
about the other pollutant categories.

Prior to the 1970s, little if any data were available on VOC in
the indoor environment. Contributing reasons for this lack of data
were the common beliefs that outdoor sources (industrial emissions
and motor vehicles) were the primary contributors of VOC in indoor
environments, and analytical techniques for monitoring VOC were
relatively insensitive. The energy crisis of 1973-74, coupled with
breakthroughs in analytical technology for VOC, such as gas chromatography coupled with mass spectrometry (GC-MS), sparked people's
interest in indoor VOC and enabled them to measure trace amounts of
a very large number of VOC with disparate individual volatility and
chemical functionality.

An early study by Yocum et al. [10] reported "enrichment factors" of benzene-soluble organics for indoor particulates as compared with outdoor particulates. These enhanced indoor organics of
lower volatility were associated with cigarette smoking and cooking.
One of the earliest studies of more volatile indoor VOC was reported
in 1978 by Moschandreas et al. [49]; the total gaseous nonmethane
hydrocarbons (NMHC), as measured indoors by flame ionization, were
higher than the NMHC outdoors in 90% of the measurements, suggesting
indoor sources. Studies by Mølhave [50, 51] demonstrated the
occurrence of solvent organic vapors in new houses and apartments in
Denmark. Mølhave also showed [52] that building and furnishing
materials were the probable sources for much of the organic emissions. Some temporary elevations in VOC concentrations of considerable magnitude are also recognized as being commonplace; Seifert
[53] reported that "during the application of one or more of the

multiple consumer products offered nowadays, it is not unusual that
the indoor air for several hours may show concentrations of organic
substances which exceed those found in ambient air by a factor of
more than 50."

A very recent and comprehensive study by Wallace et al. [54] of
IAQ in homes in Bayonne and Elizabeth, New Jersey, also showed that
these organic gases were significantly higher indoors than outdoors.
Some of the targeted VOC were two orders of magnitude higher indoors
than outdoors. A similar situation has recently been reported for
houses in northern Italy by De Bortoli et al. [55]; almost invari-
ably the indoor-to-outdoor ratio in the concentrations of individual
VOC were greater than unity. Now that researchers have established
the generality of higher indoor than outdoor concentrations of VOC,
the emphasis is on broadening our knowledge of indoor levels of VOC
and to learn more about the identity of the sources and their emission behavior.

The health effects of these VOC exposures are largely unknown
at present. Some of the individual compounds reported by Mølhave
[50-52], Seifert [53], Wallace [54], and De Bortoli [55] are known
animal carcinogens, but the carcinogenic activity has only been
demonstrated at concentrations much higher than those normally
encountered in indoor environments. Nevertheless, the chronic,
low-level exposure of building occupants to a mixture of potentially
mutagenic and carcinogenic airborne VOC is a cause for concern.

In most cases, it is also difficult to be definitive about the
acute, adverse health effects of indoor VOC exposures. Investigations of TBS complaints usually find concentrations of individual
VOC well below the threshold limit values that are established for
safe occupational exposures. This is also usually true in the
houses of individual home dwellers who perceive their ailments as
being caused by VOC exposure. In fact, occupational standards for
permissible VOC exposure seem to be ill-fitted for dealing with the
majority of IAQ health complaints. There seem to be three principle reasons for this: (1) the daily exposure to VOC of building
residents is often for periods that are much longer than the usual
8 h workday; (2) one sometimes deals with individuals who become
chemically sensitized to one or more VOC; (3) synergistic interactions exerted during these exposures to complex mixtures may exacerbate the health effects anticipated by an additivity of health
effects from exposures to the individual compounds comprising the
mixture.

A better appreciation of the potential importance of suspected
synergisms has recently been gained. Ahlström et al. [56] showed
that when formaldehyde at low concentrations (82 ppb) was mixed with
100% air from a "tight" school where complaints had been made, the
test subjects perceived a fourfold increase in the strength of odor.
A key study by Mølhave, Bach, and Pedersen [57] exposed healthy test
subjects to a "cocktail" of 20 VOC at concentrations over the range

of concentrations found in their studies of Danish homes.   Careful objective testing of the subjects' concentration and short-term memory using a digit span test showed decreasing scores (poorer performance) during VOC exposures.   Such cocktail effects could contribute to symptoms displayed by sufferers of TBS, and it may, in the future, be possible to obtain objective measures, along with the subjective complaints, to understand better the phenomena involved.

Formaldehyde

In the public's mind, the "bête noire" of VOC has been formaldehyde.   The adverse and sometimes severe, acute health effects perceived by sensitive individuals, the potential human carcinogenicity factor, the ubiquitous presence of formaldehyde in today's indoor environments, its commercial importance, and the great efforts expended by several government agencies and industry to understand and ameliorate the problems warrant our paying special attention to this important VOC.

The principal sources of indoor formaldehyde vapor are products containing urea-formaldehyde (UF) resins, the degradation of which produces free formaldehyde [58].   The most important sources of indoor formaldehyde vapor are usually pressed-wood products such as particleboard flooring, interior-grade plywood, decorative paneling, medium-density fiberboard used in furniture, and urea-formaldehyde foam insulation (UFFI).   The combined effect of tightened homes, the increased usage of these composite building materials, and the popularity of UFFI as a retrofit wall insulation led to a large increase in the number of odor complaints and formaldehyde associated health problems during the 1970s.

Formaldehyde in our homes and larger buildings is a post-World-War II phenomenon.   Particleboard bonded with UF was not invented until 1943 by Fahrui [59] and for several years was not used in large quantities.   In the United States, the commercial output of 37 wt% solution of formaldehyde [60] soared during the 1960s (Figure 11) as did the production of pressed-wood products; each year about 25% of the formaldehyde produced is used to make UF resin.   In fact, it can be said that during the last 15 years, UF-bonded particleboard, medium-density fiberboard, and plywood have essentially replaced natural whole wood as a construction material for flooring, wall paneling, cabinets, and furniture [61].

The first industrial conference dealing with formaldehyde emission problems was published in Leipzig, Germany, in 1966 [62].   The first widely publicized complaint in Europe [63] was probably in 1973 when public school teachers in Karlsruhe, Germany, refused to teach in a new classroom building because of the irritating odor. This problem episode proved to have the three causative factors involved in most formaldehyde complaints: inadequate ventilation, high-loading factors, and formaldehyde-emitting products that had high emission rates of formaldehyde.

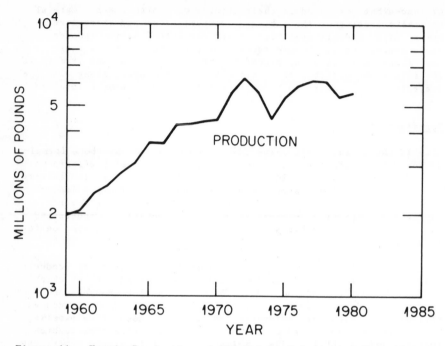

Figure 11.   Yearly Production of 37% Weight Solution of Formaldehyde
              in the U.S.

In the United States, the first published report linking formaldehyde contamination of building environments to health problems
can be traced to the study in 1961 by Breysee in Washington State
[64].   During the 1970s and early 1980s, government agencies such as
the Consumer Product Safety Commission (CPSC) and Department of
Housing and Urban Development (HUD) began to hear increased complaints about formaldehyde.   By 1982, CPSC had received over 3000
complaints, 2000 of which involved UFFI [65].   Formaldehyde levels
in a few worst instances have exceeded 1 ppm [66], while many buildings (especially newer ones) have at least occasional levels that
exceed 0.1 ppm [67, 68].   In general, however, the norm for formaldehyde levels in older buildings is in the range of 0.03 to 0.06 ppm
[68,69].

The current perception from reading the scientific literature
and the more popular news releases is that the indoor formaldehyde
problem is waning.   On the North American continent, the installation of UFFI is, to all intents and purposes, at an end [70, 71].
The rise and fall in the fortunes of UFFI are depicted in Figure 12.

With respect to pressed-wood products, during the last 5 years
industry has worked hard (and to good effect) to reduce by several-

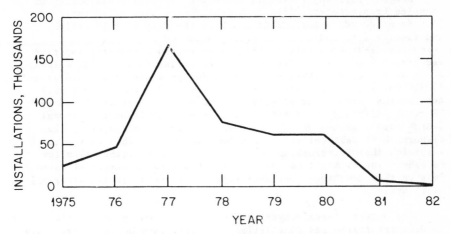

Figure 12.   Annual Installation of Urea-Formaldehyde Foam Insulation in the U.S.

fold the rates at which formaldehyde is released. There are those who now believe [72] that with best available UF resin technology, the ASHRAE standard for indoor formaldehyde of 0.1 ppm can be met in newly constructed buildings in which the ventilation rates, temperature, and humidity, as well as the load factors for the pressed-wood products, are kept within normal bounds.

Formaldehyde still poses some uncertainties with regard to the health effects associated with human exposures. Certainly formaldehyde produces irritation of the nose, eyes, and throat when inhaled, but, in most individuals, this is simple irritation and not due to an allergic reaction. Also, the formaldehyde concentration at which the acute response occurs is highly individualistic, which is why the National Academy of Sciences (NAS) [73] concluded that there is no population threshold for the irritating effects of formaldehyde. The Committee on Aldehydes of the NAS [74] further stated that persons sensitized to formaldehyde and with hyperactive airways, may respond quite severely.

The onset of adverse health symptoms is usually associated with moving into a recently acquired house, new office, or mobile home; recent remodeling or renovation; insulating with UFFI; or installation of UF-resin containing furniture or furnishings. The causal relationship between the acute illness and a formaldehyde exposure is sometimes heightened by a lessening of the symptoms' severity when the affected individual moves away from the building in question [66, 75]. Quite often, however, it turns out that the formaldehyde levels in the affected buildings are significantly below 0.1 ppm. This leads one to suspect that some other conditions within these particular indoor environments might sometimes be contributing to the person's health impairment.

Another puzzling factor is the often virtual impossibility of inducing bronchoconstriction in humans during controlled challenges with formaldehyde vapor. This has been the case with 15 healthy individuals [76] and in 13 patients suspected of having formaldehyde-induced asthma [77]. However, in a recently reported and larger study of people suspected of having formaldehyde asthma, about 5% suffered asthmatic reactions during controlled provocation [78]. One most interesting case [79] involved a woman whose child-hood asthma redeveloped after her farmhouse had been insulated with UF foam. Although she showed no reaction to challenges by formal-dehyde vapor, asthma was provoked when she was challenged by the airborne UFFI dust collected from her farmhouse. This raises the suspicion that formaldehyde-containing particles, rather than for-maldehyde vapor, might be the culprit in a sizable number of cases. Have we then sometimes been measuring formaldehyde in its wrong phy-sical state?

The data on formaldehyde-induced cancer from epidemiological studies are sparse and conflicting. As yet, they do not provide any convincing evidence that exposure to formaldehyde has produced cancer in man [80]. Nasal cancers, however, can be produced in rats exposed to high formaldehyde concentrations of 15 ppm [81].

CRYSTAL GAZING

The science of IAQ as well as the actual and the individual perception of air quality inside buildings, are each in a state of flux. The energy crises of the 1970s triggered a wave of activity and change that seems destined to continue through the rest of this century. Scientists, building engineerings, architects, utilities, and public health authorities are becoming more committed to work-ing with the occupants of buildings for optimizing their good health and comfort with the energy efficiency of the buildings. Our objec-tive is to forecast how these activities and trends will affect IAQ.

Nations such as Sweden and several utilities in countries such as the United States have policies to increase further the scale and scope of energy conservation in residences, public buildings, and offices. Every survey of the housing market shows that buyers today look for houses that conserve energy and keep heating bills low. Mortgage qualification is also easier when buying an energy-efficient residence. Media articles extol the monetary virtues of tightening up a house. These are powerful economic forces that will continue to encourage the trend towards greater numbers of tighter buildings.

In the case of homes in the United States, air infiltration is being reduced significantly by the incorporation of continuous, whole-wall plastic sheeting in new houses and to a lesser extent by undertaking economical weatherization measures in existing houses.

Modern houses in the United States now have rates of air infiltration whose median value is about 0.5 air changes per hour. This is
the value proposed as an acceptable lower limit of ventilation for
houses [82]. Further reductions in the infiltration rate of fresh
air will bring more houses into an unacceptable range of ventilation.

  Measures to counteract the associated increased pollutant levels through, for example, mechanical ventilation with heat recovery,
have lagged behind the means for achieving energy conservation. As
a specific example, air-to-air heat exchangers designed for residential use have received only limited public acceptance. Also, their
overall usefulness and cost effectiveness have been questioned [83].
Until this situation changes, tighter buildings will likely have
staler air, more pollutant buildup and more complaints from the
residents. Individual homeowners, especially those who are pollutant sensitive, will try to counteract pollutant buildup with
increasing use of new and improved types of air-cleaning devices.

  Perhaps by the turn of the century, mechanical ventilation and
energy recovery will become an integral part of space-conditioning
systems in new residences. Included, perhaps, will be the sensing
of a marker pollutant that activates an environmental conditioning
system. The ability to tailor the indoor environment to suit a
person's or family's individual needs should eventually bring considerable improvements to residential IAQ. The effort directed to
cleansing indoor air of a particular pollutant or class of pollutants will depend largely on the potential or proven adverse health
effects that are perceived. Our understanding of pollutant exposure
and health effects relationships are often the sparsest portion of
the IAQ science package. Nevertheless, we shall make some predictions about the probable trends in the types and levels of some
types of airborne pollutants.

  Radon exposures and their link to increased risk of lung cancer
pose one of the potentially gravest health threats from continuing
air-tightening measures. Indoor exposure to radon is already
estimated to cause between 1,000 and 20,000 cases of lung cancer
annually, a figure that is above par with the estimated annual
incidence of asbestos-induced mortality of 500 to 600 deaths per
year in the United States [84].

  Radon is one of those airborne pollutants that usually shows a
roughly inversely proportional increase in concentration with a
reduced rate of air exchange. Current energy conservation measures
that rely heavily on reducing the infiltration of fresh outdoor air
seem destined to increase radon levels about inversely proportionate
to the effectiveness of the tightening measures. This becomes especially serious for structures built over localized bedrocks or
porous sediments rich in radium. There is an urgency to identify
quickly these high-risk buildings and apply remedial measures to
bring the radon concentrations down to safer levels. It seems

unwise to tighten a building with no knowledge about the preexisting
levels of radon.

The oldest IAQ concern, airborne infectious agents and aeroal-
lergies, always has been and will continue to be with us. The heating, cooling, humidifying, dehumidifying, and recirculation of air
in modern buildings can, however, easily provide conditions favor-
able for the proliferation and spread of airborne pathogens. Energy
saving, by cutting back or advancing the thermostat as the seasons
change, can prevent the dry conditions necessary to kill house mites
and produce the moist and hot environment so liked by molds and
fungi. Also the water reservoirs of air conditioning systems can
become organism contaminated and produce a variety of respiratory
diseases, such as Legionella.

We anticipate that individual homeowners and the operators of
larger buildings will gradually become more aware of these problems.
People will then be better able to choose the climatic control
suited to their personal needs concerning health and allergies.

Cigarette smoking appears to be in a decline that will continue. The U.S. Public Health Service has set goals for smoking
and health for 1990 [41]. These include reducing the proportion of
adults and teenagers who smoke to below 25% (from 33% in 1980) and
6% (from 20% in 1981), respectively. Thus the amounts of second-
hand smoke available for inhalation by nonsmokers should decrease in
the future. Reduced involuntary cigarette smoking is being enhanced
by the increasing vociferation of nonsmokers, the popularity of
small air-cleaning devices that extract smoke, and the trend in
offices, public buildings, and public transport to ban smoking or
segregate smokers from nonsmokers. We anticipate that these reductions in exposure will more than counter any increased exposure
associated with more air-tight buildings.

The increasing number of people who are seeking cheaper fuel
costs are often attracted to unvented kerosene space heaters,
unvented gas appliances, home fireplaces, and wood-burning stoves.
These incomplete combustion devices can produce elevated levels of
carbon monoxide. Kerosene and natural gas burners are more important sources of NO_2 than tobacco smoke, and significant levels of
SO_2 result if sulphur-containing kerosene is burned. Escaping
woodsmoke elevates indoor concentrations of suspended particulates
and polycyclic aromatic hydrocarbons. Substitution of these sources
of heating for the often more expensive, electrically driven heating
sources will, in general, worsen the IAQ. Adequate venting of the
combustion products is a relatively easy solution to many of these
problems. Improvements are also occurring because manufacturers are
marketing devices that burn fuels more efficiently and emit lesser
amounts of noxious combustion products.

The past several decades have ushered in the era of synthetic
organic materials. Emissions from plastics, paints, sprays, pesti-

cides, automobile interior products, personal and household items, furnishings, and building materials give rise to a plethora of volatile organic compounds. Their concentrations indoors, while usually quite low, are generally well in excess of the outdoor concentrations.

There are as yet few data on the health consequences of exposures to these compounds, especially any long-term effects. Although some information is available about exposure to a few single organic compounds such as formaldehyde, next to nothing is known about human reactions to the simultaneous exposure to several organic compounds. Considering the known biological activity of many VOC, this lack of knowledge is cause for concern. Future high-priority research will be conducted to evaluate health effects, source locations, and emission strengths of VOC. For VOC in general, it remains to be seen how their concentrations and behavior will be affected by the tightening of buildings.

In the meantime, new synthetic materials, emitting as yet unidentified organic compounds, will continue to find their way indoors. It has been suggested [85] that scientists organize a global alarm system that would give an early notification when materials previously thought to be harmless are discovered to be dangerous.

People seem to be spending more time indoors than they used to. The ever increasing number of hours that TV sets are left on is a probable reflection of this trend. Certainly children are captives of their TV sets on a Saturday morning, whereas we, the authors, associate playing outside with our long past childhoods!

At work, the trend is for a higher proportion of people to be employed in office or office-like buildings; perhaps 50% of the 20 million new jobs in the next decade will be of the white-collar type [86]. Sizeable increases are also forecast in the numbers of people working at home [87], with perhaps as many as a total of 15 to 20 million people being home employed by the end of the decade.

The indoor environment and its air quality seem destined to provide an increasing share of most persons total pollutant exposure budget. It seems prudent that in our attempts to unravel the intertwining complexities of IAQ we also try to understand how we have and how we will continue to change our indoor air.

REFERENCES

[1]   "Mortality and Morbidity During the London Fog of December 1952," (London, England: Her Majesties Stationary Office. 1954).

[2]   Phair, J. J., Carey, G. C. R., Shephard, R. J., and M. L. Thomsen. "Some Factors in the Design, Organization, and Implementation of an Air Hygiene Survey," Int. J. Air Water Poll., 1: 18–30 (1958).

[3]   Biersteker, K., de Graaf, H., and A. G. Nass. "Indoor Air Pollution in Rotterdam Homes," Int. J. Air Water Poll., 9: 343–350 (1965).

[4]   Biersteker, K., and H. de Graaf. "Air Pollution Indoors: A Neglected Variable in Epidemiology," T. Soc. Geneesk, 45: 74–77 (1967).

[5]   Goldsmith, J. R., Terzaghi. J., and J. D. Hackney. "Evaluation of Fluctuating Carbon Monoxide Exposures," Arch. Environ. Health, 7: 647–663 (1963).

[6]   Benson, F. B., Henderson, J. J., and D. E. Caldwell. "Indoor-Outdoor Pollution Relationships: A Literature Review." EPAAP–112 (Research Triangle Park, NC: Environmental Protection Agency, 1972).

[7]   Langmuir, A. D. "Changing Concepts of Airborne Infection of Acute Contagious Diseases: A Reconsideration of Classic Epidemiologic Theories," in Airborne Contagion, R. B. Kundsin, Ed. (New York: Annals of the New York Academy of Sciences, 1980), 353: 35–44.

[8]   Riley, R. L., Mills, C. C., Nyka. W., Weinstock, N., Storey, P. B., Sultan, L. U., Riley, M. C., and W. C. Wells. "Aerial Dissemination of Pulmonary Tuberculosis: A Two-Year Study of Contagion in a Tuberculosis Ward," Am. J. Hyd., 70: 185–196 (1959).

[9]   "Improving Indoor Air Quality," Engineering Foundation Conference (South Berwick, ME: Berwick Academy, August 1972).

[10] Yocum, J. E., Clink, W. L., and W. A. Cote. "Indoor/Outdoor Air Quality Relationships." J. Air Pollut. Control Assoc., 21: 251–259 (1971).

[11] Cote, W. A., Wade III, W. A., and J. E. Yocum. "A Study of Indoor Air Quality," EPA–650/4–74–042 (Washington, DC: Environmental Protection Agency, September 1974).

[12]   Murature. D. A., de Kanel, J., Tang., S. Y., Steinhardt, G., and R. C. Dougherty. "Phthalate Esters and Semen Quality Parameters," presented at the Symposium on Biological Monitoring of Exposure to Organic Chemicals, 187th ACS Meeting, St. Louis, April 8-13, 1984.

[13]   Figure kindly provided by R. C. Dougherty, Department of Chemistry, Florida State University. Tallahassee, FL 32306.

[14]   Stolwijk, J. A. J. "The Tight Building Syndrome," Toxic Substances J., 5(3): 155-161 (1983/84).

[15]   Fanger, P. O. "Thermal Comfort" (Malabar, FL: Krieger Publishing Co., 1982).

[16]   Stolwijk, J. A. J. "The Sick Building Syndrome," in   Proc.   of the 3rd Int. Conf. on Indoor Air Quality and Climate, Vol. 1: 23-29, Stockholm, 1984.

[17]   Wyon, D. "Summary of Sessions on Indoor Thermal   Climate,"   in Proc. of the 3rd Int. Conf. on Indoor Air Quality and Climate, Vol. 6, to be published in 1985.

[18]   Anderson, I. "Summary of Sessions on Sick Buildings." in Proc. of the 3rd Int. Conf. on Indoor Air Quality and Climate, Vol. 6, to be published in 1985.

[19]   Tredgold, T. in: The Principles of Warming and Ventilation - Public Buildings (London, England: J. Taylor 1824).

[20]   Carnelley. T., Haldane, J. S., and A. M. Anderson, Philos. Trans. B., 178: 61 (1887).

[21]   Billings, J. S. "Ventilation and Heating." Engineering Record, New York, 120-130 (1893).

[22]   Yaglou, C. P., Riley, E. C., and D. I. Coggins. "Ventilation Requirements," ASHRAE Transactions, 42: 133-163 (1936).

[23]   "Comparison of Indoor and Outdoor Air Quality," EPRI   EA-1733, RP 1309 (Palo Alto, CA: Electrical Power Research Institute, 1984).

[24]   "Trends in Television: 1950 to Date," (Research Department, Television Bureau of Advertising, Inc., March 1984) p. 2.

[25]   Hultqvist, B., "Studies on Naturally Occurring Ionising Radiation", K. Svenska VelenskAbad. Handl. Series 4, 6(3) (1956).

[26]   O'Riordan, M. C., Duggan. M. J., Rose, W. B., and G. F. Bradford. "The Radiological Implications of Using By-Product Gypsum as a Building Material," NRPB Report R-7 (London, U.K.: Her Majesties Stationary Office, 1972).

[27]   Nazaroff, W. W., and A. V. Nero. "Transport of Radon from Soil into Residences." Proc. of the 3rd Int. Conf. on Indoor Air Quality and Climate, Vol. 2: 15-20, Stockholm, 1984.

[28]   Budnitz, R. J., Berk, J. V., Hollowell, C. D., Nazaroff. W. W., Nero, A. V., and A. H. Rosenfeld. "Human Disease from Radon Exposures: The Impact of Energy Conservation in Residential Buildings." Energy and Buildings, 2: 209-215 (1979).

[29]   Cliff, K. D. "Assessment of Airborne Radon Daughter Concentrations in Dwellings in Great Britain," Phys. Med. Biol. 23 N.4: 696-711 (1978).

[30]   Cohen, B. L. "Health Effects of Radon from Insulation of Buildings," Health Phys. 39: 937-940 (1980)

[31]   Nero, A. V. "Indoor Radon Exposures from ^{222}Rn and its Daughters: A View of the Issue," Health Phys. 45: 277-288 (1983).

[32]   Jacobi, W. "Expected Lung Cancer Risk from Radon Daughter Exposure in Dwellings," in Proc. of the 3rd Int. Conf. on Indoor Air Quality and Climate, Vol. 1: 31-42, Stockholm, 1984.

[33]   Pershagen, G., Damber, L., and R. Falk. "Exposure to Radon in Dwellings and Lung Cancer: A Pilot Study," in Proc. of the 3rd Int. Conf. on Indoor Air Quality and Climate, Vol. 2: 73-78, Stockholm, 1984.

[34]   Hildingson, O. "Radon Measurements in 12,000 Swedish Homes," Environ. Int. 8: 67-70 (1982).

[35]   Wilson, C. "Mapping the Radon Risk of Our Environment," in Proc. of the 3rd Int. Conf. on Indoor Air Quality and Climate, Vol. 2: 85-92, Stockholm. 1984.

[36]   LaFore, F. M. "Airborne Infections and Modern Building Technology." in Proc. of the 3rd Int. Conf. on Indoor Air Quality and Climate, Vol. 1: 109-127. Stockholm, 1984.

[37]   Solomon, W. R. "Assessing Fungus Prevalence in Domestic Interiors," J. Allergy Clin. Immunol. 56: 235-242 (1975).

[38]   Reed, C. E., Swanson, M. C., Lopez. M., Ford, A. M., Major, J., Witmer, W. B., and T. B. Valdes. "Measurement of IgG Antibody and Airborne Antigen to Control an Industrial Outbreak of Hypersensitivity Pneumonitis," J. Occupat. Med. 25: 207-210 (1983).

[39]   Samuelson, I. "Sick Houses - A Problem of Moisture?", in Proc. of the 3rd Int. Conf. on Indoor Air Quality and Climate, Vol. 3: 341-346, Stockholm, 1984.

[40] Reed, C. E. and M. C. Swanson. "Indoor Allergens: Identifica-
      tion and Quantification," in Proc. of the 3rd Int. Conf. on
      Indoor Air Quality and Climate, Vol. 1: 99-108, Stockholm,
      1984.

[41] Luoto, J. "Reducing the Health Consequences of Smoking - a
      Progress Report," Public Health Reports, 98(1): 34-39 (1983).

[42] Lebowitz, M. D. "The Potential for Lung Cancer from Passive
      Smoking," in Proc. of the 3rd Int. Conf. on Indoor Air Quality
      and Climate, Vol. 1: 59-70, Stockholm, 1984.

[43] Repace, J. L. and A. H. Lowrey. "A Proposed Indoor Air Quality
      Standard for Ambient Tobacco Smoke," in Proc. of the 3rd Int.
      Conf. on Indoor Air Quality and Climate, Vol. 5: 235-239,
      Stockholm, 1984.

[44] Weber, A "Environmental Tobacco Smoke Exposure: Acute   Effects
      - Acceptance Level - Protective Measures," in Proc. of the 3rd
      Int. Conf. on Indoor Air Quality and Climate, Vol. 2: 297-301,
      Stockholm, 1984.

[45] Matsukura, S., Taminato, T.,   Kitano,   N.,   Seino,   Y.,
      Hamada. H.,   Uchihashi, M.,   Nakajima, H.,   and Y. Hirata.
      "Effects of Environmental Tobacco Smoke on Urinary Cotinine
      Excretion in Nonsmokers," New England J. Med. 311 (13): 828-
      832 (1984).

[46] Smith, K. R., Apte, M., Menon, P. and M. Shrestha. "Carbon
      Monoxide and Particulates from Cooking Stoves: Results from a
      Simulated Village Kitchen," Proc. of the 3rd Int. Conf. on
      Indoor Air Quality and Climate, Vol. 4, 389-395, Stockholm,
      1984.

[47] Kim, Y. S. and W. Kreisel. "Effects of Household Exposures to
      Carbon Monoxide Poisoning in Korea," in Proc. of the 3rd Int.
      Conf. on Indoor Air Quality and Climate, Vol. 4: 117-122,
      Stockholm, 1984.

[48] Goldstein, I., Hartel, D., and L. Andrews. "Indoor Exposure of
      Asthmatics to Nitrogen Dioxide," in Proc. of the 3rd Int.
      Conf. on Indoor Air Quality and Climate, Vol. 2: 269-274,
      Stockholm, 1984.

[49] Moschandreas, D. J., Stark, J. W. C., Morse, S. E., and S. M.
      Bromberg.   "Indoor Air Pollution in the Residential Environ-
      ment," Report 68-02-2294 (Research Triangle Park, NC: Environ-
      mental Protection Agency, August 1978).

[50] Mølhave. L. "The Atmospheric Environment in Modern Danish
      Dwellings - Measurements in 39 Flats," in Proc. First Intl.
      Indoor Climate Symp. P. O. Fanger and O. Valbjorn, Eds.

(Copenhagen, Denmark: August 31–September 1, 1978), pp. 171–186.

[51]   Mølhave. L., Moller, J., and I. Anderson. "Air Concentrations of Gases, Vapours and Dust in New Houses," Ugeskr. f. Laeger, 141: 956–961 (1979).

[52]   Mølhave, L. "Indoor Air Pollution Due to Organic Gases and Vapors of Solvents in Building Materials," Environ. Int. 8: 117–127 (1982).

[53]   Seifert, B. "Relationship Between Indoor and Outdoor Concen-trations of Inorganic and Organic Substances," in Indoor Air Quality,   K. Aurand,   B. Seifert,   and   J. Wegner.   Eds. (Stuttgart/New York; Gustav Fischer Verlag, 1982), p. 41.

[54]   Wallace. L., Pellizzari, E., and T. Hartwell. "Analyses of Exhaled Breath of 355 Urban Residents for Volatile Organic Compounds," in Proc. of the 3rd Int.   Conf. on Indoor Air Quality and Climate, Vol. 4: 15–20, Stockholm, 1984.

[55]   De Bortoli, M., Knöppel, H., Pecchio, E., Peil. A., Rogora, L.,   Schauenburg, H.,   Schlitt, H.,   and H. Vissers. "Integrating Real Life Measurements of Organic Pollution in Indoor and Outdoor Air of Homes in Northern Italy," in Proc. of the 3rd Int. Conf. on Indoor Air Quality and Climate, Vol. 4: 41–26, Stockholm, 1984.

[56]   Ahlström, R., Berglund, B., Berglund. U., and T. Lindvall. "Odor Interaction between Formaldehyde and the Indoor Air of a Sick Building," in Proc. of the 3rd Int. Conf. on Indoor Air Quality and Climate. Vol. 3: 461–466, Stockholm, 1984.

[57]   Mølhave, L., Bach. B., and 0. F. Pederson. "Human Reactions During Exposures to Low Concentrations of Organic Gases and Vapors Known as Normal Indoor Air Pollutants." in Proc. of the 3rd Int. Conf. on Indoor Air Quality and Climate, Vol. 3: 431–436, Stockholm, 1984.

[58]   Allan, G. G. "Long–Term Stability of Urea – Formaldehyde Foam Insulation," Environ. Sci. Technol. 14: 1235–1240 (1980).

[59]   Fahrni. F. "Procede pour la Fabrication des Plaques Artifi-cielles en Bois Comprime," French Patent 881,781 (1943).

[60]   "Formaldehyde Production and Sales," Chemical Economics Hand-book (Stanford, CA: SRI International, January 1983).

[61]   Meyer, B., and K. Hermanns. "Formaldehyde Release from Pressed–Wood Products." in Formaldehyde, V. Turoski, Ed. (ACS Advances in Chemistry Series No. 210), in press.

[62]   Thomas, M. "Zur Problematik des Freiwerdens von Formaldehyd von Hamstoff-Formaldehyd Harzen." Holztechnologie 5: 79 (1964).

[63]   Deimel, M. "Formaldehyd Belastung in Schulban," in Organische Verunreinigungen in der Umwelt, K. Auraud, Ed. (Hamburg, W. Germany: E. Schmidt Publishers, 1978) pp. 416–427.

[64]   Breysee, P. A. "Formaldehyde in Mobile and Conventional Homes," Environ. Health and Safety News, 26: 19 pp. (1977).

[65]   Gupta, K. C., Ulsamer, A. G., and P. W. Preuss. "Formaldehyde in Indoor Air: Sources and Toxicity." Environ. Int. 8: 349–358 (1982).

[66]   Breysee, P. A. "Formaldehyde Levels and Accompanying Symptoms Associated with Individuals Residing in Over 1000 Conventional and Mobile Homes in the State of Washington," in Proc. of the 3rd Int. Conf. on Indoor Air Quality and Climate, Vol. 3: 403–408, Stockholm, 1984.

[67]   Wanner, H. U., and M. Kuhn. "Indoor Air Pollution by Building Materials." in Proc. of the 3rd Int. Conf. on Indoor Air Quality and Climate, Vol. 3: 35–40, Stockholm. 1984.

[68]   Gammage, R. B. and A. R. Hawthorne. "Current Status of Measurement Techniques and Concentrations of Formaldehyde in Residences." in Formaldehyde, V. Turoski, Ed. (ASC Advances in Chemistry No. 210), in press.

[69]   "Final Report of the Canadian National Testing Survey." (Place du Centre, 4th Floor, 200 Promenade du Portage. Hull. Quebec KIA OC9: UFFI Centre, September, 1983.

[70]   Hanson, D. J. "Effects of Foam Insulation Ban Far Reaching," Chem. & Eng. News, pp. 34–37. March 29, 1982.

[71]   "Gulf South Insulation, et al, vs U.S. Consumer Product Safety Commission," 701 F. 2d 1137 (5th Cir. 1983).

[72]   Meyer, B. "Formaldehyde Release from Building Products," in Proc. of the 3rd Int. Conf. on Indoor Air Quality and Climate, Vol. 3: 29–34, Stockholm, 1984.

[73]   "Formaldehyde, An Assessment of its Health Effects," (Washington, DC: National Academy of Sciences, March 1980).

[74]   "Formaldehyde and Other Aldehydes," National Academy Press (Washington, DC: National Research Council, 1981).

[75]   Godish. T. "Formaldehyde and Building-Related Illness," J. Environ. Health 44(3): 116–121 (1981).

[76]   Witek, T.J., Schachter, E. N., and T. Tosun. "Acute Pulmonary Effects from Exposure to Low Concentrations of Formaldehyde." in Proc. of the 3rd Int. Conf. on Indoor Air Quality and Climate, Vol. 3: 41-45, Stockholm, 1984.

[77]   Frigas, E., Filley, W. V., and C. E. Reed. "Bronchial Challenge with Formaldehyde Gas: Lack of Bronchoconstriction in 13 Patients Suspected of Formaldehyde-Induced Symptoms," Mayo Clin. Proc. 59: 295-299 (1984).

[78]   Nordman, H., Keskinen, H., and M. Tuppurainen. "Asthma Caused by Formaldehyde," in Proc. of the 3rd Int. Conf. on Indoor Air Quality and Climate, Vol. 3: 217, Stockholm, 1984.

[79]   Frigas, E., Warren, F. V., and C. E. Reed. "Asthma Induced by Dust from Urea-Formaldehyde Foam Insulating Material" Chest 79(6): 706-707 (1981) and 82(4): 511-512 (1982).

[80]   "Epidemiology Panel Report," Deliberations of the Consensus Workshop on Formaldehyde (Little Rock, AR: National Center for Toxicological Research, October 1983), p. 46.

[81]   Albert. R., Sellakumar, A. R., Laskin, S., Kushner, M., Nelson N., and C. A. Snyder. "Gaseous Formaldehyde and Hydrogen Chloride Induction of Nasal Cancer in the Rat," J. Nat. Cancer Inst. 68: 597-603 (1982).

[82]   "Ventilation for Acceptable Indoor Air Quality." ASHRAE Standard 62-1981 (American Society of Heating, Refrigerating, and Air Conditioning Engineers, 1981).

[83]   Lannus, A. "Control of the Indoor Environment in Buildings: Problems and Research Needs" presented at the Conference on Environmental Research Needs in the Tennessee Valley Region, Knoxville. TN, September 26-28, 1984, to be published in the Proceedings.

[84]   Ross, M. "A Survey of Asbestos-Related Disease in Trades and Mining Occupations and in Factory and Mining Communities as a Means of Predicting Health Risks of Nonoccupational Exposures to Fibrous Minerals," in Definitions for Asbestos and Other Health-Related Silicates, B. Lavadie, Ed. (Philadelphia, PA: ASTM Special Publication 834, July 1984).

[85]   Holm, L. "Future Building and Building Hygiene in a Historical Perspective." in Proc. of the 3rd Int. Conf. on Indoor Air Quality and Climate, Vol. 1: 19-21, Stockholm, 1984.

[86]   McLellan, R. K. "The Health Hazards of Office Work," Toxic Substances J. 5(3): 162-181 (1983/84).

[87]   Goerth, C. R. "Trend of Home Work Poses New Health, Safety Challenges," Occup. Health & Safety. p. 31, September 1984.

Part One

Radon

3.

Part One: Overview

Wayne M. Lowder

Environmental Measurements Laboratory
U.S. Department of Energy
New York, New York 10014

Naturally occurring radioactive gas radon is always present in the air that man breathes. Indoor concentrations are typically 2 to 10 times those in outdoor air. Inhalation of the decay products of radon attached to particulates results in an alpha dose to the critical cells of the respiratory tract that has been shown to produce lung cancer in miners. Current estimates of radon exposure and the consequent risk indicate that about 10% of nonsmoking-related lung cancers may be produced by radon. There are also hints in the current literature that the respective roles of radon and smoking in the induction of lung cancer may not be unrelated. Thus, radon has been generally recognized as one of the most significant indoor pollutants in terms of potential human health effects, considering both current exposure levels and the trend toward increased exposures induced by the application of advanced energy conservation technology in new housing.

The reports in this section treat current knowledge concerning both radon exposure and risk and the implications of the uncertainties in this knowledge for future research. The uncertainty in the risk is particularly critical, since the multiplication of our current "best" estimates of lung cancer risk per unit exposure and average exposure yields a lifetime risk of a few tenths of a percent. A significant number of people may be living in environments where radon exposure produces a lifetime lung cancer risk of about 10%. Thus, the radon problem is of sufficient magnitude to render the risk uncertainties described in these reports somewhat uncomfortable. There also remains the still-open question concerning the role of radon exposure in the induction of those lung cancers attributed to smoking.

However, despite this situation, it is clear that our knowledge in this area has increased dramatically in the past few years, and this knowledge has enabled us to scope the problem, define future information needs, and at least get a glimmer of what the possible

39

solutions might be. This is being done at the federal level through
the activities of the Federal Committee on Indoor Air Quality, co-
chaired by representatives from EPA, DOE, CPSC, and HHS. At the
request of Congress, a federal research strategy document is now in
preparation that will define the overall research needs and direc-
tions and possible ways in which various levels of government and
the private sector can cooperate in responding to these needs.

Our state of knowledge can be briefly summarized in a manner
that is only slightly modified from material submitted verbally at
this symposium by A. C. James, U.K. National Radiological Protection
Board. Our statements illustrate that viewpoints on both sides of
the Atlantic are quite consistent.

1.   We know what to measure as an index of population risk and also
     how to measure population exposure.

2.   Techniques for carrying out large-scale surveys of population
     exposure to radon are available and are being used in a number
     of countries; for example, passive monitors can be placed in
     homes to integrate exposure for days, months, or over the whole
     year.

3.   Therefore, statistically valid information on the distribution
     of indoor exposure is emerging.

4.   There is a general understanding of the effect of reduced ven-
     tilation in increasing population exposure and the risk of lung
     cancer.

5.   There is also a consensus that the main cause of high indoor
     exposures is the ingress of radon in soil gas. Research work
     should therefore be directed towards understanding the physical
     processes involved, developing economical means of control, and
     identifying geographical regions at high risk.

6.   To avoid problems of high radon exposures in future housing
     stock, there is an urgent need to develop criteria for screen-
     ing building land and to establish minimum standards of build-
     ing practice in high-risk areas.

7.   Although we are almost certain that exposure to radon at
     environmental levels is associated with increased risk of lung
     cancer, the numerical value of risk per unit exposure must
     always remain uncertain. The range of uncertainty is narrow-
     ing.

8.   A risk estimate can be obtained by direct consideration of lung
     cancer incidence among uranium miners, although it must be
     recognized that the individual exposures of miners in the popu-
     lations studied are all highly uncertain, and that the transla-

tion of miner risks to the general population is not necessarily simple.

9. The EPA evaluation of uranium miner data implies that 1 in 200
    of the U.S. population will die of lung cancer caused by radon
    exposure in the home. This represents half of the lung cancer
    risk for the nonsmoking population and appears somewhat high
    when all of the other potential carcinogens in the environment
    are considered.

10. The NCRP analysis, which is based on a different interpretation
    of the time course of cancer incidence following exposure to
    radon daughters, gives a risk estimate about a factor of 3
    lower (i.e., about 20% of the lung cancers in nonsmokers are
    related to radon exposure).

11. The OECD/NEA and UNSCEAR have estimated risk by dosimetric
    analysis, applying the ICRP concept of summing risks to indivi-
    dual organs (effective dose equivalent). The best estimate of
    risk to the general population obtained by this method is
    slightly lower (i.e., about 10% of the nonsmokers' risk of lung
    cancer).

12. The overall range of risk estimates is therefore about a factor
    of 6. Whichever estimate is adopted, however, the risks from
    environmental exposures to radon are significant and of public
    health concern.

13. Although the public health impact of environmental (mainly
    indoor) radon exposure is considerable and is being enhanced by
    energy conservation measures that reduce indoor-outdoor air
    exchange rates, the problems are likely to be controllable in a
    cost-effective manner. Changes in building construction prac-
    tices and the application of control technology now being
    developed should have a significant impact on future radon
    exposure levels.

4.

Indoor Concentrations of Radon-222 and Its Daughters: Sources, Range, and Environmental Influences

Anthony V. Nero, Jr.

Co-Leader, Building Ventilation and Indoor Air Quality Program, Lawrence Berkeley Laboratory, University of California, Berkeley, California 94720

INTRODUCTION

The radiation dose from inhaled daughters of ^{222}Rn constitutes about half of the total effective dose equivalent that the general population receives from natural radiation. A variety of results from the United States suggests that ^{222}Rn concentrations in residences average about 1 pCi/l (37 Bq/m^3). Estimation of the incidence of lung cancer due to the daughters associated with this much ^{222}Rn yields thousands of cases per year among the U.S. population.

It is also clear that indoor levels are sometimes an order of magnitude or more higher than average: it is the common experience of the community performing measurements in homes that ^{222}Rn concentrations in the range of 10 to 100 pCi/l occur with startling frequency. And, whereas the risk associated with even 1 pCi/l is very large compared with many environmental insults of concern, living for prolonged periods at the higher concentrations observed leads to estimated individual lifetime risks of lung cancer that exceed 1%. For the extreme concentrations that have been found, risks appear to approach that from cigarette smoking.

Not surprisingly, the early work on indoor concentrations has given rise to a broad range of research characterizing ^{222}Rn and its daughters indoors. This work has included significant monitoring programs in homes (although, as noted below, not in a statistical sampling of U.S. homes), investigation of the sources of indoor radon, examination of the factors affecting indoor concentrations, study of the behavior of ^{222}Rn daughters, and - of course - development of techniques to control indoor concentrations. In addition, radiobiologists and epidemiologists have begun to apply dosimetric and dose-response data to the problem of environmental exposures to daughters of ^{222}Rn.

The international research effort in this area has been very substantial, beginning in the 1970's. For detailed information, one must turn to the very large literature, which has – for example – recently culminated in two substantial collections of research papers on indoor radiation exposures (1,2). The purpose of the present paper is to distill the growing understanding of indoor concentrations and the factors – sources, ventilation rates, and daughter reactions – that affect them.

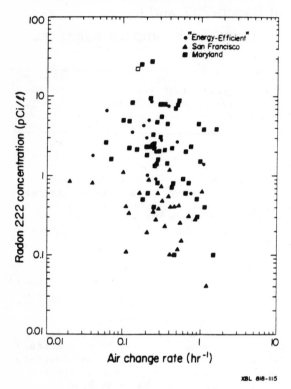

XBL 818-1115

Figure 1. Radon Concentrations and Air Change Rates Measured in 98 U.S. Residences.

The results shown are from three survey groups: "energy-efficient" houses in the United States and (one) in Canada; conventional houses in the San Francisco area; and conventional houses in a community in rural Maryland.

OVERVIEW OF IMPORTANT FACTORS

The initial observation of significant concentrations of radon (taken to be ^{222}Rn unless otherwise stated) in ordinary U.S. homes (3) occurred at about the same time that programs to increase the efficiency of energy use were hitting their stride. Reducing ventilation rates in buildings can be a cost-effective, and hence attractive, component of such programs. However, given the expectation in first order that indoor pollutant concentrations equal the ratio of the entry rate per unit volume to the ventilation rate, energy-conservation programs were thought to have the potential to exacerbate the indoor air quality problem significantly. This expectation still has some merit, although –

for every major class of airborne pollution from indoor sources -
it has become clear that the presence or absence of substantial
emission rates is the major determinant of whether or not a
building has excessive indoor concentrations.

The importance of differences in source strength became clear
in initial investigations of the dependence of indoor
concentrations on ventilation rate. It was found that, with supply
of differing amounts of mechanical ventilation in a given house,
the indoor radon concentration varied as the inverse of the
ventilation rate, as expected (4). However, at about the same
time, paired measurements of radon concentration and ventilation
rate in several housing samples showed no apparent correlation
between these two parameters (5). As shown in Figure 1, for a
given sample, the radon concentration and ventilation rates showed
an approximately order-of-magnitude range; for the combined
samples, the concentration showed a significantly larger
variability than the ventilation rate, suggesting that the source
strength was the dominant determinant of the wide range of
concentrations observed in U.S. housing.

These indications have prompted substantial work in
understanding the size and variability of radon entry rates, as
discussed below. However, it is important to emphasize that other
factors still play an important role: ventilation rates vary
substantially within the building stock, which - after all - is one
major incentive for instituting major energy conservation programs
(the adjunct incentive being that the average ventilation rate in
U.S. buildings is rather high as compared with rates in some
countries). And, even for a given radon concentration, the
concentrations and physical state of the daughters - which account
for the health effects of interest - can vary significantly.

It is worth noting at this point the substantial tendency of
the research community to measure radon concentrations in survey
efforts, rather than daughter concentrations. This tendency
arises, of course, from the availability of a reasonably reliable
and very simple integrating radon monitor (6), a significant
contrast to the state of daughter monitoring. And yet, given a
reasonable understanding of the relationship between radon and its
daughters and an awareness of the fact that the daughter-to-radon
ratio does not vary as widely as radon concentrations, measurement
of radon concentrations is a reasonable indicator of daughter
concentrations and is certainly a very effective tool in survey
efforts. This is entirely analogous to the situation for many
other pollutants: for example, although any health effects
associated with NO_2 exposures may have a substantial dependence on
peak (as opposed to average) concentrations, an integrating sampler
can be a very effective survey instrument, provided associated
studies examine relationships between average and peak
concentrations under well-characterized conditions.

As seen below, another incentive for emphasizing the radon concentration per se is that the very fact that this parameter shows the widest variability suggests clues to identifying and controlling excessive concentrations. Considering the origin of this wide variability, it will not be surprising that attention to radon sources and entry modes appears to have the greatest potential as a basis for control strategies.

SOURCES OF INDOOR RADON

Radon arises from trace concentrations of radium in the earth's crust, and indoor concentrations depend on access of this radon to building interiors. Radon can enter directly from soil or rock that is still in the crust, via utilities such as water (and in principle natural gas) that carry radon, or from crustal materials that are incorporated into the building structure in the form of concrete, rock, and brick. The relative importance of these pathways depends on the circumstances, but - for U.S. single-family homes - it has become clear that the first dominates the indoor concentrations that are observed.

Indications of this arose in early investigations, when it was found that measurements of radon emanating from structural materials could not account for observed indoor concentrations, based on estimates of the air exchange rate (3). A clearer picture emerged from the distribution of entry rates inferred from direct measurements of radon concentration and ventilation rate (such as those shown in Figure 1). Figure 2 shows such entry-rate distributions from various countries, as well as indicating the potential contribution of various sources. Although building materials were first suspected as the major source, based on experience in Europe (7), the initial U.S. results (5) strongly suggest that the soil must be the major source. Understanding how the rate of radon entry could be approximately equal to the unimpeded flux from the ground (i.e., in the absence of the house) has been a major focus of research on radon entry, both in the United States and Europe.

Soil and Building Material

Establishing a radon mass-balance for a building requires consideration of the various sources. As indicated in Figure 2, a median (or geometric mean) entry rate for U.S. single-family homes is in the vicinity of 0.5 pCi l^{-1} h^{-1} (18 Bq m^{-3} h^{-1}). Based on emanation rate measurements from concretes (17), one might expect emissions from this source to account for a median of about 0.07 pCi l^{-1} h^{-1}, far below the total observed. On the other hand, the potential contribution from unattentuated soil flux, a median of

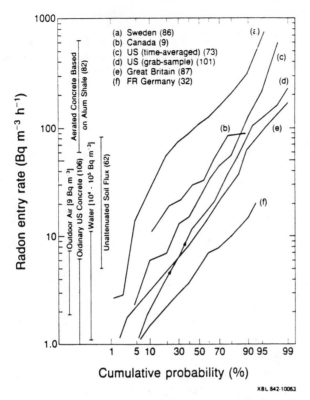

Radon entry rate (Bq m⁻³ h⁻¹) — graph axes

(a) Sweden (86)
(b) Canada (9)
(c) US (time-averaged) (73)
(d) US (grab-sample) (101)
(e) Great Britain (87)
(f) FR Germany (32)

XBL 842-10063

Figure 2. Cumulative Frequency Distributions of Radon Entry Rate
Determined in Dwellings in Several Countries as the Product of
Simultaneously-Measured Ventilation Rate and Radon Concentration.

The number of residences in each sample is indicated in
parentheses; the sources of these results are a) (7), b) (8),
c) (9, 4, 10-13), d) (5), e) (14), f) (15). The bars at the
left indicate the range of contributions expected from a variety
of sources, with assumptions indicated in brackets. For each source
we have assumed a house having a single story of wood-frame construction with a 0.2-m-thick concrete-slab floor. The floor area and
ceiling height are assumed to be 100 m² and 2.4 m, respectively;
water usage is assumed to be 1.2 m³ per day, with a use-weighted
transfer efficiency for radon to air of 0.55; the ventilation rate
is assumed to be in the range of 0.2-0.8 h⁻¹. (References for
source contribution estimates: outdoor air (16): U.S. concrete (17);
alum-shade concrete (18); water (19); soil flux (20).) 37 Bq m⁻³
equals 1 pCi l⁻¹.

0.7 pCi l⁻¹ h⁻¹ (based on ref. 20), corresponds well with the
indoor observations. However, houses have understructures that
might be expected to impede substantially the ingress of radon, at
least by diffusion. In fact, as discussed in more detail in a
recent review of the source of radon indoors (19), although

transport via diffusion accounts well for observed fluxes from building materials and exposed soil, and could account for comparably small fluxes from the soil through some understructure materials (such as concrete), diffusion cannot account for the total entry rates observed specifically in single-family houses. Another mechanism must account for the efficiency with which radon from soil enters homes. It appears that this mechanism is bulk flow of soil gas driven by small pressure differences between the lower part of the house interior and the outdoors.

As it turns out, such pressure differences are precisely the cause of ventilation in homes during seasons when the windows are closed. These pressure differences arise from two environmental factors. First, the difference in temperature between indoors and outdoors causes small pressures across the building shell, with pressures at the bottom pointing toward the higher temperature (e.g., the heated interior) and pressures at the top causing air flow toward the colder temperature. This "stack" effect causes a convection pattern that exchanges indoor air with outdoor, with outdoor air being drawn from the vicinity of the understructure during the heating season. The second factor of importance is wind, which causes a depressurization of the house interior, as well as strong differences in pressure across the different walls that depend on the wind direction. The pressure differences caused by temperature differences and winds are roughly comparable in size, averaging on the order of a few pasqual (with higher values in relatively severe climates). These extremely small pressure differences account for air infiltration through walls (and other components of the house), the dominant contributor to home ventilation during heating season. These same pressure differences can, in principle, drive the small flows of soil gas that can account for the observed rate of radon entry into homes: soil gas contains enough radon that only 0.1% of infiltrating air would have to be drawn from the soil (19). Before proceeding to the studies specifically addressing this possibility, I note that it is often very useful to parameterize infiltration rates in terms of the temperatures differences and winds that drive them, and a very simple and useful model for doing so is now in wide use (21).

Recent work has begun to characterize directly the potential for pressure differences to cause entry of radon via soil gas, probably through imperfections and penetrations in the house understructure that permit passage of the relatively small amount of soil gas required. A study of radon entry in a single-family house with a basement analyzed the entry rate versus the ventilation rate, measured over a period of months, and concluded that entry could usefully be represented by a sum of two components: one - the smaller - independent of ventilation rate, much as diffusion would be, and a larger term that is proportional to ventilation rate, as pressure-driven flow might be (13). Moreover, the authors concluded that the observed pressure and soil parameters were consistent with the soil-gas flow rate that was implied by the measured concentrations and ventilation rates. More detailed theoretical work (22) is helping to formulate a fundamental picture of the pressure and velocity fields in the soil surrounding homes with basements. Finally, recent experiments have directly observed, in two houses with basements, the underground depressurization implied by this picture, and have measured underground soil-gas movement by injecting and monitoring tracers (23). It is interesting to note that these results may also have significant implications for entry of other pollutants from the soil.

In respect to other housing types, the basement studies have given results that might also be expected to apply in large part to slab-on-grade structures, where the pressure difference generated can still draw soil gas through any penetrations in or around the slab. However, direct measurements in such structures have not been performed. The other understructure type of substantial importance is the crawl-space, which to some extent isolates the interior from the soil - at least in respect to pressure-driven flow between the two. Limited measurements of the transport efficiency of radon through crawl-spaces yield the result that a substantial portion of the radon leaving uncovered soil manages to enter the interior, even if vents are open to permit natural ventilation of the space (12).

In retrospect, this is not entirely surprising, since the stack effect will still tend to draw infiltrating air into the home from the crawl-space, which can retain radon from the soil in conditions where winds are not sufficient to flush it to the outdoors via the vents. Furthermore, for structures where the vents are sealed shut, e.g., to save energy, it is conceivable that the crawl-space still provides sufficient connection between the house interior and the soil that pressure-driven flow can enhance the flux from the soil above levels associated merely with diffusion; the work reported in ref. (12) may have observed this effect. Another result of this study is that energy conservation efforts that focus on tightening the floor above a crawl-space can significantly reduce infiltration rates, while reducing radon entry a corresponding amount, as a result of which indoor concentrations are little affected.

Thus sufficient mechanisms exist to account for the substantial amount of radon that appears to enter homes from the soil, apparently without great regard to understructure type. However, this does not imply that other sources of radon are unimportant. It is clear that materials utilized in a building structure can contribute substantial indoor concentrations, although this is not usually the case (even for natural stone that is higher than average in radium content). Moreover, in buildings that are relatively isolated from the ground, such as multi-story apartment buildings, indoor concentrations are expected to be lower than average - as is often the case in central European dwellings -

and to arise primarily from the building materials and, for typical U.S. infiltration rates, from radon in outdoor air.

Water

    Probably more important than building materials, as a source of radon in certain parts of the housing stock, is domestic water drawn from underground sources. Surface waters have radon concentrations too small to affect indoor concentration when used indoors, but ground water is in a good position to accumulate radon generated within the earth's crust. As a result, very high radon concentrations can be found in associated water supplies: as an example, concentrations exceeding 100,000 pCi/l (3.7×10^6 Bq/m^3) have been found in wells in Maine (24). With normal water use, if the radon from such (admittedly rare) water enters the air of a typical house, an indoor concentration of about 10 pCi/l would result, among the higher levels observed.

    Past examinations of the overall potential contribution from water supplies have been little more sophisticated than the estimate just given, which corresponds to a ratio of radon in air to radon in water of 10^{-4}, comparable to estimates and direct observations made by a number of authors (e.g., ref. 24). However, substantial data have recently become available on concentrations of radon in public water supplies drawn from ground water (25), indicating that the majority of such supplies have concentrations below 1000 pCi/l, but that a very small percentage has concentrations exceeding 100,000 pCi/l. Moreover, data are available to assess the effect of radon release to indoor air in a more comprehensive way: a very recent analysis has combined information on water use rate, efficiency of radon release from domestic water used in various ways, house volumes, and ventilation rates, to yield a frequency distribution of air-to-water ratios that is approximately lognormal and that has an arithmetic mean of about 1.1×10^{-4}, close to the value cited above (26). The importance of these developments is that they permit quantitative assessment of the contribution of public water supplies to indoor radon concentrations. The preliminary result of such assessment is that such supplies contribute an average of approximately 0.03 pCi/l in homes served by ground water, about 3% of the average indoor radon concentrations in U.S. homes (see below). However, the very high water-borne concentrations that are <u>sometimes</u> found will contribute much larger airborne concentrations in the homes affected.

    This distribution for the air-to-water ratio may also be used for assessing the contribution from private wells, to the extent that concentrations in water are known. Using data that are available for the approximately 18% of the population using private wells, ref. 26 calculates the indoor radon concentration from water for this segment of the housing stock to average about 0.4 pCi/l. Moreover, about 1% of the entire housing stock would be expected to have indoor concentrations from water of about 1 pCi/l, due primarily to concentrations from private wells in high activity areas (such as Maine). The authors emphasize that these estimates for private-well contributions cannot be regarded to be reliable, but it is significant that, if they were approximately correct, the portion of the population using private wells would be experiencing significantly higher radon exposures than average, particularly in high-activity areas.

DISTRIBUTION OF INDOOR CONCENTRATIONS

Despite a broad range of efforts to characterize indoor radon, and a significant number of studies that have included measurements in existing U.S. homes, no unequivocal estimates may be made of the concentrations to which the U.S. population is exposed. The reason for this difficulty is that the studies that have been performed have varied significantly in incentives, scientific objectives, selection of homes, and measurement procedures. The results, not surprisingly, vary significantly, as may the conclusions that can be drawn from them. Thus, although the community has a general appreciation that the average indoor concentrations is in the vicinity of 1 pCi/l and that a notable number of homes exceed 10 pCi/l, no useful quantitative appreciation of the actual distribution in U.S. homes has been available. Knowledge of this distribution is essential in formulating a strategy for controlling excessive concentrations, as well as for making a reliable estimate of even the average population risk.

An obvious solution to this difficulty is to carry out measurements in a valid statistical sampling of U.S. homes. Given our current appreciation of typical concentrations and the incidence of high levels, it is thought that monitoring of perhaps 1000 to 2000 homes would determine mean concentrations very accurately and ascertain the fraction of homes at high concentrations (e.g., 10 times the mean) to a reasonable accuracy. However, although the Federal agencies interested in indoor air quality have been seriously considering the potential for a national survey of radon and other pollutants, this will not occur very rapidly, if at all. The main effect of the Federal evaluation may be to formulate a design that enhances the potential for aggregating results from smaller regional efforts.

Regardless of such efforts, the data already available are quite substantial and deserve careful evaluation, if only as a basis for proper design of subsequent monitoring efforts. In particular, although past studies have not been conducted with a consistent approach, the number of such studies is substantial, yielding some tens of data sets (with the precise number depending

on criteria for consideration). For this reason, a systematic
analysis of U.S. results has recently been undertaken, explicitly
considering the differences between studies and using lognormal
representations as a basis for aggregating the various data sets to
yield a nominal distribution for the United States. The results
are quite robust, i.e., they have little dependence on selection of
data sets, on normalizations having to do with season of
measurements, and on weighting of the data (27). Figure 3 shows
the result of <u>direct</u> aggregation of 19 of the data sets that are
available as individual data, totalling 552 houses. Because of the
lack of proper normalization and weighting, no general conclusion
may be drawn from this specific aggregation, aside from the
substantial conformance to a lognormal representation, a result
that has been observed in many individual studies.

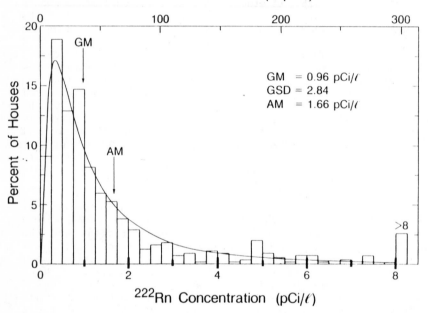

Figure 3. Probability Distribution from Direct Aggregation of the
552 Individual Data in 19 Sets.

The smooth curve is the lognormal functional form corresponding to
the indicated parameters, calculated directly from the data.

To lay the basis for citing a principal result, I note that
the analysis of ref. 27 utilized from 22 to 38 sets, corresponding
to different areas (usually a state or urban area) of the country,
with the larger number including monitoring efforts that were

prompted by some prior knowledge of a potential for elevated concentrations. Characterizing the results of each set in terms of a lognormal function, the geometric mean radon concentrations from these sets range from 0.3 to 5.7 pCi/l (11 - 210 Bq/m^3), with geometric standard deviations ranging from 1.3 to 4. As just noted, the results of aggregating these sets are quite robust, the main differentiation among different aggregations being that including the full 38 sets yields somewhat higher results than including only the 22 "unbiased" samples. The overall result, relying primarily on the 22-set aggregations and including a renormalization of data taken only during heating season, is a geometric mean of about 0.9 pCi/l and a geometric standard deviation of 2.8, implying an average concentration of 1.5 pCi/l and 1 to 3% of houses exceeding 8 pCi/l. This result can only be associated with the portion of the housing stock consisting of single-family houses, since 99% of the data are drawn from such houses. However, this is the dominant element in the housing stock, and the results of this analysis suggest that of the order of a million houses have annual-average concentrations averaging 8 pCi/l or more. This corresponds to exposures approaching the 2 WLM/yr remedial-action limit recently recommended by the National Council on Radiation Protection and Measurements (NCRP) (28). Another interesting observation from this analysis is that the geometric means of the 22 sets are themselves lognormally distributed, with a geometric standard deviation of 2.0. This index demonstrates the substantial variability in mean radon concentration from one area to another and indicates the potential value of strategies that locate homes with high concentrations by first trying to identify areas that have unusually high mean concentrations.

We are still faced with difficulty in estimating concentrations in apartments (and, of course, in other types of buildings). Indications are, based on a few measurements (such as a single third-floor apartment in ref. 3), that concentrations are typically a few tenths of a pCi/l, substantially lower than concentrations in single-family houses. In fact, for an apartment or other building where the average space is relatively well-isolated from the ground, the major contributions to indoor concentrations may be expected to be outdoor air and building materials. The few measurements made to date in the United States are consistent with this expectation, which is also confirmed by the much larger European efforts monitoring concentrations in apartments.

Overall, the approximate contribution of various sources to U.S. residences may be summarized as in Table 1. There it is seen that the dominant contributor to indoor concentrations in single-family houses is the soil, with water and building materials contributing only a few percent. In contrast, for large buildings, the main contributors appear to be the building material and the outdoor air, which also contributes significantly to concentrations in single-family housing. However, for the portion of residences served by private wells in high activity areas, the contribution from water may be much large than in ordinary circumstances.

Table 1.

Approximate Contributions of Various Sources to

Observed Average Indoor Radon Concentrations

	Single-Family Houses (pCi/l)	Apartments (pCi/l)
Soil potential (based on flux measurements)	1.5	< 1
Water (public supplies)	0.01*	0.01*
Building materials	0.05	0.1^{+}
Outdoor air	0.2	0.2
Observed indoor concentrations	1.5	(0.3)

*   Applies to 80% of population served by such supplies; contribution from water may average about 0.4 pCi/l in homes using private wells, with even higher contributions in high-activity areas.

+   A higher contribution to apartment air is suggested on the presumption that, on the average, apartments have a higher amount of radon-bearing building materials per unit volume than do single-family houses.

BEHAVIOR OF RADON DAUGHTERS INDOORS

Basic Considerations

Even for a given indoor concentration of radon, the concentrations of its daughters (or "progeny") and their physical state can vary substantially. The behavior of the daughters is determined by their fundamental physical and chemical characteristics. Their chemical activity is what distinguishes the daughters substantially from radon itself: the daughters are chemically active and can therefore attach to airborne particles, to indoor (macroscopic) surfaces, and - indeed - to the human tracheobronchial tract, where they can deposit either directly or after attachment to airborne particles. On the other hand, the detailed behavior and health significance of the daughters is influenced greatly by their half lives and decay modes, indicated in Figure 4. The two alpha decays that impart the radiation dose of greatest significance are shaded in the figure and are both the results of decay of polonium isotopes. The amount of (polonium) alpha energy that will ultimately be emitted from an arbitrary mixture of radon daughters in air is uniquely specified by the concentrations of the first three daughters, ^{218}Po, ^{214}Pb, and ^{214}Bi (known as Radium A, B, and C in earlier years). The next daughter, ^{214}Pb, has such a short half life that its activity concentration is, for practical purposes, identical to that of its parent, ^{214}Bi. Finally, ^{210}Pb has such a long half life that it is effectively cleared from indoor air - as well as from the lung - in contrast to the earlier daughters, whose 3, 27, and 20 minute half lives permit their accumulation in buildings with typical ventilation rates and their decay in the lung before they are cleared.

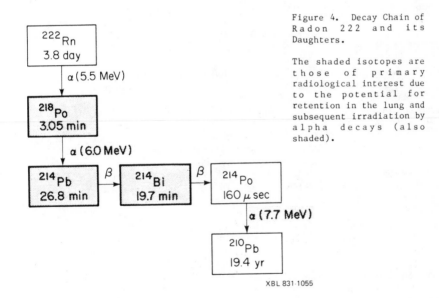

Figure 4. Decay Chain of Radon 222 and its Daughters.

The shaded isotopes are those of primary radiological interest due to the potential for retention in the lung and subsequent irradiation by alpha decays (also shaded).

XBL 831-1055

A useful measure of daughter concentration, the potential alpha energy concentration (PAEC), is therefore determined from the individual daughter concentrations by the expression, PAEC = $K(0.10 \times I_1 + 0.52 \times I_2 + 0.38 \times I_3)$, where I_i is the activity concentration of the ith daughter and the constant K has a value that yields a PAEC of approximately 1 WL (working level) - equal to 1.3×10^5 MeV/l - when all of the daughter concentrations are equal to 100 pCi/l. Alternatively, one can measure the concentration in terms of an equilibrium equivalent daughter concentration (EEDC), which is the concentration that - if attributed to every daughter - yields the same PAEC as the mixture that is actually present. In fact, equilibrium - a condition where all the daughters have the same concentration as radon - would only be attained if daughters were only removed by radioactive decay. It is useful, therefore, to define an equilibrium factor equal to the ratio of EEDC to radon concentration. In the real world, daughters are removed from the indoor air by several mechanisms, so that the equilibrium factor is always less than 1, most frequently in the range 0.3 to 0.7.

Were it not for the chemical activity of the daughters, the departure from equilibrium would depend solely on the ventilation rate. But the fact that the daughters can attach to particles or to surfaces, and that these attachment rates can vary with conditions, makes general characterization of the state of the daughters - and of its dependence on ventilation rate, particle concentrations, and other factors - exceedingly complex. However, since we are dealing only with a few species, whose rate of production from early members of the decay chain is determined solely by known half lives, it is possible to specify a relatively straightforward framework for considering the behavior of the daughters.

Figure 5 illustrates, for an unspecified daughter, various of the mechanisms for changing the state (or presence) of the daughter, other than radioactive decay itself. Because the deposition rates for the daughters depend strongly on whether or not they are attached to particles - and indeed on the particle characteristics - airborne particles play a crucial role in determining the concentrations that are present in the air, and potentially on the radiation dose that results from a given concentration. Given the parameters that are indicated in Figure 5, one can write down a system of mass-balance equations, following Jacobi (29), that determine the concentrations, based on given rate constants, or - conversely - that can determine specific rate constants on the basis of experiments that measure individual daughter concentrations. Practical application of such a theoretical approach usually requires assumptions that simplify the picture. One of the usual simplifications is consideration only of a single well-mixed space. Another is the lack of differentiation of rate constants on the basis of particle size or chemical composition.

These simplifications aside, key issues of interest are the rate of attachment of radon daughters to particles, as well as the rate at which free and attached daughters deposit on walls. (In many cases, deposition is parameterized in terms of the "deposition velocity", which - for a given space - is proportional to the deposition rate.) By way of perspective, in contrast to typical ventilation rates on the order of 1 h^{-1} and daughter radioactive decay constants that are similar (or, in the case of ^{218}Po, one order of magnitude higher), rates of attachment to particles, for typical particle concentrations, appear to be on the order of 50 h^{-1}, with slightly lower rates - perhaps 15 h^{-1} - for plateout of unattached daughters onto interior surfaces. In contrast, rates for deposition onto walls of airborne particles (and of any daughters attached to them) are very low, on the order of 0.1 h^{-1}. As a result, an atmosphere with low particle concentrations tends to have a higher overall rate of deposition onto the walls - because a higher proportion of the daughters are unattached - and a lower equilibrium factor. This condition can, of course, be attained by use of particle-cleaning devices. However, the advantage indicated by the lower equilibrium factors (and hence lower PAEC) may be balanced by the fact that the detailed behavior of unattached daughters in the lung may cause a more significant radiation dose than that associated with attached daughters.

Other Removal Processes:

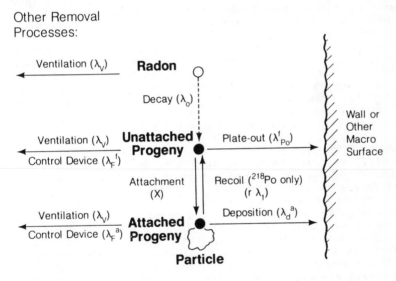

XBL 8311-647

Figure 5. Daughter Removal Mechanisms (Other Than Radioactive Decay) and Associated Rate Constants.

Once created by decay of its parent, a daughter may attach to airborne particles, a process that is usually considered to be reversible for ^{218}Po because of the substantial recoil energy associated with alpha decay. Whether attached or not, a daughter can be removed from the indoor air by plateout/deposition on indoor surfaces, by ventilation, or by processing of the air through an air-cleaning system.

Recent Results

The complexity and importance of radon daughter behavior, as well as the potential interest in air-cleaning as a control technique, has given rise to a substantial amount of work - both experimental and theoretical - on characterization of radon daughters. A great deal of such work is reported in papers in references (1) and (2), as well as in other publications. References 30 - 32 constitute effective reviews. Here it is worth mentioning a few examples of important progress over the last several years.

Experiments in small and room-sized chambers, and related analysis in terms of the Jacobi model, have suggested the values for deposition rates indicated above (33). This and other work has demonstrated that the rate at which "unattached" daughters plate out, while very high as compared with particle deposition rates, is smaller than would occur if the daughters were present in the form of single unattached atoms (which would have a very high diffusion constant). The implication that an unattached daughter is actually a cluster of atoms including a daughter atom appears to be confirmed in experiments that measure the size distribution of daughters: the daughters appear to divide into two regimes, one mode having a size median diameter of about 100 nm, as might be expected based on the size distribution of particles typically present in the room, and a smaller fraction with median diameter in the vicinity of 10 nm, perhaps an order of magnitude greater than the size expected of a single ^{218}Po atom (34).

Considering what is known about the behavior of radon daughters, estimates have also been made of the effect of air cleaning techniques on the radiation dose to the lung. Such estimates suggest that the radical reduction in PAEC that is possible by air cleaning may not cause a corresponding decrease in estimated health effects; it is even possible that there is no decrease at all (35). On the other hand, a detailed review of dosimetric models yields the result that, although the PAEC is an imperfect measure of dose and - ultimately - of health effects, it is still a reasonably good indicator, assuming that parameters are in the normal range (36). These results would seem to suggest that - to the extent that air-cleaning devices result in particle concentrations outside the normal range - there is the potential that PAEC is no longer a good indicator of dose. On a related matter, the effectiveness of generally available particle-cleaning devices is highly variable, as indicated in Figure 6, from ref. 32, where it is found that - although systems based on HEPA filters and electrostatic precipitation can have a substantial particle cleaning rate (and an attendant substantial effect on PAEC) - many devices, especially the small and inexpensive ones that have recently become very popular, can have very small removal rates.

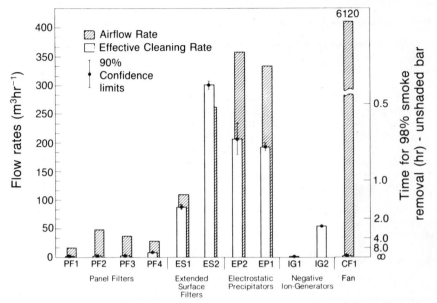

Figure 6. Performance of Various Unducted Air-Cleaning Devices for Removal of Airborne Particulates (generated in these experiments by a cigarette-smoking machine).

Bars give flow rates: shaded is measured air flow rate, unshaded is effective rate at which air is cleaned of particles, calculated as the inflow of particle-free air required to produce the observed decay rate of cigarette smoke. (The right-hand vertical scale indicates 98% clearance times for each device.)

Finally, Figure 5 does not explicitly indicate one of the potentially substantial influences on daughter behavior, i.e., the fact that air within a room moves and that the pattern and rate of air movement can significantly affect plateout rates. Recent advances have made it possible to simulate air movement in an enclosure, thereby removing the simplifying assumption ordinarily used for simulation of radon-daughter behavior, that of a well-mixed room. The more detailed formulation permits treatment of the boundary layer more realistically, thereby providing a basis for determining the manner in which the plateout rate (or deposition velocity) depends on conditions in the room and, particularly, near the wall (37).

The importance, not only of ventilation rate, but also of other measures of air movement, indicates the need for a more complete understanding of the manner in which buildings operate if we are to understand how radon daughters behave. A similar conclusion arises from considering the manner in which radon enters buildings, where it has become clear that the building is not a passive object into which radon diffuses, but actively contributes to the entry of radon indoors.   And, indeed, both of these issues - i.e., radon entry and daughter behavior - are linked to the question of ventilation rate in a much more subtle way than was initially envisioned.   Whereas the ventilation rate might be thought to influence indoor concentrations primarily in providing a means for removal of radon from the building interior, it is now clear that the same factors that account for infiltration affect radon entry decisively and - indeed - comparable factors drive air movement indoors, which has its influence on the behavior of the radon daughters.   In a similar way, consideration of how to control excessive indoor concentrations requires attention to the several factors that determine indoor concentrations and that therefore offer the potential for reducing those that are deemed excessive.

STRATEGIES FOR CONTROLLING INDOOR CONCENTRATIONS

    There is ample evidence that indoor radon concentrations constitute a significant portion of natural radiation exposures. As determined either by direct measurement of the radon daughters or by measurement of radon and assumption of typical equilibrium factors, even average concentrations yield estimated lung cancer rates that equal a substantial portion of the lung cancers that are not associated with cigarette smoking.   Thus the average radon concentration in U.S. dwellings appears to account for thousands of cases of lung cancer annually.   And, considering the apparent distribution of indoor concentrations, a substantial number of homes have very high concentrations.   As noted above, perhaps a million U.S. homes have concentrations causing exposures that exceed the remedial action criterion recently recommended by the NCRP; occupants of these million homes have exposures comparable to those now received by uranium miners.   Faced with a large number of homes apparently in need of help, we must consider how effectively to find these homes and to reduce concentrations significantly.   A related question is how to lower the average exposure of the population, if this seems desirable.

    Control strategies entail two basic elements: 1)   the specific techniques for controlling concentrations, and 2) the framework within which such techniques are applied to homes found to be in need of action.   The control techniques available correspond quite closely to the factors found to affect indoor concentrations.   For several of the major classes of indoor pollution - whether radon, combustion products, or airborne chemicals - these factors may include the source strengths for the pollutants of interest, the ventilation rate (and ventilation effectiveness), and reactions of the pollutants with each other and with the buildings or its contents.   For each pollutant class, concentrations appear to be distributed approximately lognormally, and the largest contributor to the width of the distribution is typically the source strength, but - in each case - variability in ventilation rates and in reaction rates contribute significantly. Thus, the apparent ranking indicated above for the influence of these factors on radon concentrations is not anomalous.

Given our current understanding of the strong variability in radon entry rates and its origin, one technique for controlling indoor concentrations is to minimize entry rates, particularly in cases where they are unusually large. Considering the importance of pressure-driven flow of soil gas into houses through their understructures, substantial attention has been given in recent years to the potential for reducing this flow. It is clear that use of better barriers, sealants, and construction techniques can have a significant effect on the radon entry rate, but this potential appears - in most cases - to be limited, one reason being that slight movements of the house understructure can create small imperfections that appear to be adequate pathways for the entry of soil gas.

An alternative approach that has the potential for reduction of entry rates by large factors is to apply some technique that flushes radon from the region immediately below the house understructure, effectively presenting an alternative pathway for the radon flux from the ground. In certain cases, where the main entry route is highly localized, as through a drain tile and sump system, provision of local venting is highly effective. In the more general case of a basement or slab-on-grade, one or a few pipes that use a small fan to depressurize the soil (preferably gravel) immediately below the concrete floor can strongly reduce the radon entry rate, in effect by reversing the pressure differential that would ordinarily draw soil gas into the home. In the case of crawl spaces, active ventilation of the space below the house can easily be accomplished, although - as noted above - careful sealing of the floor may be quite practical in this case.

For situations where large entry rates are responsible for excessive concentrations, such entry reduction techniques appear to have the greatest potential effect. However, there are also circumstances where increases in ventilation rates are appropriate, whether because the ventilation rate in question is unusually low, because source reduction techniques do not appear effective for the case at hand, or because - in rare cases of extremely high concentrations - an immediate, if only temporary, solution is required. The primary limitation of increased ventilation rates, especially in homes that have very high concentrations, is that reduction of indoor concentrations by large factors will require increases in ventilation rate by large factors, which is often impractical, uncomfortable, or too expensive, at least for the long term. For homes where only modest reductions are sought, ventilation increases are quite practical, including systems that recover energy that would otherwise be lost - either by incorporation of an air-to-air heat exchanger between incoming and outgoing air streams or by recovery of heat from an exhaust ventilation system. (However, an exhaust ventilation system must be used with some caution, since it may result in an increased depressurization of the house that leads to higher radon entry rates.)

An alternative means of control is, of course, use of air cleaning systems to remove radon daughters. The most common of these, as suggested above, employ particle removal techniques such as filtration and electrostatic precipitation. However, although such techniques can substantially reduce the daughter concentrations as measured by the potential alpha energy concentration, their effect on the actual dose to the lung is far from clear. As a result, the radon research community as a whole presently favors employment of source-reduction techniques as the first choice and use of increased ventilation rates as the next option.

The other element in a control strategy, formulation of a framework within which concentrations are controlled, is - if anything - more difficult to define. Even given an adequate understanding of the general occurrence of radon in homes (and other buildings) and of the attendant health implications, such a framework itself involves several interconnected elements. One is some agreement on objectives of the control strategy, including specification of concentration limits or guidelines. A related issue is allocation of responsibility, both for locating houses with excessive concentrations and for implementing the appropriate control techniques. Next is formulation and implementation of the actual scheme for identifying areas and individual houses with high concentrations. And finally is their a logical structure that indicates what control approach should be used in each situation?

Formulating this framework will be no easy task. Parts of it are already being attacked, often helter skelter, but it is important to appreciate that no satisfactory and systematic treatment of the problem of indoor radon will be possible without conscious attention to each of these elements and to their interconnections. Some of the elements, e.g., the formulation of a scheme for finding cases in need of help, depend largely on our growing understanding of the factors that affect concentrations - a primarily scientific question. But other elements, such as that of responsibility, have very substantial social and even political components. As a result, attacking the problem of indoor radon goes far beyond the purely scientific or technical.

CONCLUSIONS

Due to work of recent years, the research community has made substantial progress in understanding the factors that affect indoor concentrations of radon 222 and its daughters. A very wide range of concentrations is present in homes, e.g., in the United States, and these are found to depend, as expected, on source strengths, ventilation rates, and daughter reactions. The highest variability is found in source strengths, which - for single-family homes - are contributed primarily by radon from the soil, but with substantial contributions in some cases from water and building materials. Ventilation rates also affect indoor concentrations directly, although they do not appear to vary as much from house to house as do entry rates. Perhaps the more interesting aspect of ventilation rates is their indirect influence on source strengths and daughter removal because of common factors - such as temperatures and pressures - that affect source strengths, ventilation rates, and daughter behavior. As to the behavior of daughters themselves, considerable effort is being devoted to an understanding of the rates at which they attach to particles or to interior surfaces and to the behavior of "unattached" versus attached daughters.

Corresponding to the relative influence of these factors on indoor concentrations, we have available to us an array of techniques for controlling excessive concentrations, including reduction of the source strength, provision of more ventilation (including use of energy-efficient techniques), and removal of the daughters using systems that clear particles from the air. It appears that source reduction, followed by increased ventilation, has the greatest potential effectiveness. But even more challenging than the development of specific control techniques may be the formulation of an overall strategy within which they may effectively be employed.

This work was supported by the Director, Office of Energy Research, Office of Health and Environmental Research, Human Health and Assessments Division and Pollutant Characterization and Safety Research Division, and by the Assistant Secretary for Conservation and Renewable Energy, Office of Building Energy Research and Development, Building Systems Division, of the U.S. Department of Energy under Contract No. DE-AC03-76SF00098.

REFERENCES AND NOTES

1.   Nero, A.V., and W.M. Lowder, Eds. Indoor Radon, special issue of Health Phys., vol. 45, no. 2, Aug. 1983.

2.   Clemente, G.F., H. Eriskat, M.C. O'Riordan, and J. Sinnaeve, Eds. Indoor Exposure to Natural Radiation and Associated Risk Assessment (Proc. of conf., Anacapri, Italy, Oct. 3-5, 1983), special issue of Rad. Prot. Dos., vol. 7, 1984.

3.   George, A.C., and A.J. Breslin. "The Distribution of Ambient Radon and Radon Daughters in Residential Buildings in the New York - New Jersey Area," in Natural Radiation Environment III, T.F. Gesell and W.M. Lowder, Eds. (Technical Information Center/U.S. Department of Energy Rep. CONF-780422, 1980), pp. 1272-1292.

4.   Nazaroff, W.W., M.L. Boegel, C.D. Hollowell, and G.D. Roseme. "The Use of Mechanical Ventilation with Heat Recovery for Controlling Radon and Radon-Daughter Concentrations in Houses," Atmosph. Env. 15: 263-270 (1981).

5.   Nero, A.V., J.V. Berk, M.L. Boegel, C.D. Hollowell, J.G. Ingersoll, and W.W. Nazaroff. "Radon Concentrations and Infiltration Rates Measured in Conventional and Energy-Efficient Houses", Health Phs. 45: 401-405 (1983).

6.   Alter, H.W., and R.L. Fleischer. "Passive Integrating Radon Monitor for Environmental Monitoring," Health Phys. 40: 693-(1981).

7.   Hildingson, O. "Radon Measurements in 12,000 Swedish Homes," Environment Intl. 8: 67-70 (1982).

8.   Smith, D. "Ventilation Rates and Their Influence on Equilibrium Factor," in Second Workshop on Radon and Radon Daughters in Urban Communities Associated with Uranium Mining and Processing, report AECB-1164 (Ottawa: Atomic Energy Control Board, 1979).

9.   Doyle, S.M., W.W. Nazaroff, and A.V. Nero. "Time-Averaged Indoor Radon Concentrations and Infiltration Rates Sampled in Four U.S. Cities," Health Phys. 47: 579-586 (1984).

10. Nazaroff, W.W., M.L. Boegel, and A.V. Nero. "Measuring Radon Source Magnitude in Residential Buildings", in Radon-Radon Progeny Measurements, report EPA 520/5-83/021 (Washington, D.C.; U.S. Environmental Protection Agency, Office of Radiation Programs, 1983), pp. 101-124.

11. Nazaroff, W.W., F.J. Offermann, and A.W. Robb. "Automated System for Measuring Air-Exchange Rate and Radon Concentration in Houses," Health Physics 45: 525-537 (1983).

12. Nazaroff, W.W., and S.M. Doyle. "Radon Entry Into Houses Having a Crawl Space", Lawrence Berkeley Laboratory report LBL-16637, Berkeley California, 1983, to be published in Health Physics.

13. Nazaroff, W.W., H. Feustel, A.V. Nero, K.L. Revzan, D.T. Grimsrud, M.A. Essling, and R.E. Toohey. "Radon Transport Into a Single-Family House With a Basement," Lawrence Berkeley Laboratory report LBL-16572, Berkeley, California,

1984, to be published in Atmospheric Environment.

14. Cliff, K.D. "Assessment of airborne radon daughter concentrations in dwellings in Great Britain". Phys. Med. Biol. 23: 696-711 (1978).

15. Wicke, A. "Untersuchungen zur Frage der Naturlichen Radioaktivitat der Luft in Wohn- and Aufenthaltstraumen", Ph.D. Thesis, Justus Liebig Universitat, Giessen (1979), in German.

16. Gesell, T.F. "Background Atmospheric ^{222}Rn Concentrations Outdoors and Indoors: A Review". Health Phys. 45: 289-302 (1983).

17. Ingersoll, J.G. "A Survey of Radionuclide Contents and Radon Emanation Rates in Building Materials Used in the U.S." Health Phys. 45: 363-368 (1983).

18. UNSCEAR. Ionizing Radiation: Sources and Biological Effects (New York: United Nations, 1982).

19. Nero, A.V., and W.W. Nazaroff. "Characterising the Source of Radon Indoors," Radiation Prot. Dos. 7: 23-39 (1984).

20. Wilkening, M.H., W.E. Clements, and D. Stanley. "Radon-222 Flux Measurements in Widely Separated Regions," in J.A.S. Adams, W.M. Lowder and T.F. Gesell (eds.), Natural Radiation Environment II (Springfield: NTIS report CONF-720805, 1972), pp. 717-730.

21. Grimsrud, D.T., M.P. Modera, and M.H. Sherman. "A Predictive Air Infiltration Model - Long-Term Field Test Validation," ASHRAE Trans. 88 (Part 1): 1351-1369 (1982).

22. DSMA Atcon Ltd. "Review of Existing Instrumentation and Evaluation of Possibilities for Research and Development of Instrumentation to Determine Future Levels of Radon at a Proposed Building Site," report INFO-0096 (Ottawa, Canada: Atomic Energy Control Board, January 18, 1983).

23. Nazaroff, W.W., S.R. Lewis, S.M. Doyle, B.A. Moed, and A.V. Nero. "Migration of Air in Soil and Into House Basements: A Source of Indoor Air Pollutants," Lawrence Berkeley Laboratory report LBL-18374, in draft.

24. Hess, C.T., C.V. Weiffenbach, and S.A. Norton. "Environmental Radon and Cancer Correlations in Maine," Health Phys. 45: 339-348 (1983).

25. Horton, T.R. "Methods and Results of EPA's Study of Radon in Drinking Water," report EPA 520/5-83-027 (Montgomery, Al.:

U.S. Environmental Protection Agency Eastern Environmental Radiation Facility, 1983).

26. Nazaroff, W.W., S.M. Doyle, and A.V. Nero. "Potable Water as a Source of Airborne Radon-222 in U.S. Dwellings: A Review and Assessment," Lawrence Berkeley Laboratory report LBL-18154, to be submitted to Health Phys.

27. Nero, A.V., M.B. Schwehr, W.W. Nazaroff, and K.L. Revzan. "Distribution of Airborne ^{222}Radon Concentraions in U.S. Homes," Lawrence Berkeley Laboratory report LBL-18274, November 1984, submitted to Science.

28. National Council on Radiation Protection and Measurements (NCRP). Exposures from the Uranium Series with Emphasis on Radon and its Daughters (Bethesda, Md.: NCRP, March 15, 1984).

29. Jacobi, W. "Activity and Potential Alpha Energy of Radon 222 and Radon 220 Daughters in Different Air Atmospheres," Health Phys. 22: 441 (1972).

30. Bruno, R.C. "Verifying a Model of Radon Decay Product Behavior Indoors," Health Phys. 45: 471-480 (1983).

31. Porstendofer, J., "Behavior of Radon Daughter Products in Indoor Air," Radiation Prot. Dos. 7: 107-113 (1984).

32. Offermann, F.J., R.G. Sextro, W.J. Fisk, W.W. Nazaroff, A.V. Nero, K.L. Revzan and J. Yater, report LBL-16659 (Berkeley, Cal.: Lawrence Berkeley Laboratory, February 1984).

33. George, A.C., E.O. Knutson, and K.W. Tu. "Radon Daughter Plateout - I. Measurements," Health Phys. 45: 439-444 (1983); E.O. Knutson, A.C. George, J.J. Frey, and B.R. Koh. "Radon Daughter Plateout - II. Prediction Model," Health Phys. 45: 445-452 (1983).

34. Knutson, E.O., A.C. George, R.H. Knuth, and B.R. Koh. "Measurements of Radon Daughter Particle Size," Radiation Prot. Dos. 7: 121-125 (1984).

35. Jonassen, N. "Removal of Radon Daughters by Filtration and Electric Fields," Radiation Prot. Dos. 7: 407-411 (1984).

36. James, A.C. "Dosimetric Approaches to Risk Assessment for Indoor Exposure to Radon Daughters," Radiation Prot. Dos. 7: 353-366 (1984).

37. Schiller, G.A., A.V. Nero, K.L. Revzan, and C.L. Tien. "Radon Decay-Product Behavior Indoors: Numerical Modeling of

Convection Effects," presented at the Annual Meeting of the
Air Pollution Control Association, San Francisco, 24-29 June
1984.

5.

Comparing Radon Daughter Dosimetric and Risk Models

Naomi H. Harley

New York University School of Medicine, Institute of Environmental
Medicine, 550 First Avenue, New York, NY 10016

INTRODUCTION

Only within the past few years have the data become available
to evaluate the significance of indoor levels of radon.
Occupational exposure standards were established in the 1950's and
were unrealistically high by present day understanding of lung
cancer risk per unit exposure. The usual unit of exposure is the
working level month (WLM) and is still as commonly a reported value
as radon air concentration. There are now four large
epidemiological studies of underground miners which provide the
basic data on lung cancer mortality for particular underground
exposure conditions. The most relevant quantity for assessing any
exposure to radon daughters is the lifetime risk of lung cancer per
unit exposure and this value must be derived from the existing
studies through risk projection models since none of the mining
studies have gone to closure. Also, environmental atmospheric
characteristics differ from those underground and environmental
exposures affect the general population rather than working males.
Therefore, dosimetric models must be considered since the dose is
the unifying factor for comparing effects upon different
populations.

This paper attempts to evaluate the existing dosimetric and
risk projection models which have been devised by radiation
protection authorities in various countries and to comment on their
biological integrity.

DOSIMETRIC MODELING

There are three primary models utilized to determine the alpha
dose delivered by the short-lived daughters of radon. The dose
must be calculated using data concerning atmospheric and biological
parameters. The free-ion fraction and particle size for radon

daughters attached to aerosols are major atmospheric factors and breathing rate and target-cell depth are biological factors.   The three models are described below.

ICRP Model.   The International Commission on Radiation Protection has developed three models which are used in its publication 32 (1). They are, the ICRP Task Group (TG) model, the Jacobi-Eisfeld (JE) model and the James-Birchall (JB) model.   The TG model uses the single tracheobronchial compartment developed for other inhalation exposures and delivers the energy into the entire 45 gram bronchial tree for the calculation of absorbed dose. Interestingly, although it is known that the target cells for cancer induction lie in the few grams of bronchial epithelium which line the airway surfaces, the TG dose is numerically close to the other two model values.

     The JE model utilizes Weibel (2) lung morphometry and deposition fraction calculated with Gormley-Kennedy diffusion equations (3) corrected for turbulent diffusion in the first few upper airways, with factors determined by Martin and Jacobi (4). Mucociliary clearance times are estimated using a mass-balance approach as described by Altshuler et al. (5).   The physical dosimetry utilizes average basal cell depth in six bronchial regions reported by Gastineau et al.(6) and energy deposition from measurements by Harley and Pasternack (7).   It was assumed that both RaA ions and RaA desorbed from particles cross the epithelium to blood.

     The JB model uses the same deposition calculation but without the correction for higher deposition values in the first few airways due to turbulence. The more detailed lung morphometry reported by Yeh-Schum (8) was used for the JB calculations in ICRP 32. The JB model also uses physical dosimetry for average depths of basal cells in six airway groups reported by Gastineau. et al. (6). Energy deposition was calculated from theoretical range energy curves.   Transfer of the daughters to epithelium and blood is assumed.

     Finally, the reported dose for either the JE or JB models is an average dose, with the averaging performed over airway generations 2 to 16.

     The dose factors are calculated for an average particle size of 0.2 - 0.3 μm AMAD and for a fraction of the potential energy, f-pot, existing as free ions of from 0 to 0.05.

     The average bronchial dose per WLM for this range of f-pot is shown in Table 1 for each of the models presented in ICRP 32.

OECD Nuclear Energy Agency.   This group (9) utilized the same JE

Table 1. Alpha Dose Conversion Factors for Occupational and
Environmental Exposures.

		Occupational		Environmental	
ICRP *	JE	0.38 - 0.55	f-pot= 0 - 0.05	–	
	JB	0.30 - 0.90			
	TG	0.32 - 0.42			
OECD *		0.63	f-pot= 0.03	0.41	f-pot= 0.02
NCRP +		0.50	F= 0.04	0.70	F= 0.07

and JB dosimetric models. Weibel morphometric data were also added
in an additional JB model to supplement the original JB Yeh-Schum
model given in ICRP 32.

Several characteristic atmospheres were treated with the JE
and JB models to evaluate the dose for 4 types of mining situations
- low ventilation, medium ventilation, high ventilation and open
pit. Averaging of the 4 mine atmospheres results in an effective
f-pot of 0.03. This is not stated in the report but can be
calculated from the information given. Three particle sizes, 0.1,
0.2 and 0.3 µm were analyzed and the final dose factor is again an
average for these sizes.

In addition to the mine atmospheres, three environmental
atmospheres were treated. These atmospheres had an effective f-pot
of 0.02. Again, this value was not reported but can be inferred
from the calculations.

The average bronchial dose resulting from these calculations
is shown in Table 1.

NCRP. The National Council on Radiation Protection and
Measurements has reported on the dosimetry of mine and
environmental atmospheres (10). The dose is calculated using the
Weibel model for mines and both the Weibel (2) and Yeh-Schum (8)
morphometric model for environmental exposures.

Particle deposition is calculated using Gormley-Kennedy (3)
deposition equations corrected with factors developed by Martin-
Jacobi (3) for turbulance in the upper airways. Mucus clearance
follows values calculated by Altshuler et al. (5). The dose is to
shallow basal cells (22 µm below the epithelial surface) and is
detailed airway by airway. The measurements of transfer to blood
from carrier-free Pb-212 and Bi-212 (11) indicate that removal of

Pb-212 is negligible and an effective half-life for Bi-212 of about 20 minutes. This would decrease the bronchial dose by about 20% and so is not considered important. For purposes of indicating a pertinent cancer-causing dose, airway 4 (generation 4) is considered typical since the majority of lung cancers appear in the first few airway generations. The fraction of potential energy associated with free ions is not used. Instead, the original approach of using the measured free ion fraction, F, was continued (activity of unattached RaA expressed as a fraction of radon activity). For mines an average value of F of 0.04 was used in the modeling based upon measurements of George et al. (12) as was the particle size of 0.18 μm median AMD.

The relationship between free ion fraction, F, and f-pot is shown in Table 2. There is a currently a trend to use f-pot as though it were numerically identical with F. This is not the case and although the exact ratio depends upon daughter equilibrium, F is in general a factor of 4 higher than f-pot.

Table 2. Relationship Between Free ion Fraction (F) and Fraction of Potential Energy as Free Ions (f-pot).

	Values of f-pot in Percent for F = 10%		
	---------------- RaB/RaA ----------------		
RaC/RaA	1	0.6	0.4
1	1.6	–	–
0.6	1.8	2.4	–
0.4	2.0	2.8	3.4

NCRP calculates environmental dose in a manner similar to that in mines except that the median particle size is smaller and free ion fraction larger, 0.12 μm and 0.07 respectively (13).

The doses per WLM for mining and environmental atmospheres are given in Table 1.

DOSE SUMMARY

Although the numerical results of the various calculations are similar, a very disparate set of input parameters are used to arrive at the end points. The dose averaging over the bronchial tree preferred by ICRP and OECD is discouraged by NCRP since lung cancer is primarily bronchial in origin and specific to the first few airway generations.   In fact, it is reported to be specific to the right upper lobe of the bronchial tree (14).

The free ion fractions used in the models are decidedly different.   Measured values of F show a relationship with condensation nuclei concentration (10).   Mines should be dustier in general than home indoor environments and thus averages for this parameter in modeling should reflect this with a lower free ion concentration (or f-pot) for mines than homes.   A median value for F in mines is reported to be 0.04 (f-pot=0.01)and 0.07 (f-pot=0.017) for homes.   It is surprising that the ICRP and OECD models prefer higher values and especially that a higher value of f-pot for homes versus mines can be inferred. The dose factor in mines appears higher than that for homes which is not the case. There is certainly little justification for these assumptions.

The use of the ICRP weighting factor, Wt, to calculate effective dose equivalent is introduced to allow adding different exposures for radiation protection purposes .   Thus, it is not relevant for assessing risk in underground mines.   In particular, the use of equal weighting factors for the tree and pulmonary region is clearly inappropriate.

RISK PROJECTION MODELS

There are several risk projection models that may be used to predict lifetime risk of lung cancer following an exposure to radon daughters.   Various groups including ICRP 32, the Environmental Protection Agency (EPA), UNSCEAR, the BEIR III Committee and NCRP have developed models using the same existing epidemiological studies.   The UNSCEAR and BEIR III models are similar conceptually to the ICRP model, so only ICRP, EPA and NCRP will be described.

ICRP Risk Model.   ICRP 32 calculates that the observed annual excess lung cancer rate (lung cancers per year/WLM per million persons exposed) leads to values of 5 to 15 per million per year per WLM.   This they regard as "the most probable range".   ICRP assumes a 30 year manifestation period for lung cancer and thus their model for total lifetime risk is

30x(5 to 15/1000000)=1.5 to 4.5/10000 persons per WLM

This model is unrealistic in that lung cancer expression does not manifest itself for 30 years but can appear from age 40 to end of life, which is usually taken as 85 years for modeling purposes. Lung cancer must follow a minimum latent interval (the shortest interval between exposure and frank cancer that has been observed) and this is normally considered to be from 5 to 10 years.   The ICRP model projects a uniform rate of cancer appearance and is therefore an absolute risk model. No minimum latent interval is considered.

EPA Risk Projection model. EPA concluded from three studies which appear to show that the temporal pattern of appearance of lung cancer follows the normal occurrance of the disease that a relative risk model is appropriate (15). A relative risk model assumes that the excess lung cancer mortality will increase the natural mortality of the disease in direct proportion to exposure (constant fraction per WLM x age specific mortality). They arbitrarily selected 0.03 as the constant, that is, a 3% increase per WLM exposure. Parenthetically, this is a factor of 10 higher than the best estimate of the relative risk coefficient obtained upon reanalysis of the U.S. underground mining cohort (16).

The results of the EPA model predict 8.6 excess lung cancers/10000 persons per WLM.

NCRP Risk Projection Model. The NCRP developed a modified absolute risk projection model which empirically fits the epidemiological studies (10). The model follows the temporal pattern of appearance of lung cancer with appearance only after age 40, a minimum latent interval of 5 years before potential tumor appearance and an exponential loss of effect with a 20 year half-life from the year of exposure. The base rate is 10 excess lung cancers/1000000 person year per WLM. This model results in 1.0 to 1.5 excess lung cancers/10000 per WLM depending upon age at first exposure. This risk is for occupational exposure to allow comparison with the other two models. Whole life exposure under environmental conditions yields 1.1 excess lung cancers/10000 persons per WLM.

All of the lifetime risk values are summarized in Table 3.

Table 3. Lifetime Lung Cancer Risk Projections Models.

	Lifetime Lung Cancers/10000 Persons. WLM	Model Type
EPA	8.6	Relative Risk
ICRP	1.5 – 4.5	Absolute Risk
NCRP	1.3 – 2.1	Modified Absolute

RISK PROJECTION MODEL SUMMARY

It will not be known for perhaps 20 years which projection models provide suitable numerical values for extrapolation to environmental exposures. Even then this extrapolation may not be verified directly. The lowest exposures reported to produce excess lung cancer in underground miners are about 40 to 80 WLM (17,18).

Average environmental exposure in the U.S. is about 10 to 20 WLM per lifetime.

Model validation is difficult but at least one way in which these values can be anchored to reality is to compare environmental exposure with the known lung cancer mortality in nonsmokers. Whenever more lung cancer is calculated than is observed in nonsmokers, such a model can be considered unrealistic. Choosing the EPA relative risk model, for example, and an average environmental exposure of 0.2 WLM/year yields a lifetime risk of lung cancer in nonsmokers of 0.5% for males and 0.25% for females or at least 1/2 of that actually observed. The NCRP model would predict 0.18% for both sexes or about 1/5 of that actually observed. Consideration of the other carcinogens known to be in the environment, radon alone should not be a sole source of lung cancer. Therefore, the EPA estimates border on fiction.

Unfortunately, the ICRP model does not give equivalent lifetime risk estimates.

A few comments are in order on the biological integrity of the models in general. It is not known whether a relative or absolute type of model is appropriate. However, if a relative risk model were appropriate, the histological type of radiogenic lung cancer should be in the same ratio as in the base population used. This is known not to be the case in radon daughter induced tumors. Simple absolute risk models do not follow the observed temporal pattern of radon daughter induced lung cancer and are therefore inappropriate.

The various models seem to compare reasonably well for the rad/WLM factor. This agreement can only be described as fortuitous, and does not really prove the validity of any particular model. Naturally, I lean to the empirical approach of selecting most probable values for model parameters exemplified by the NCRP models.

ACKNOWLEDGEMENTS

The author would like to thank Aimee Miranda for preparation and typing the manuscript. Support from Core Grants ES 00260 and CA 13343 from the National Institute of Environmental Health Sciences and National Cancer Institute and from Grant AC02 80EV10374 from the USDOE is gratefully acknowledged.

REFERENCES

1. Limits for Intake of Rn-222 and Rn-220 and their Short-Lived Daughters. International Commission on Radiological Protection ICRP Publication 32 (New York: Pergamon Press, 1981)

2. Weibel ER, Morphometry of the Human Lung. (New York: Academic Press Inc 1963)

3. Gormley PC and Kennedy M. "Diffusion from a Stream Through a Cylindrical Tube." Proc. R. Ir. Acad. Sect A 52: 163 (1949)

4. Martin D. and Jacobi W. "Diffusion Deposition of Small Sized Particles in the Human Lung." Health Physics 23: 23 (1972)

5. Altshuler B., Nelson N., Kuschner M. "Estimation of Lung Tissue Dose from Inhalation of Radon Daughters." Health Physics 10: 1137 (1964)

6. Gastineau RM., Walsh PJ., Underwood N., "Thickness of Bronchial Epithelium with Relation to Exposure to Radon." Health Physics 23: 857 (1972)

7. Harley NH and Pasternack BS. "Alpha Absorption Measurements Applied to Lung Dose from Radon Daughters." Health Physics 23: 771 (1972)

8. Yeh HC and Schum M. "Models of Human Airways and their Application to Inhaled Particles Deposition." Bull. Math. Biol. 42: 461 (1980)

9. Dosimetry Aspects of Exposure to Radon and Thoron Daughters. Nuclear Energy Agency/Organization for Economic Cooperation and Development June 1983.

10. Evaluation of Occupational and Environmental Exposures to Radon and Radon Daughters in the United States. Report 78 (Bethesda, MD National Council on Radiation Protection and Measurements, 1984)

11. James AC., Greenhalgh JR. and Smith H. "Clearance of Pb-212 Ions from Rabbit Bronchial Epithelium to Blood." Phys. Med. Biol. 22: 932 (1981)

12. George AC. and Hinchliffe L. and Sadowski R. "Size Distribution of Radon Daughter Particles in Uranium Mines." Am. Ind. Hyg. Assn. J. 36: 4884 (1975)

13. George AC. and Breslin AJ. "The distribution of Ambient Radon and Radon Daughters in residential Buildings in the New York, New Jersy Area."In The Natural Radiation Environment III Gesell TF. and LowderWM. Eds. (Springfield, VA. National Technical Information Service, (1980)

14. Byers TE., Vena JE. and Rzepka TF. "Predilection of Lung Cancer for the Upper Lobes:An Epidemiologic Inquiry." JNCI 72: 1271 (1984)

15. Draft Environmental Impact Statement for Standards for the Control of Byproduct Materials from Uranium Ore Processing (40 CFR 192) EPA 520/1-82-022 (1983)

16. Whittemore AS. and McMillan A. " Lung Cancer Mortality Among U.S. Uranium Miners; A Reappraisal." J. Nat. Can. Inst. 71:489 (1983)

17. Muller J., Wheeler WC., Gentleman JF., Suranyi G., Kusiak R. " Study of Mortality of Ontario Miners." Int. Conf. on Occuptaional Radiation Safety in Mining, Toronto, Canada, October 1984

18. Radford EP., and Renard KG. St. Clair. "Lung Cancer in Swedish Iron Miners Exposed to Low Doses of Radon Daughters." N. Engl. J. Med 310: 1485 (1984)

6.

Epidemiology and Risk Assessment: Testing Models for Radon-Induced Lung Cancer

William H. Ellett

Chief, Bioeffects Analysis Branch, Office of Radiation Programs, Environmental Protection Agency, Washington, D.C.

and

Neal S. Nelson

Senior Radiobiologist, Office of Radiation Programs, Environmental Protection Agency, Washington, D.C.

INTRODUCTION

Modeling the risk of lung cancer due to the inhalation of radon progeny is not as simple as current models imply. Although a number of epidemiological studies of underground miners show a strong association between exposure to radon progeny and excess lung cancer, these data are not directly applicable to a general population. Male miners are not fully analogous to a population composed of both sexes, all age groups, and smokers, never smokers, and former smokers. Given this complexity, it is important to identify the best type of model for the testing of hypotheses on the risks due to radon exposures. A secondary aim of risk models is to estimate the outcome of remedial options for controlling radon levels both in the workplace and in the general indoor environment. As shown below, some models show more promise than others of being worthy of further development.

In this paper, we examine a number of current risk models and their underlying assumptions. We then compare results obtained with these models to epidemiological data on radiogenic lung cancer obtained since the models were developed to test their ability to predict the temporal distribution of lung cancer death observed in underground miners exposed to relatively low levels of radon progeny. Finally, we address some of the difficulties inherent in modeling a general population and suggest some tests that could be applied to the dosimetric models that are being developed by the NCRP and ICRP.

PROJECTING RISKS BEYOND THE PERIOD OF DIRECT OBSERVATION

The manner in which radiogenic lung cancers are distributed in time, after a minimum induction period, is a crucial factor in numerical risk estimates. For radiation-induced leukemia and bone cancer, the period of risk expression is relatively brief; most occur within 25 years of exposure. However, for other radiation-induced cancers, including lung cancer, it appears that people are at risk for the remainder of their lives [1]. None of the epidemiological studies of underground miners provides information on lifetime expression; indeed the majority of those in the study populations are still alive and still at risk. Because lifetime risks cannot be estimated only on the basis of observations to date, a model is needed to project the risk beyond the period of direct observation. As discussed in the 1980 NAS BEIR report, there are two basic models of risk projection; the absolute risk projection model in which it it assumed that the annual numerical excess cancer per unit exposure (or dose) continues throughout life, and the relative risk projection model in which it is assumed that the observed percent increase of the base line cancer risk per unit exposure (or dose) is constant with time [1].

In the case of lung cancer, and most other solid cancers, a relative risk model leads to larger estimated risks because of the high prevalence of this disease at old age. Figure 1 shows the number of lung cancer deaths that occurred in the 1970 U.S. population as a function of age. The decrease in the number of deaths for ages greater than 65 years is due to depletion of the population by competing risks, not a decrease in the age-specific incidence of lung cancer mortality which is relatively constant until age 95 [2]. In this regard, we note that estimates of lung cancer risks should not be based on an average life span, e.g. 70 years. State-of-the-art estimates of the risk of radiogenic cancer should allow for both competing risks and for a full lifetime of risk expression by means of a life table based risk analysis, as outlined in [3].

TYPICAL RISK MODELS

Table 1 illustrates the range of published risk estimates for radon progeny [1, 4-7]. All of these risk estimates are based on a consideration of the same epidemiological studies and a linear non-threshold response. While there may be minor differences in the way a working level month (WLM) is used to characterize the exposure (or the dose), the major difference between these estimates is in the underlying modeling assumptions, not data selection or technical factors. With the exception of the NCRP model, most of the differences between the risk estimates listed in Table 1 are due to the way in which risks are projected beyond the period of direct observation.

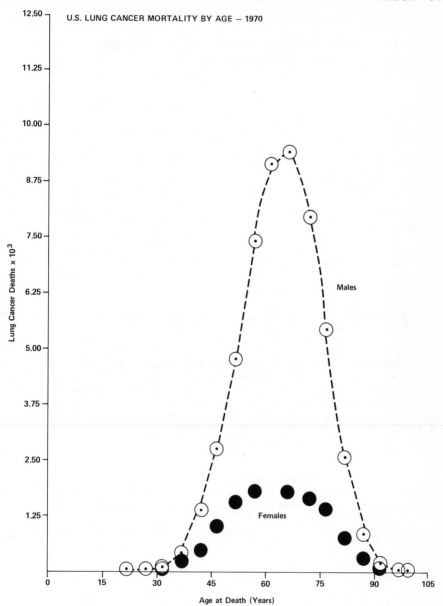

Fig. 1.   The occurrence of lung cancer (ICDA-162) by age and sex
          in 1970.

Table 1

MODELS AND RISK ESTIMATES FOR RADON PROGENY
GENERAL POPULATION

Model		Exposure Period	Duration of Expression	Fatalities per 10^6 person WLM
BEIR-3	Absolute*	Lifetime	Lifetime**	730
AECB	Relative	Lifetime	Lifetime	600***
UNSCEAR	Absolute	Lifetime	40 Years	200-450
ICRP	Absolute	Working Lifetime	30 Years	150-450
NCRP	Absolute	Lifetime	Lifetime****	130

    * Age-dependent - see text.
   ** No expression before age 35.
  *** Adjusted to reflect the sex ratio of the 1970 U.S. population.
 **** No expression before age 40.

    The BEIR-3 Committee's risk model is unique among absolute
risk models in that it takes into account the increase that
occurs in the number of cases of radiogenic lung cancer per
person year at risk as a population ages [8]. The age-dependent
parameters for the BEIR-3 risk model for radon are listed in
Table 2. Unlike the BEIR-3 Committee's absolute risk estimates

Table 2

THE BEIR-3 RADON MODEL

Age-Dependent Parameters for Absolute Risk Projection [1]

Age at Diagnosis (y)	Risk Coefficient (Cases/10^6 WLM-Py)	Minimum Induction Period (y)
≤ 34	0	25 - 15
35 - 49	10	10
50 - 64	20	10
≥ 65	50	10

for low-LET radiation, where the risk coefficient depends on the recipients age at exposure, coefficients for their radon risk model vary with "Age at Diagnosis," page 327 in NAS80. This risk model also assumes age-dependent minimum induction periods. The inclusion of long minimum induction periods before age 35 insures that few radiogenic lung cancers are expressed before age 40. It should be noted that the zero risk shown in Table 2 for those under 35 years of age at exposure does not mean no harm occurs but rather that it is not expressed until the person is more then 35 years old, i.e., only after the minimum induction period. After age 35, the risk coefficient increases at an accelerated rate so that after age 64 it is five times larger than at the age interval 35 to 50. This leads to a comparatively large number of fatalities per 10^6 person WLM, as shown in Table 1. A relative risk model with a risk coefficient of 3% per WLM yields nearly identical estimated risks, (760 fatalities per 10^6 person WLM) [9].

The Atomic Energy Control Board (AECB) relative risk model, listed in Table 1, is based upon a thorough analytical study of the reported lung cancer among five groups of underground miners [4]. This is the only review we are aware of that treated each data set in a consistent fashion and used modern epidemiological techniques, such as controlling for age at exposure and duration of followup. A number of risk models were applied to all of the epidemiological studies that contained enough data to define a dose response function. The authors of the study concluded that a relative risk projection model was more consistent with the excess lung cancer mortality observed in underground miner groups than the other models tested. This study clearly shows that the results for U.S. uranium miners are inconsistent with those obtained in four other studies i.e., underground miners in Czechoslovakia, Newfoundland, Sweden, and Ontario. A regression on combined data from these four studies yielded a relative risk coefficient and standard error of 2.28 ± 0.35 percent excess per WLM. For the U.S. miners the calculated excess per WLM was about half a percent [4].

The AECB estimate for lifetime exposure was for Canadian males, 830 fatalities per million person WLM [4]. In Table 1 this estimate has been adjusted for the 1970 U.S. general population (males and females) in order to make it comparable to the other estimates listed.

Unlike the BEIR-3 and AECB risk estimates, the next two models in Table 1, UNSCEAR and ICRP, are not based on a life table analysis that explicitly considers the full duration of cancer expression. Although the UNSCEAR model is supposedly for lifetime exposure, it assumes that risk expression is limited to a 40-year period [5]. The UNSCEAR estimated, from miner data, that absolute risk coefficient for radon progeny exposure ranges

from about 5 to 10 cases per 10^6 person year WLM [5], hence
the range in Table 1. The ICRP estimate is for occupational
exposure with an expression period of 30 years. Based on
epidemiological data, the ICRP estimated that the absolute risk
coefficient ranges from 5 to 15 cases per 10^6 person year WLM
[6].

The NCRP model has been described in detail by Harley and
Pasternack [10] as well as in an NCRP report [7]. This risk
model uses a life table analysis which explicitly considers
lifetime exposure and lifetime expression. Unlike the other
models listed in Table 1, it is a dosimetric model in which the
dose to the bronchial epithelium is used as the independent
variable in calculating response. This has the advantage that
differences in dose as a function of age, sex, occupational
status, and anatomical site of origin (see below) can be
considered explicitly as additional physiological data become
available. The NCRP uses an absolute risk coefficient of 10
cases per 10^6 person year WLM and a 5-year minimum induction
period. It also assumes that no cases occur before age 40.

A unique assumption in the NCRP model is that the
effectiveness of a given dose of alpha particle radiation
diminishes exponentially with time (half time = 20 years).
While this is not a controlling factor in estimates of occupational risk, it is important for estimating the risk due to
lifetime exposure. Our analyses with the NCRP model indicate
that the estimated lifetime risk is reduced by a factor of 2.4
when the dose accumulated over a lifetime is discounted in this
manner. We are unaware of any evidence that the risk of radiogenic lung cancer decreases with time. A reduction in the risk
of lung cancer as a function of time has not been observed in
the A-bomb survivors. Rather, these data show that the absolute
risk of lung cancer increases with time, post exposure [8]. In
contrast, the relative risk remains fairly constant.
Nevertheless, it should be appreciated that the exponential
reduction in risk with time proposed in the NCRP model is not an
arbitrary assumption but rather an attempt to account for the
large risk that is observed in miners who start underground work
at middle age [10]. This age-dependence can also be accounted
for by a risk coefficient that increases with age, as in the
BEIR-3 model, or by a relative risk model.

A COMPARISON BETWEEN OBSERVATION AND PREDICTION

All of the models in Table 1 provide reasonable agreement
with the lung cancer risk observed, to date, among underground
miners. However, the proper test of a model is how well it
predicts the results of observations that are independent of the
data used in its development.

Recently, Radford and St. Clair Renard have reported the results of a study of iron miners in Malmberget, Sweden which we believe provides useful data for testing models of the type listed in Table 1. The study is described in reference [11]. The Malmberget miner population has a mean follow-up period of 44 years which is much longer than in other studies of underground miners. All of these iron miners were born between 1880 and 1920 and have been followed until 1977. In addition to the long follow-up time and old age of this population, ascertainment of cause of death is close to 100 percent as is information on smoking habits for those dying of lung cancer. In Appendix I, we present the data on the Malmberget lung cancer cases that we have used in this analysis, as supplied by Drs. Radford and St. Clair Renard.

While the Radford - St. Clair Renard study has many useful attributes, it is not ideal. The number of miners at risk is only 1415 of whom about half were still alive in 1977. Fifty-three lung cancer deaths had occurred by 1977; 18 among nonsmokers. Since then, another 12 miners have died of lung cancer. We hope that this study will be continued until almost all of the miners have died as it is the only study we are aware of that can provide information on lifetime risk within the foreseeable future.

A realistic risk prediction model should be able to predict the age-dependence observed in the occurrence of radiogenic cancer due to radon progeny exposures. The analysis we have used is like that of Land and Norman who showed that the temporal pattern of radiogenic lung cancer death among highly irradiated A-bomb survivors was the same as that for nonradiogenic lung cancer and concluded that this was indicative of a relative risk model [12]. Most lung cancers normally occur very late in life. The probablity of death due to this disease varies with age as is shown in Figure 2 for the 1970 U.S. male population dying at the 1970 age specific lung cancer mortality rates, c.f. Figure 1.

Figure 3 shows the probability of death due to lung cancer before a given age for the Swedish iron miners described in Appendix I. Data for smokers and nonsmokers are shown separately. While there is a suggestion that lung cancer deaths among smokers occur at an earlier age, application of a two tailed Kolmogirov-Smirnov (K-S) test, indicated the difference shown in Figure 3 for smokers and nonsmokers is not significant. We have combined these two set of data in the comparisons described below.

Using the life table analysis described in [3], we have calculated the pattern of lung cancer deaths for each of the risk models listed in Table 1. We have modeled the first 51

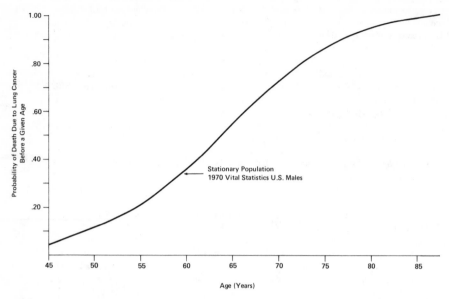

Fig. 2.   The probability that a lung cancer death occurs before a
given age in a stationary population of U.S. males – 1970
U.S. Vital Statistics.

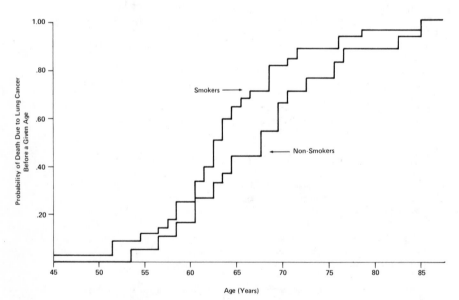

Fig. 3.   The mortality experience of the Malmberget iron miners
who died of lung cancer before 1977.

cases in Appendix I by assuming a cohort of miners who started
underground work at age 24 and were exposed, through age 51, to
a uniform concentration of radon progeny corresponding to the
Malmberget miners average annual exposure. For the lung cancer
cases, Appendix I indicates the mean starting age was 25.3
years, first, second, and third quartiles are 21, 24.0, and 30
years respectively. The mean duration of exposure for these
cases was 26.9 years, median, 28.5 years. Relative risk
estimates were calculated using U.S. data. We hope to repeat
these calculations at a later date when we receive Swedish vital
statistics. However, we do not expect much change in the
temporal distribution of lung cancer deaths since the ratio of
age-specific lung cancer rates in the two countries are similar
[13]. (See note A in Appendix I.)

Figure 4 compares the probability of death before a given
age, assuming death from lung cancer as calculated with a
relative risk model, to that observed in the Malmberget underground miners. We have tested the null hypothesis that there is
no difference between these data sets by means of the K-S test.
The difference between the two cumulative distributions shown in
Figure 4 is not significant.

Figure 5 shows a similar comparison using the BEIR-3
age-dependent absolute risk model outlined in Table 2. Although

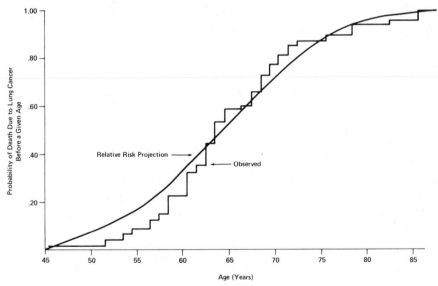

Fig. 4.   A comparison of the Malmberget miner's lung cancer
mortality experience with that predicted by a relative
risk projection model.

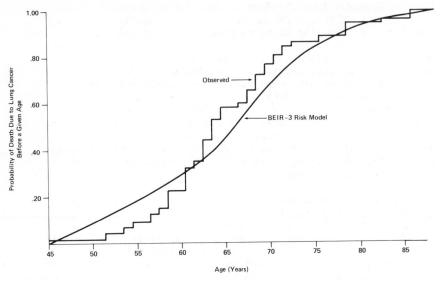

Fig. 5.   A comparison of the Malmberget miner's lung cancer
mortality experience with that predicted by the BEIR-3
age-dependent absolute risk projection model.

this model appears to account reasonably well for the observed
data, the K-S test indicates that if the null hypothesis is
true, differences as large as those shown in Figure 5 would
occur with a probablity of less than 0.05.

The temporal pattern of predicted lung cancer deaths is
quite different for absolute risk models. In Figure 6, we have
compared the Malmberget observations to results obtained with an
absolute risk model having a risk coefficient of 10×10^{-6}
lung cancer deaths per person year WLM, a 5-year minimum induction period, no deaths before age 40 and, unlike the NCRP model,
no reduction of dose effectiveness with time. In this case,
there is less than a one percent chance the null hypothesis is
true. While an absolute risk model with a 10-year minimum
induction period provided a slightly better fit, it also fails
at the one percent level.

The poorest congruence between observations and calculated
results is found for the NCRP risk model, Figure 7. The
exponential repair assumed in this model does not improve the
fit but rather has increased the difference between observed and
calculated results, c.f. Figure 6. Again, the probability that
the null hypothesis could be correct is less than 1%.

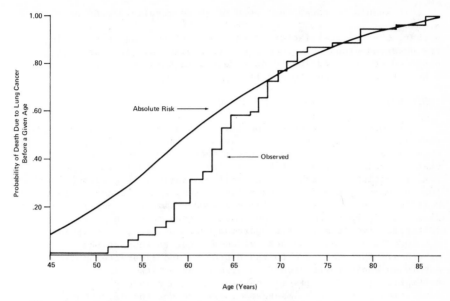

Fig. 6.   A comparison of the Malmberget miner's lung cancer
           mortality experience with that predicted by an absolute
           risk projection model.

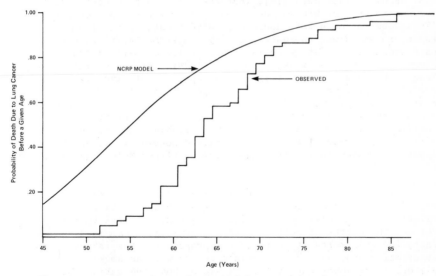

Fig. 7.   A comparison of the Malmberget miner's lung cancer
           mortality experience with that predicted by the NCRP
           risk model.

It should be noted that none of these results depend on the accuracy of the estimated exposures to the Swedish iron miners. Rather, the critical factor is that most of the lung cancers in the observed group were indeed due to radiation. The ratio of observed to expected lung cancer deaths in this group of Swedish miners is nearly four to one [11].

CONFOUNDING DUE TO CIGARETTE SMOKING

Although these results indicate that a relative risk model provides a satisfactory match between the temporal pattern of observed and calculated lung cancer deaths, they do not mean that the relative risk estimate in Table 1 is wholly adequate for estimating the risk to a general population. Neglecting, for the moment, any differences due to sex, a single coefficient relative risk model would provide a good estimate of population risks if cigarette smoking and radon progeny acted only as multiplicative (synergistic) carcinogens. However, this appears not to be the case. Current evidence tends to support the hypothesis that exposure to radon progeny and cigarette smoking act in an additive fashion to a greater extent than as multiplicative, synergistic, carcinogens. Absolute risk coefficients for the Malmberget miners are not too different, 22 versus 16 cases per 10^6 person year WLM for smokers and nonsmokers, respectively [11]. This result appears to confirm the earlier observation by Lundin, based on an analysis of the U.S. uranium miner data [14], that the relative risk coefficient for nonsmokers per WLM is considerably larger than for smokers. (This does not mean that nonsmokers have greater risks from radon progeny since the baseline lung cancer mortality rate of male nonsmokers relative to male smokers is so small that a nonsmokers numerical risk, as opposed to his percentage risk, is less.) Radford and St. Clair Renard report a relative risk coefficient for male smokers of 2.4% per WLM and for male nonsmokers, 10.7% per WLM. As mentioned above, their sample of nonsmokers is small so that this estimate of the relative risk coefficient for nonsmokers may not be reliable. Ascertainment of lung cancer mortality among long-lived nonsmokers is another important reason for continuing follow-up of the Malmberget iron miners.

A relative risk coefficient for radon progeny that is dependent on smoking history complicates the application of relative risk models. Age-specific lung cancer mortality rates reflect the deaths of both smokers and nonsmokers combined. Factoring this mortality data into appropriate rates is not straightforward, particularly since sex as well as smoking history is an important factor. Table 3 shows the ratio of lung cancer mortality for U.S. smokers relative to nonsmokers [15]. The ratio increases with the amount smoked per day but is apparently different for males and females.

Table 3

LUNG CANCER MORTALITY RATIOS
SMOKERS RELATIVE TO NONSMOKERS

Males			Females	
Cigarettes/day	(V.A.)	(ACS)	Cigarettes/day	(ACS)
<10	3.89	4.62	<10	1.3
10 - 20	9.63	8.62	10 - 19	2.4
21 - 39	16.70	14.69	20 - 39	4.9
40+	23.70	18.71	40+	7.5

(V.A.) Veterans Administration, (A.C.S.) American Cancer Society,
data reprinted from [15]

Moreover, there are strong cohort effects in cigarette use, as shown in Figure 8, that are sure to be reflected in age specific cancer mortality rates [16]. Currently, about 60% of the adult male population are nonsmokers, compared to over 50% in 1964 and less than 30% among the U.S. uranium miners studied by Lundin, Archer, and Wagoner [17]. In contrast, cigarette smoking among young women is increasing and now exceeds the rates for males of a comparable age [16]. Obviously, projecting deaths that will occur in the future due to exposures to radon progeny is not a simple task.

Another major problem is the lack of any data on the sensitivity of women and children to radon progeny. The A-bomb survivor data give some indication that for external low-LET radiation, there may not be a large sex-related difference [18]. We do not know if this is actually the case for inhaled particulates. Table 3 indicates less difference in lung cancer mortality between smoking and nonsmoking females than for their male counterparts, who smoke the same amount. This could be due to a greater amount of cancer initiation occurring in males due to their occupational exposures to other inhalable carcinogens as well as differences in smoking habits. A comparison of the sex-specific information in Figures 1 and 8 also leads to the conclusion that differences in smoking history alone cannot account for the much lower incidence of lung cancer in females. Moreover, among never smokers, the age-specific lung cancer mortality of females is less than half that of males [19].

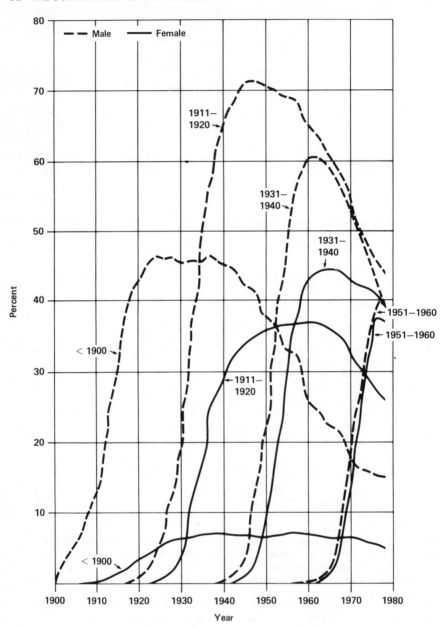

Fig. 8.   Prevalence of Cigarette Smoking in birth cohorts from
1900 to 1978 (GS80).

Projecting the risk to females due to radon on the basis of observations on males is unlikely to be as accurate as one would like.

DOSIMETRY AND RISK ASSESSMENT

Another problem that needs more consideration is the difference in the dose distribution in the various airways of the lung. As mentioned above, all of the risk estimates in Table 1, except that of the NCRP, are based on exposure to radon progeny in WLM rather than the dose in rads. The ICRP, OECD, and NRPB as well as the NCRP are working towards making the alpha dose to basal cells in the respiratory epithelium a useful basis for assessing radon risks. Assuming, for the moment, that basal cells are the correct target cells for induction of radio-genic cancer, the calculation of dose should include: the depth of basal cells, i.e. how far they are from the lumen of the airway; fractional deposition of radon progeny in airways, by amount, location, and interaction with mucous; amount of radon progeny inhaled, considering respiration rate, tidal volume, mouth or nose breathing; the fate of unattached ions in the nasal passages; and the distribution of radon progeny in the mucous and in the epithelium itself. Given the choice of variables and recalling that pertinent physiological data in this research area is scarce, it is not surprising that estimates of the dose per WLM show considerable variation, Table 4.

The conversion factors listed in Table 4 are based on such a wide variety of modeling assumptions (the original papers should be consulted for details) that results are not fully comparable [10, 20-23]. In spite of this, there is a common thread in the calculated results: (1) most models yield a dose in the main stem bronchi which is less than the dose in the subsequent bronchi, (2) in general, the dose in segmental bronchi is as high or higher than doses in other bronchi and (3) peripheral bronchi and bronchioles have relatively high doses compared to the main stem dose.

If the calculated dose conversion factors actually reflected the potential risk per WLM exposure, then we would expect the greatest number of radon-progeny-induced lung cancers to occur in the segmental bronchi with lobar bronchi having a similar or perhaps fewer number. Peripheral bronchi would also have an appreciable number of cancers and main stem bronchi the smallest number. These predictions are at variance with the reported sites of origin of lung cancers in underground miners. Although these data are rather sparse, the distribution of reported sites of origin, as illustrated in Figure 9, indicates that the majority of radiogenic lung cancers occur in the main stem and lobar bronchi with slightly more in main stem. In any

Table 4

ESTIMATED DOSE CONVERSION FACTORS FOR EXPOSURE
TO INHALED RADON PROGENY

| | Dose to Bronchial Basal Cell Nuclei (Rad/WLM)* | | | |
| | Conducting Airways | | | |
	Main Stem	Lobar	Segmental	Peripheral
Harley and Pasternack [10]				
Working Males	0.36-0.41	0.16-0.95	0.24-0.58	0.09-0.55
James et al. [20]**				
Miners-				
Moderate Work	0.09	0.5-0.6	0.7-0.9	0.4-0.9
Heavy Work	0.1-0.2	0.9-1.4	1.3-1.5	0.8-1.5
Working Males	0.2	0.8	1.1	1.2
Jacobi and Eisfeld [21]**				
Working Miners	0.3	0.4-0.6	0.4-0.50	0.1-0.6
Harley [22]				
Working Miners	0.43-0.58	0.18-1.10	0.29-0.57	0.06-0.37
Hofmann [23]**				
Males				
Moderate Work	0	0.02-0.07	0.2-0.4	0.3-0.5
Heavy Work	0	0.04-0.1	0.4-0.9	0.5-1.1

 * Range is for different airways in same generation.
** Numerical values estimated from graphs.

case, the segmental bronchi are not a prominent site of origin
even though the estimated dose is highest in this region.

The data on the site of origin of lung cancers in
underground miners are primarily from autopsy reports, with
x-ray confirmation of diagnosis in most cases and biopsy and
resection in a few cases [24-26]. In Table 5, the observations
in underground miners are compared with results in selected
reports from the medical literature with complete or nearly
complete ascertainment of the site of origin for nonradiogenic
cancers [27-30]. There is apparently little difference in site
prevalence between radiogenic and nonradiogenic lung cancers
when both estimates are derived from autopsy data.

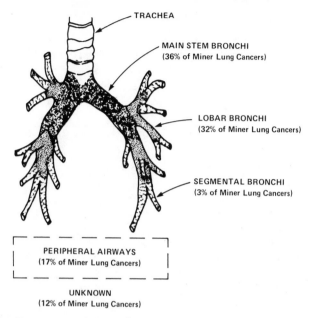

Fig. 9.   The reported location of lung cancers in underground uranium miners.

It should be noted that the cases identified after lung resection, Table 5, are not fully representative. Lung cancer of the more central bronchi, especially main stem bronchi, is usually less amenable to resection and would not be included in the series. In addition, patients undergoing resection are selected on clinical grounds [30]. The bronchographic series are also less than completely representative. There are a number of clinical circumstances in which production of a bronchogram is contraindicated; e.g., severe respiratory impairment, severe recent hemoptysis, fever, acute pneumonia, severe heart failure [27]. To what extent these exclusions may have influenced the results is unknown. Therefore, we believe the best group in Table 5 for comparison with the miner data is the autopsy data.

The data described above suggests that either the dosimetric model or the target cell for which the dose is calculated, or perhaps both, need further consideration. There are many questions about dosimetric models for lung tissue. D.R. Fisher, In Search of the Relevant Lung Dose, pp. 29-36; and W. Hofmann, Dosimetric Concepts for Inhaled Radon Decay Products in the Human Lung, pp. 37-43; in Current Concepts in Lung Dosimetry, PNL-SA-11049, Pacific Northwest Laboratory, U.S. DOE,

Table 5

REPORTED SITES OF ORIGIN OF LUNG
CANCERS IN SELECTED STUDIES
% Distribution

Site	Underground Miners	Diagnosis Based on				
		A [29]	A [28]	B [27]	R [28]	B*
Main Stem Bronchi	36	32	29	7	3	15
Lobar Bronchi	32	46	55	46	23	55
Segmental Bronchi	3		0	39	54	23
Peripheral	17	12	16	8	20	7
Unknown	12	9	0	0	0	0
Number of Cases	59	956	206	565	66	388

A - Autopsy
B - Bronchography
R - Resection

---

* Molnar, W. "Bronchography in Bronchogenic Cancer" (cited by
Garland et al., [30]).

1983 have provided recent reviews. Attenuation of the alpha
particles by the mucous in the airways may be particularly
important [31]. Smoking, particularly heavy smoking, is
associated with chronic bronchitis and increased mucous
production. Although, in this paper, we have compared risk
coefficients for smokers and nonsmokers in terms of WLM, the
dose per WLM may be quite different in these two cases. Animal
as well as human studies suggest the dose is less per WLM in
smokers [11, 14, 32, 33].

In Table 6, we have examined the depth of basal cell nuclei
in different airways [34, 35]. Again, we note that the cells
receiving the highest dose are in regions where few radiogenic

Table 6

FRACTION OF BASAL CELLS IN VARIOUS AIRWAYS AS A
FUNCTION OF DEPTH OF BASAL CELL NUCLEI IN THE TISSUE [34]

	Distance from Surface of Mucous (micron)			
	< 30 μ	< 50 μ	< 70 μ	> 70 μ
Main Stem	-	-	-	1.00
Lobar	.10	.40	.90	.10
Segmental	.06	.41	.79	.21
Peripheral	.50	1.00	-	-

n.b.
Range of Radium A, alpha is ∿47 μ.
Range of Radium C, alpha is ∿71 μ [35].

lung cancers are reported. It is of interest that the main stem basal cells are at a depth greater than the maximum range of alpha particle from radon-222 progeny, unless these progeny diffuse rather rapidly through the mucous or even into the epithelium before they decay. For these reasons, we believe that calculations of dose should not be restricted to basal cells alone. The list of potential target cells now includes mucous cells, neurosecretory cells and Clara cells as well as basal cells [36]. Recently, it has been shown that in normal hamster tracheal epithelium, dividing mucous cells are twice as prevalent as dividing (mitotic) basal cells and after mechanical injury about 14 times more prevalent [37]. It may be that the basal cell is one of the lessor contributors to lung cancer induced by radon progeny. The pattern of radiogenic cancer in the different airways, Figure 9, and the distribution basal cells in tissue lead to the conclusion that the dose estimates in Table 4 are of a preliminary nature.

CONCLUSION

In this paper, we have tried to develop two complementary ideas. First, both dosimetric and risk projection models should be compared with biological and clinical observations that are independent of the model's origins. Secondly, the partial failure of a proposed model to predict observations should be viewed as an opportunity to improve the model.

The current interest in developing realistic dosimetric
models for inhaled radon progeny is long overdue. It is obvious
that the working level and working level month are of limited
usefulness for characterizing the insult to a diverse population. Moreover, an adequate evaluation of the dose from alpha
particles in bronchial tissue has the potential for predicting
an RBE for human lung cancer. Until this is accomplished, the
large body of experience gained from observations on underground
miners cannot be integrated into the information base derived
from the A-bomb survivors and other groups exposed to low-LET
radiations. Finally, we believe a new generation of relative
risk models that estimate risk on the basis of sex specific
incidence data and which also take into account smoking history,
has the best promise of providing good predictive models for the
general population.

The authors would like to acknowledge the technical
assistance of James T. Walker, Robert E. Sullivan, and
You-yen Yang, and the careful preparation of the manuscript
by Rose M. Henderson

APPENDIX I - NOTE A

In this paper we have examined the intrinsic consequences of differing modeling assumptions and not commented on the absolute value of the risk estimates in Table 1. It is obvious that absolute and relative risk models will yield the same numerical results (in terms of fatalities per 10^6 person WLM) when based on observations for the entire life span of a cohort. The relevant question is, given the information available now, which model (and risk coefficients) provides a better estimate of lifetime risk? In this regard, it should be noted that male lung cancer incidence in Sweden is considerably smaller than that in the U.S., presumably because of different smoking habits. Radford and St. Clair Renard report a relative risk coefficient of 3.6% increase per WLM for the Malmberget miners (smokers and nonsmokers) compared to Swedish national lung cancer mortality rates [11]. In the U.S., the national rates are considerably larger and the ratio of observed to expected cases would be smaller. This leads to a smaller relative risk coefficient for the U.S. male population if it is assumed, as in BEIR-3 [1], that the number of excess cases is transferable between populations having differing cultural characteristics.

Unfortunately, we do not have Swedish age-specific lung cancer mortality rates available to us, but we do have the 1980 Swedish age-specific incidence data, courtesy of the Attache for Work Environment at the Swedish Embassy in Washington, D.C. [13]. Assuming lung cancer incidence is the same as mortality, it is only slightly larger, we have calculated a relative risk coefficient based on U.S. male lung cancer mortality rates and the observed and expected excess in the Malmberget miners. The increase per WLM for a U.S. male population is 2.1%, which is about the same as the 2.4% per WLM reported in Th82. We note, however, that the mixture of smokers and nonsmokers and, hence, the relative risk for combined groups will differ between populations. Therefore, a risk coefficient for U.S. males of 2.1% per WLM is at best only a first order approximation. As indicated in the main text, specific consideration of sex and smoking history is likely to provide better estimates.

APPENDIX I (cont.)

E.P. Radford and K.G. St. Clair Renard
Detailed information on lung cancer cases -- 1951-76
as described in the N. Engl. J. Med. 310:1485-94, 1984

Case No.	Yr. of Death	Age at Death	Yr. Began	Total Years	Chief Mine[1]	Cum. Dose WLM	Dose Excl[2] WLM	Smoking Status[3]
				Underground				
1	1952	62	1920	27	Ko	47	0	Cur.
2	1955	71	1912	29	He	159	0	Cur.
3	1955	67	1912	17	O	75	0	Never
4	1956	69	1919	32	Ka	135	5	Ex 1909
5	1957	53	1925	33	Ko	48	10	Never
6	1959	66	1914	9	Ti	43	0	Cur.
7	1960	61	1919	24	Ti	137	0	Cur.
8	1960	60	1917	15	He	89	0	Ex 1923
9	1960	56	1936	23	Ko	34	9	Never
10	1963	61	1930	6	Ti	36	0	Cur.
11	1963	60	1927	34	He	186	11	Cur.
12	1964	69	1918	13	Ti	62	0	Ex 1935
13	1964	67	1923	38	Ka	191	14	Ex 1930
14	1965	72	1920	29	Ko	50	0	Ex 1930
15	1965	60	1927	35	Ti	210	11	Cur.
16	1965	54	1937	27*	Ti*	120*	16*	Cur.
17	1965	51	1937	26	Ti	159	21	Cur.
18	1966	56	1937	30	Ko	45	11	Cur.
19	1966	51	1936	27	Ti	157	19	Cur.
20	1968	71	1927	33	Ti	179	0	Cur.
21	1968	63	1928	6	He	32	0	Cur.
22	1968	62	1927	33	Ko	53	10	Never
23	1968	60	1928	30	Ti	171	16	Cur.
24	1970	63	1937	31	Ko	52	6	Ex 1939
25	1970	58	1932	24	O	118	14	Cur.

1. Code: Ti, Tingvallskulle; He, Hermelin; Ka, Kapten; Ko, Koskoskulle; O, Other (no specific mine, e.g., timberman, railroad worker, maintenance worker).
2. Dose excluded from exposure less than five years prior to death.
3. Smoking status: Cur., current smoker at death; Ex, gave up smoking in year shown; Never, never smoked.

*   Left Malmberget after 9 yrs. underground and worked 18 yrs. underground in Grangesberg Exportfalt mine. Dose for this period assumed to be 72 WLM (4WLM/yr), of which 16 were excluded.

APPENDIX I (Cont.)

Case No.	Yr. of Death	Age at Death	Yr. Began	Total Years Underground	Chief Mine[1]	Cum. Dose WLM	Dose Excl[2] WLM	Smoking Status[3]
26	1971	82	1925	28	Ti	186	0	Ex 1920
27	1971	76	1927	9	He	54	0	Ex 1940
28	1971	75	1917	44	He	296	0	Never
29	1971	68	1924	23	Ko	40	0	Cur.
30	1971	64	1956	13	Ti	58	16	Cur.
31	1972	85	1917	29	Ti	164	0	Never
32	1972	64	1929	33	Ti	189	15	Cur.
33	1972	63	1937	28	Ti	169	13	Cur.
34	1972	63	1937	34	Ka	156	22	Cur.
35	1973	85	1927	14	Ti	70	0	Cur.
36	1973	70	1929	28	He	199	0	Never
37	1973	67	1920	46	Ti	281	0	Ex 1970+
38	1973	62	1927	32	Ti	186	0	Ex 1971+
39	1973	62	1929	32	Ti	162	9	Cur.
40	1973	57	1951	22	O	87	14	Cur.
41	1974	68	1928	28	Ti	188	0	Cur.
42	1974	64	1937	28	Ti	190	0	Never
43	1974	58	1949	23	Ka	82	19	Ex 1956
44	1974	58	1937	33	Ti	209	4	Cur.
45	1975	76	1928	16	O	74	0	Cur.
46	1975	68	1928	28	He	200	0	Ex 1970+
47	1975	60	1938	34	O	166	4	Ex 1939
48	1975	58	1939	35	Ko	77	8	Ex 1971+
49	1976	70	1925	42	Ko	77	0	Cur.
50	1976	68	1925	36	O	203	0	Cur.
51[#]	1964	45	1955	9	Ti	26	26	Cur.
52[#]	1967	62	1936	29	Ti	159	14	Cur.
53[#]	1976	78	1919	37	He	253	0	Cur.

+   Recent ex-smokers. Counted as current smokers.
#   Cases 51, 52, and 53 are excluded by Radford and St. Clair Renard.

APPENDIX I (cont.)

Case No.	Silicosis Grade[4]	Diagnosis
1	±	Bronchial ca, left, no biopsy.
2	3	Bronchial ca, RLL, biopsy: polymorphic cells.
3	0	Oat cell ca.
4	1	Undiff. large cell ca.
5	1	Undiff. large cell ca.
6	0	Ca of lung. Bronchoscopy: "malignant cells."
7	0	Pleural cancer (coded ICD 163).
8	0	Ca mediastinum, metast. to lung (coded ICD 163).
9	0	Oat cell ca.
10	0	Cancer, left lung, metastatic.
11	1	Undiff. ca of bronchus.
12	2	Mod. diff. epidermoid ca.
13	±	Poorly diff. epidermoid ca.
14	0	Poorly diff. epidermoid ca.
15	2	Poorly diff. oat cell ca.
16	0	Slightly diff. epidermoid ca.
17	0	Poorly diff. possible oat cell ca.
18	0	Mod. diff. epidermoid ca.
19	0	Undiff. oat cell ca.
20	0	Poorly diff. epidermoid ca.
21	1	Undiff. oat cell ca.
22	1	Undiff. oat cell ca.
23	0	Poorly diff. large cell epidermoid ca.
24	1	Bronchial ca. Bronchoscopy: atyp. squam. epith.
25	0	Ca of lung. Sputum: "malignant cells."

4. Silicosis grades by x-ray opacities; ±: questionable; 1: diffuse 1.5 mm; 2: diffuse 1.5-3.0 mm; 3: conglomerate nodules.

APPENDIX I (Cont.)

Case No.	Sili- cosis Grade[4]	Diagnosis
26	0	Poorly diff. epidermoid ca.
27	0	Anaplastic oat cell ca.
28	0	Mod. diff. epidermoid ca., metastatic.
29	1	Poorly diff. adenocarcinoma.
30	0	Oat cell ca.
31	0	Poorly diff. epidermoid ca.
32	0	Mod. diff. adenocarcinoma.
33	0	Bronchia ca, right. Sputum: malig. squam. cells.
34	1	Anaplastic oat cell ca.
35	0	Mod. diff. epidermoid ca. Healed tuberculosis.
36	1	Epidermoid ca. Inactive tuberculosis.
37	1	Mod. diff. epidermoid ca., metastatic.
38	0	Poorly diff. oat cell ca.
39	0	Anaplastic oat cell ca. metastatic.
40	0	Oat cell ca.
41	$\overline{+}$	Mod. diff. epidermoid ca.
42	0	Undiff. oat cell ca.
43	0	Undiff. oat cell ca.
44	0	Poorly diff. oat cell ca, metastatic.
45	$\overline{+}$	Poorly diff. epidermoid ca. Also ca of stomach.
46	2	Mod. diff. epidermoid ca.
47	0	Mod. diff. epidermoid ca, metastatic.
48	$\overline{+}$	Fusiform small cell ca, WHO group 2.
49	0	Mod. diff. epidermoid ca.
50	0	Poorly diff. epidermoid ca.
51[#]	0	Mod. diff. epidermoid ca. Died 10 yrs. after start.
52[#]	0	Bronchial ca. Path: metastatic from intest. (coded ICD 199).
53[#]	1	Oat cell ca of lung, metastatic (coded ICD 199).

\# Cases 51, 52, and 53 are excluded by Radford and St. Clair Renard.

References

[1]   National Academy of Sciences - National Research Council,
       "The Effects on Populations of Exposure to Low Levels of
       Ionizing Radiation," Committee on the Biological Effects
       of Ionizing Radiation, (Washington, D.C.: National
       Academy of Sciences, 1980).

[2]   National Center for Health Statistics. Public Use Tape,
       Vital Statistics - Mortality, Cause of Death Summary -
       1970, PB80-133333, National Technical Information
       Service, Washington, D.C., 1973

[3]   Bunger, B., J. R. Cook, and M. K. Barrick. "Life Table
       Methodology for Evaluating Radiation Risk: An
       Application Based on Occupational Exposure," Health
       Physics 40:439-455 (1981).

[4]   Thomas, D. C. and K. G. McNeill. "Risk Estimates for the
       Health Effects of Alpha Radiation," Report INFO-001,
       (Ottawa, Canada: Atomic Energy Control Board, 1982).

[5]   United Nations Scientific Committee on the Effects of
       Atomic Radiation. "Sources and Effects Ionizing
       Radiation," Report to the General Assembly UN Publication
       E.77 IX.1., (New York: United Nations, 1977).

[6]   International Commission on Radiological Protection.
       "Limits for Inhalation of Radon Daughters by Workers,"
       ICRP Publ. 32, (Oxford, England: Pergamon Press, 1981.)

[7]   National Council on Radiation Protection and
       Measurements, "Evaluation of Occupational and
       Environmental Exposures to Radon and Recommendation,"
       NCRP 78, (Bethesda, MD: National Council on Radiation
       Protection and Measurements, 1984).

[8]   Kato, H. and W. J. Schull. "Studies of the Mortality of
       A-bomb Survivors, 7:Mortality, 1950-1978: Part I, Cancer
       Mortality," Rad. Research 90:395-432 (1982). (Also
       published by the Radiation Effect Research Foundation
       as: RERF TR 12-80, "Life Span Study Report 9, Part 1.")

[9]   Environmental Protection Agency. "Radionuclides
       Background Information Document for Final Rules,
       Volume I," EPA 520/1-84-022-1, Washington, D.C. 1984.

[10]   Harley, N. H. and B. S. Pasternack. "Environmental Radon
       Daughter Alpha Dose Factors in a Five-Lobed Human Lung,"
       Health Physics 42:789-799 (1982).

[11]   Radford, E. P. and K. G. St. Clair Renard. "Lung Cancer
        in Swedish Iron Miners Exposed to Low Doses of Radon
        Daughters," New Eng. J. Med. 310:1485-1494 (1984).

[12]   Land, C.E. and J. E. Norman. "Latent Periods of
        Radiogenic Cancers Occurring Among Japanese A-bomb
        Survivors," in Late Biological Effects of Ionizing
        Radiation, Volume 1, (Vienna: IAEA, 1978), pp. 29-47.

[13]   Lagerlof, E. Swedish Cancer Incidence Rates 1980 by
        Site, Sex and Age. Personal communication.

[14]   Lundin, F. E., V. E. Archer, and J. K. Wagoner. "An
        Exposure-Time Response Model for Lung Cancer Mortality in
        Uranium Miners: Effects of Radiation Exposure, Age, and
        Cigarette Smoking," in Energy and Health, Proc. of the
        2nd Conf. of the Society for Industrial and Applied
        Mathematics, N. E. Breslow and A. S. Whitemore, Ed.
        (Philadelphia, PA: SIAM, 1979), pp. 243-264.

[15]   Surgeon General. "The Health Consequences of Smoking -
        Cancer - A Report of the Surgeon General," DHHS (PHS)
        82-50179, (Rockville, MD: U.S. Public Health Service,
        Office on Smoking and Health, 1982).

[16]   Surgeon General. "The Health Consequences of Smoking for
        Women - A Report of the Surgeon General," DHHS (PHS)
        (Rockville, MD: U.S. Public Health Service, Office on
        Smoking and Health, 1980).

[17]   "Report on a Review and Evaluation of the PHS
        Epidemiological Study, August 1967," Appendix 21, in
        Radiation Exposure of Uranium Miners, Hearings before the
        Joint Committee on Atomic Energy, (Washington, D.C.:
        USGPO, 1967), pp. 1261-1267.

[18]   Prentice, R. L., Y. Yoshimoto, and M. W. Mason.
        "Relationship of Cigarette Smoking and Radiation Exposure
        to Cancer Mortality in Hiroshima and Nagasaki," J. Natl.
        Cancer Inst. 70:611-622 (1983).

[19]   Enstrom, J. E. and F. H. Godley. "Cancer Mortality Among
        a Representative Sample of Nonsmokers in the United
        States During 1966-68," J. Natl. Cancer Inst.
        65:1175-1183 (1980).

[20]   James, A. C., J. R. Greenhalgh, and A. Birchall. "A
        Dosimetric Model for Tissues of the Human Respiratory
        Tract at Risk from Inhaled Radon and Thoron Daughters,"
        in Radiation Protection, Vol. 2, Proceedings of the 5th
        Congress of the International Radiation Protection
        Society. (New York: Pergamon Press, NY, 1980),
        pp. 1045-1048.

[21]   Jacobi, W. and K. Eisfeld. "Internal Dosimetry of
        Inhaled Radon Daughters," in Radiation Hazards in
        Mining: Control, Measurement, and Medical Aspects,
        M. Gomez, Ed. (New York: Society of Mining Engineers,
        1981), pp. 31-35.

[22]   Harley, N. H. "Comments on the Proposed ICRP Lung Model
        as Applied to Occupational Limits for Short Lived Radon
        Daughters: A Comparison with Epidemiologic and Dosimetry
        Models," presented at the Berlin Colloquium, October 7-9,
        1980.

[23]   Hofmann, W. "Cellular Lung Dosimetry for Inhaled Radon
        Decay Products as a Base for Radiation-Induced Lung
        Cancer Risk Assessment. I. Calculation of Mean Cellular
        Doses," Radiat. Environ. Biophys. 20:95-112 (1982).

[24]   Pirchan, A., and H. Sikl. "Cancer of the Lung in the
        Miners of Jachymov (Joachimstal)," Amer. J. Cancer
        16:681-722 (1932).

[25]   de Villiers, A. G. and J. P. Windish. "Lung Cancer in a
        Fluorspar Mining Community, I. Radiation, Dust, and
        Mortality Experience," Brit. J. Indust. Med. 21:94-109
        (1964).

[26]   Wagoner, J. K., V. E. Archer, F. E. Lundin, D. A. Holaday
        and J. W. Lloyd. "Radiation as the Cause of Lung Cancer
        Among Uranium Miners," New Eng. J. of Med. 273:181-188
        (1965).

[27]   Christoforidis, A. J. and J. C. Johnson.
        "Bronchography," Curr. Problems in Radiol. 111(4).
        July-August 1973.

[28]   Rosenblatt, M. B. and J. R. Lisa. Cancer of the Lung
        Pathology, Diagnosis and Treatment, (New York: Oxford
        University Press, 1956).

[29]   Herman, D. L. and M. Crittenden. "Distribution of
        Primary Carcinomas in Relation to Tissue as Determined by
        Histochemical Techniques," J. Natl. Cancer Inst.
        27:1227-1271 (1961).

[30]   Garland, L. H., R. L. Beier, W. Coulson, J. H. Heald, and
       R. L. Stein. "The Apparent Sites of Origin of Carcinomas
       of the Lung," Radiology 78:1-11 (1962).

[31]   National Academy of Sciences - National Research Council,
       "The Effects on Populations of Exposure to Low Levels of
       Ionizing Radiation," Committee on the Biological Effects
       of Ionizing Radiation, (Washington, D.C.: National
       Academy of Sciences, 1972).

[32]   Axelson, O., and L. Sundell. "Mining, Lung Cancer and
       Smoking," Scand. J. Work, Environ. & Health 4:46-52
       (1978).

[33]   Cross, F. T., R. F. Palmer, R. E. Filipy, R. H. Busch,
       and B. O. Stuart. "Study of the Combined Effects of
       Smoking and Inhalation of Uranium Ore Dust, Radon
       Daughters and Diesel Oil Exhaust Fumes in Hamsters and
       Dogs," PNL-2744. (Richland, Washington: Battelle
       Pacific Northwest Laboratory, 1978.)

[34]   Gastineau, R.M., P. J. Walsh, and N. Underwood.
       "Thickness of Bronchial Epithelium with Relation to
       Exposure to Radon," Health Physics 23:857-860 (1972).

[35]   Harley, N. H. and B. S. Pasternack. "Alpha Absorption
       Measurements Applied to Lung Dose from Radon Daughters,"
       Health Physics 23:771-782 (1972).

[36]   McDowell, E. M., J. S. McLaughlin, D.K. Merenyl,
       R. F. Kleffer, C. C. Harris, and B. F. Trump. "The
       Respiratory Epithelium. V. Histogenesis of Lung
       Carcinomas in the Human," J. Natl. Cancer Inst.
       61:587-606 (1978).

[37]   Keenan, K. P., J. W. Combs, E. M. McDowell.
       "Regeneration of Hamster Tracheal Epithelium After
       Mechanical Injury. Parts I, II, and III," Virchow's
       Arch. [Cell Pathol.] 41:193-252 (1982).

7.

European Radon Surveys and Risk Assessment

Fritz Steinhausler

Division of Biophysics, University of Salzburg
Salzburg, Austria

INTRODUCTION

For many years the international scientific community has recognized the significance of doses resulting from the exposure of the general public to the natural radiation environment (NRE) [1,2]. However, although the risk for the public due to NRE exposure is many times larger than that from man-made practices, the system of dose limitation recommended by the International Commission on Radiological Protection (ICRP) does not apply to natural radiation [3]. Up until 1984 no dose-equivalent limits were recommended by international regulatory agencies for the NRE-component, which causes by far the largest contribution to the total dose to the public, specifically by indoor exposure to radon and thoron daughters. It took the publication of recommendation 39 in 1984 by ICRP to acknowledge the importance of issuing internationally coherent principles for limiting exposure of the public to natural sources of radiation [4].

Due to the lack of coordinated actions by international organizations, such as the International Atomic Energy Agency or the World Health Organization, on this important issue, it was largely left to individual national or regional agencies to address the problem of indoor radon daughter exposure.

At the level of European regional coordination, it was particularly the Commission of the European Communities (CEC) that declared its interest in the issue in the 1980-1984 Radiation Protection Research Programme by providing extensive support for experimental and theoretical research in the various member states [5]. The activities sponsored covered basic research tasks, such as the attachment of radon daughter products on aerosol particles, as well as applied research on technical countermeasures for the development of radon barriers.

Transcontinental coordination of research tasks between member states is carried out by the Nuclear Energy Agency (NEA) of the Organization for Economic Cooperation and Development (OECD). This agency is actively engaged in providing technical guidelines with the assistance of an international task group for a variety of topics ranging from radon daughter dosimetry to international intercomparison programs on radon and thoron daughter measuring methods and instrumentation; the latter is carried out in close collaboration with the CEC.

The following is an overview on CEC-NEA coordinated and sponsored activities, as well as on current national research programs related to the field of indoor exposure of the public to radon and thoron daughters.

Individual European researchers and their affiliations engaged in the various topics are identified by numbers in brackets in the text and listed at the end.

EUROPEAN RESEARCH ON RADIUM CONCENTRATION AND RADON EXHALATION OF BUILDING MATERIALS

Large-scale investigations of commercially available material and manufactured building materials are carried out in Austria, Denmark, (West) Germany, Greece, Hungary, and Sweden [9,10,11,12, 13,14,15,16]. Useful criteria for the decision on the suitability of a given material are the recommendations issued by the OECD-NEA on the specific activity of masonry defined as [6]:

$$\frac{C_K}{1500} + \frac{C_{Ra}}{150} + \frac{C_{Th}}{100} = 1$$

where   C = specific activity in Bq/kg.

Material complying with this equation is capable of causing an annual increment in dose equivalent of 0.1 mSV (for apartments) and 0.03 mSv (for single family houses) to the whole body through living in masonry buildings.

It was generally found that all natural materials, except granite, are below the OECD limit. However, manufactured materials, particularly if using industrial wastes (chemical gypsum, fly ash), can show Ra 226 levels considerably higher than the OECD limit. Risk-benefit analysis shows that as long as only small amounts of waste material (about 10% of total material) are added, e.g., in the cement production, this results in only small dose increments which are outweighed by the benefits of using an otherwise unwanted waste product. In view of the amounts of industrial waste material used already in the cement production, e.g., in Denmark 250,000 tons/year of fly ash, the radiological impact of unsuitable materials used as building material can have a significant impact on the collective dose equivalent of populations.

The radium and thorium content of the source term "building material" describes rather the potential of a given material to supply gaseous radon or thoron to the indoor atmosphere, while the actual amount of radon and thoron released depends on the exhalation rate. This parameter is extensively studied in several countries [10,11,13,16,21]. Figure 1a shows a three-chamber system for the determination of radon and thoron exhalation from building materials. It is possible to test simultaneously sandy materials (e.g., cement) in chamber V_1, building blocks (e.g., prefabricated concrete blocks) in V_1, and pre-built and plastered test walls with chamber V_3.

The exhalation rate is determined to a large extent by the moisture and temperature of the material.

Figure 1b shows (as an example) the influence of moisture on radon exhalation for concrete. The experimental results indicate that up to a certain level of moisture radon exhalation of most materials increases with increasing moisture due to increased "direct recoil fraction," reduced gas adsorption on internal surfaces of the material, and active radon transport on water molecules. After an optimum moisture content of about 5%, the diffusion is greatly impaired and the exhalation is reduced. Exhalation also increases with temperature due to decreased physical Van der Waal adsorption of radon on solids.

The influence of the physical structure of concrete as a function of the type of aggregates used, water/cement ratio (W/C ratio), and curing conditions have also been studied. While the radium activity concentration follows a linear correlation with the radon exhalation rate for the same type of material, concrete samples prepared with different W/C ratios and cured under different conditions show no significant difference in the exhalation rate. Due to the fact that about 30% of the total dose from inhaled short-lived radon and thoron daughters is caused by thoron daughters, it is important also to quantify the thoron exhalation rate. Good correlation has been found between porosity of the material (p) and the corresponding diffusion length (R) for radon as well as thoron (Figure 2) and can be described by the equation:

$$R = ap^b$$

Table 1 summarizes the characteristic values for the frequency distribution of radon and thoron exhalation rates measured from different building materials. The highest mean rates are generally found for limestone, gypsum, and pumice, while marble and slagstone have mostly very low exhalation rates.

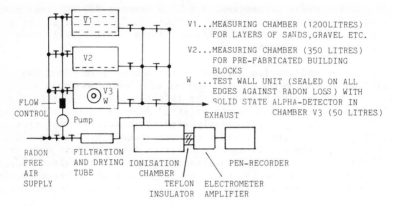

V1...MEASURING CHAMBER (1200LITRES)
      FOR LAYERS OF SANDS,GRAVEL ETC.

V2...MEASURING CHAMBER (350 LITRES)
      FOR PRE-FABRICATED BUILDING
      BLOCKS

W ...TEST WALL UNIT (SEALED ON ALL
      EDGES AGAINST RADON LOSS) WITH
      SOLID STATE ALPHA-DETECTOR IN
      CHAMBER V3 (50 LITRES)

Figure 1a.   Measuring of all-sided (V1,V2) and one-sided (V3) exhalation (Ref. 10)

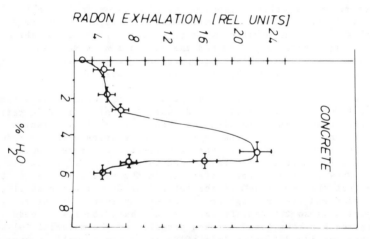

Figure 1b.   General trends of the influence of moisture on radon-exhalation for concrete. The exhalation rate is given relative to the exhalation rate for dry samples. Moisture is quantified in (%) H_2O. (Ref. 16)

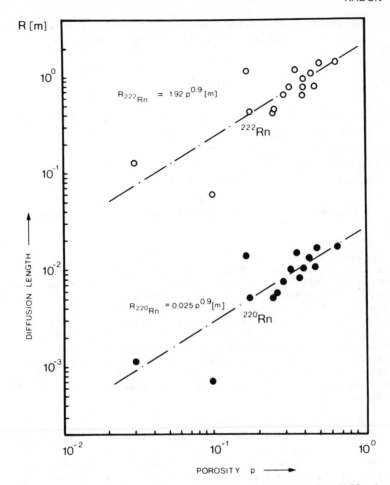

Figure 2. Correlation between the porosity and the diffusion
length of the building materials investigated
(Ref. 11)

CURRENT EUROPEAN RESEARCH IN MEASUREMENT TECHNIQUES FOR
RADON, THORON, AND DAUGHTERS

Generally, research interests in this field are focusing on the
development of low-cost screening devices and improved continuous
radon and daughter monitors, international intercomparison, and
intercalibration of measuring devices [17,18,19,20,21,22,23].

Whenever there is no need for high accuracy of radon daughter
measurements, e.g., in large-scale regional screening programs, a

Table 1.   Characteristics Values for the Frequency Distribution of the Rn 222 and Rn 220 Exhalation Rates   (Ref. 11)

N = 70 Parameter	^{222}Rn 10^{-3} [Bq.m^{-2}.s^{-1}]	^{220}Rn 10^{-3}[Bq.m^{-2}.s^{-1}]
Median value	0.49, σ = 2.9	45, σ = 2.3
95% confidence range	0.13 − 1.92	14.5 − 140
Arith. mean	0.86	64
Most freq. value	0.16	22
Min − Max	0.05 − 4.50	10.0 − 565

(σ = scattering factor, N = Number of measurements).

simple method can combine cost-effectiveness with the requirement for fast data acquisition. Figure 3 shows the detector arrangement for such an instrument. It consists of three GM tubes suitable for beta particle detection (inside metal shielding for background reduction). Beta radiation from RaB, RaC, and background is detected by the first tube; the second tube has a filter from absorption of the beta-contribution from RaB; and the filter on the third tube absorbs all beta particles and measures background only. A filter strip in front of the detectors moves during the measurements and thereby shields against plate-out. Typically, the statistical uncertainty corresponding to one standard deviation is 30% at 100 Bq/m^3.

In the area of continuous radon monitoring, a compact portable system based on an ionization chamber has been developed (Figure 4). It uses a 10-liter chamber connected to an electrometer amplifier with a lower limit of detection down to 10 Bqm^{-3}. The chamber is filled with air by a small pump as part of the system and can be used for radon measurements in soil gas, air, and other gases (e.g., natural gas of public networks).

For the continuous registration of radon daughters, a portable battery-powered, two-filter detector assembly for alpha spectroscopy with integrated data processor has been suggested (Figure 5). Alpha spectra are collected continuously from both filters, while both filter devices take air samples alternately (cycle time: two hours). At a resolution of 10 minutes the detection limit is about 5 mWL (flow rate: 0.2 m^{3}h^{-1}. By adding diffusion plates to one of the samplers it is possible to determine also the unattached fraction.

In order to ensure the comparability of results obtained in different regional surveys, international intercalibration and intercomparison programs have been initiated by the CEC for European member states and the OECD/NEA for the western industrialized coun-

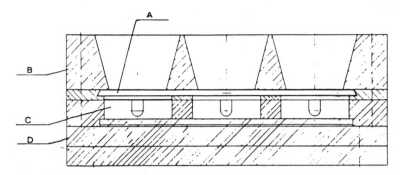

Figure 3.   Detector arrangement: A) sutter, B) collimater,
            C) GM tube, D) background shielding material (Ref. 21)

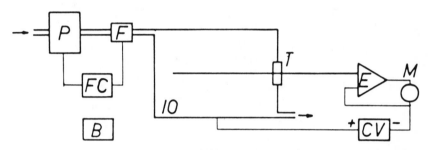

Figure 4.   Schematic diagram of the instrument (P...pump, F...flow
            meter, FC...flow control, B...battery, IO...ionization
            chamber, T...Teflon isolator, E...electrometer amplifier,
            M...display/recorder, CV...chamber voltage) (Ref. 23)

tries. For increased cost effectiveness, both programs are coordinated and basically use a framework of designated "regional reference laboratories": the UK National Radiological Protection Board
for Europe; the US DOE Environmental Measurements Laboratory and the
US Bureau of Mines Denver Research Center for the North American
continent; and the Australian Radiation Laboratory for the Pacific
area. After these four laboratories have standardized their
calibration procedure, they will serve as centers for regional
intercomparisons among interested institutions in each region.

    An essential component for such interregional comparisons was
the installation of radon chambers in various European countries
[10,13,18,21). Figure 6 shows a typical design for a chamber consisting of an anteroom with sampling port holes and the actual test
chamber with an optional closed loop or open air circulation system.

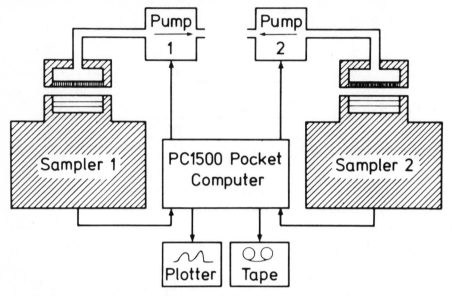

Figure 5.   Schematic diagram of the twin channel device for
            continuous radon daughter spectrometry (Ref. 22)

EUROPEAN REGIONAL SURVEYS ON RADON EXPOSURE INDOORS

     Two different types of radon surveys have been carried out:
detailed studies with emphasis on the urban environment [10,25,28]
and large-scale nationwide surveys [16,18,21,24,26,27].

     Since the major part of the population in Central Europe lives
in cities and towns, it is important to obtain frequency distributions of the mean radon activity concentration in such areas (Figure
7).   It is also possible to correlate this information with the
building material used, the age of the building, the floor number of
the dwelling, and the season (Figure 8).

     National surveys often have to compromise between the large
number of measurements necessary for statistical significance and
the financial, temporal, and manpower restraints.

     Mostly integrating measuring methods based on solid-state
nuclear-track detectors are used for the detection of radon gas, and
values of the equilibrium factor (F-value) indoors are assumed to be
in the order of 0.3 to 0.5.   Figures 9-12 and Table 2 show different
national approaches for obtaining the necessary information.

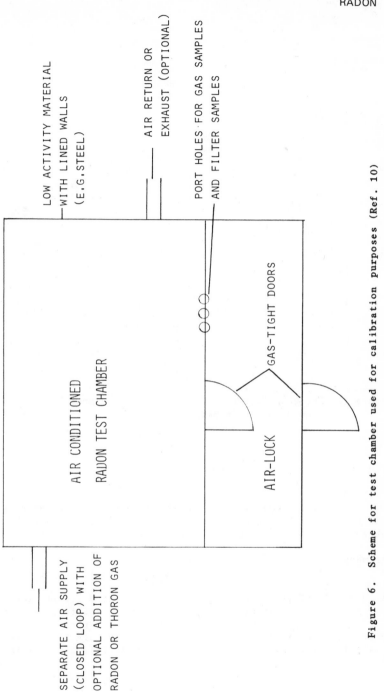

LOW ACTIVITY MATERIAL
WITH LINED WALLS
(E.G.STEEL)

AIR RETURN OR
EXHAUST (OPTIONAL)

PORT HOLES·FOR GAS SAMPLES
AND FILTER SAMPLES

AIR CONDITIONED

RADON TEST CHAMBER

GAS-TIGHT DOORS

AIR-LUCK

SEPARATE AIR SUPPLY
(CLOSED LOOP) WITH
OPTIONAL ADDITION OF
RADON OR THORON GAS

Figure 6.   Scheme for test chamber used for calibration purposes (Ref. 10)

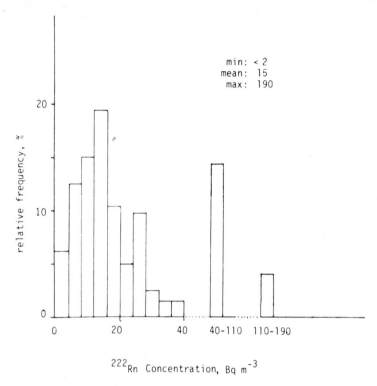

Figure 7.   Frequency distribution of the mean annual Rn222
concentration in Salzburg city (Ref. 10)

RESEARCH ON TECHNICAL COUNTERMEASURES

Due to the magnitude of the problem, it is mainly the Scandina-
vian countries that have concentrated on research concerning technical countermeasures [13,21,30].

In many cases various methods for modifying the existing ventilation system are used or new systems are installed (Figure 13).

In view of efforts to conserve energy, increased ventilation is
not always feasible in Nordic climates. As an alternative, electrostatic precipitators have been found to lower considerably the relative concentration of unattached decay products (Figure 14). Also,
radon daughters can be removed effectively by electric fields (Figure 15).

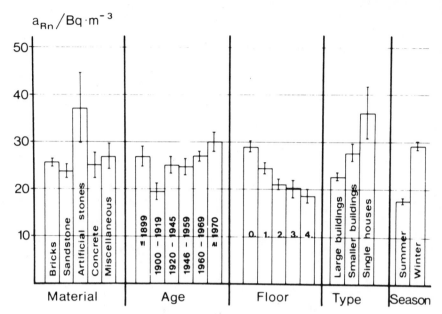

Figure 8.   Mean values of the radon activity concentration in
             dwellings, dependent on material and age of building,
             floor number of the dwelling, and season (Ref. 25)

RISK ASSESSMENT

     The assessment of the risk to the public for lung cancer induc-
tion due to radon daughter exposure using epidemiology requires the
availability of radiological, demoscopical, and medical data (Figure
16).   Unfortunately, large uncertainties are associated with the
available data, thereby limiting the usefulness of risk factors
derived from such studies, particularly for low-level indoor exposure [7].   However, it is possible to obtain a range of probable
values by combining information derived from epidemiological data
with dosimetric calculations.   In that manner the fraction of the
normal lung cancer frequency associated with a typical mean indoor
exposure can be estimated (Figure 17).

     Using these risk factors as best estimates it can be concluded
that -- related to the total reference population -- about 5% of the
observed lung cancer frequency may be due to the indoor exposure to
radon daughters (in any case less than 2% of the observed lung
cancer frequency for non-smokers).   However, it should be noted that
other atmospheric pollutants that may also represent a carcinogenic
risk can be found in the indoor environment.

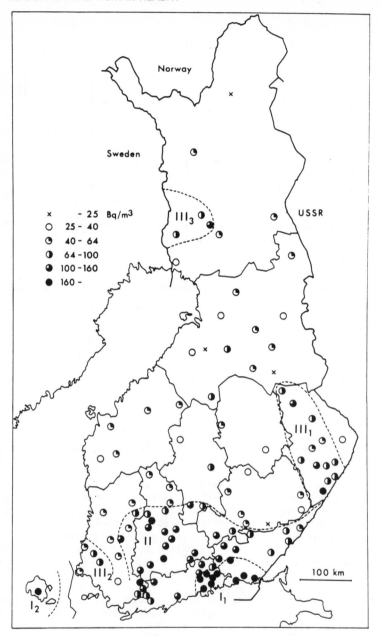

Figure 9. Median of the radon concentration in indoor
air of small houses (Finland) (Ref. 29)

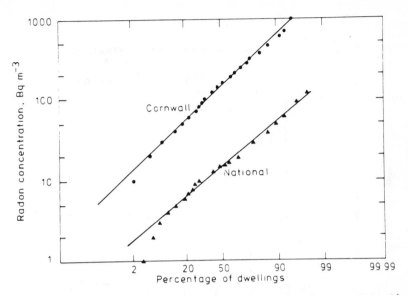

Figure 10.   Cumulative frequency distribution of radon concentration
in air in the living areas of some UK dwellings (Ref. 18)

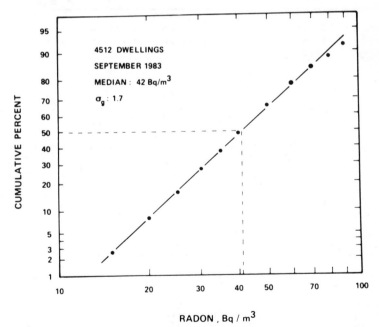

Figure 11.   Cumulative distribution of radon levels indoors
in Germany (Ref. 27)

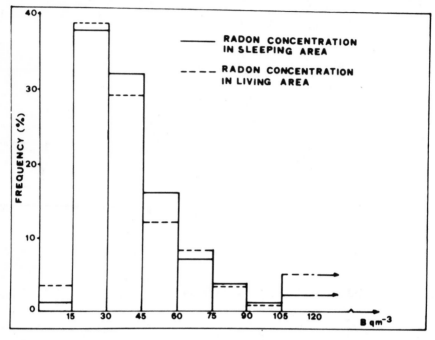

Figure 12.   Distribution of radon concentration in sleeping and
living areas in Italian dwellings (April—July 1983)
(Ref. 24)

Table 2.   Distribution of Dwellings Measured by the Local Authorities by Region, Type of House, and the
Radon Daughter Concentrations in Bqm^{-3} During the Period from July 1, 1979 to June 30, 1982 in
Sweden (Ref. 21)

Regions	Total number of measured dwellings		Radon daughter concentration in Bq.m⁻³					
			> 200		> 400		> 800	
	detached houses	multi-fam houses	detached houses	multi-fam-h.[a]	detached houses	multi-fam-h.[a]	detached houses	multi fam-h.[a]
The County of Stockholm								
alum shale based concrete	9,258	2,010	3,797	75	1,074	11	195	3
other materials	1,490	56	537	9	207	4	72	2
Alum Shale Regions								
alum shale based concrete	1,387	408	788	98	290	12	59	2
other materials	1,098	6	290	0	138	0	57	0
Other Regions								
alum shale based concrete	11,345	3,821	5,384	732	1,441	102	162	6
other materials	957	144	183	10	67	2	17	2
All Regions								
– alum shale based concrete	21,990	6,239	9,969	905	2,805	125	416	11
– other materials	3,545	206	1,010	19	412	6	146	4

[a]Number of dwellings.

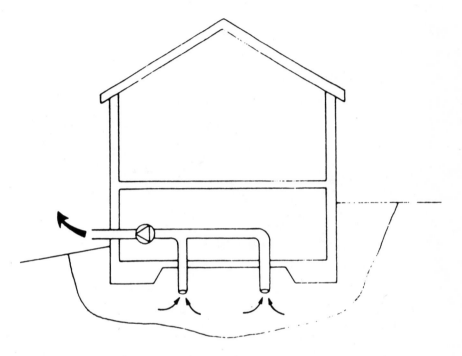

Figure 13.   Ventilating the drainage to evacuate the radon
             under the slab (Ref. 42)

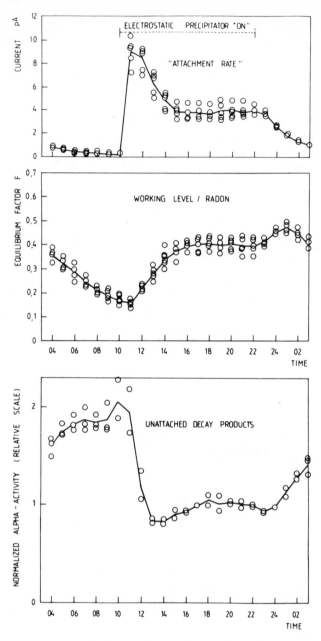

Figure 14. Upper graph shows the reading of an aerosol monitor; the equilibrium factor is shown in the middle graph: in the lower graph, the normalized concentration of unattached decay products during the same period is shown (Ref. 30).

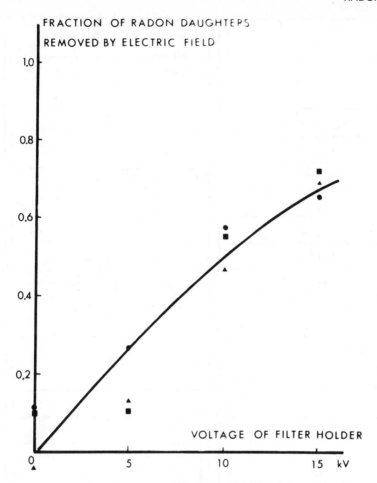

Figure 15.   Removal of radon daughters by electric field.
● RaA ▲ RaB ■ RaC.   Aerosol concentration 10^9
to 10^{10} m^{-3}.   No filtration. (Ref. 24)

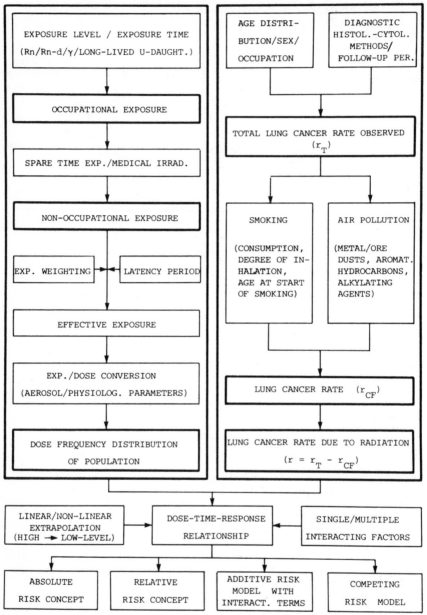

Figure 16.   Scheme of radiological, demoscopical, and medical data
needed for control and test populations for risk
assessment of lung cancer induction due to radon
daughter exposure (Ref. 10)

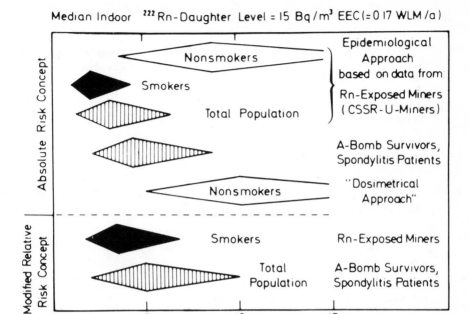

Figure 17.   Estimated fraction of the normal lung cancer frequency
             associated with the mean indoor exposure to Rn222
             daughters (EEC = $15Bqm^{-3}$) (Ref. 8)

REFERENCES

[1]   Adams, J.A.S., Lowder, W. M., and Gesell, T. F. (Eds.). "The
      Natural Radiation Environment II," Symposium Proceedings, ERDA
      – Report CONF–720805 (1972).

[2]   Gesell, T. F. and W. M. Lowder (Eds.). "Natural Radiation
      Environment III," DOE Symposium Series, CONF–780422 (1980).

[3]   International Commission on Radiological Protection, "Recom-
      mendations," ICRP Publication 26 (1977).

[4]   International Commission on Radiological Protection, "Princi-
      ples for Limiting Exposure of the Public to Natural Sources of
      Radiation," ICRP Publication 39, Ann. ICRP 14, No. 1 (1984).

[5]   Commission of the European Communities, "Radiation Protection
      Programme 1980–1984: Research Priorities and Scientific Docu-
      mentation" (1979).

[6]   OECD–NEA, "Exposure to Radiation from the Natural Radioac-
      tivity in Building Materials," Report by a Group of Experts
      (1979).

[7]   Steinhausler, F., "Possible Lung Cancer Risk from Indoor Expo-
      sure to Radon Daughters," Rad. Protection Dosimetry Vol. 7,
      389–394 (1984).

[8]   Jacobi, W., "Possible Lung Cancer Risk from Indoor Exposure to
      Radon Daughters," Rad. Protection Dosimetry Vol. 7, 395–401
      (1984).

      LIST OF EUROPEAN INSTITUTIONS ACTIVELY ENGAGED IN RADON RELATED
      RESEARCH (CONTACT MEMBERS IN BRACKETS):

[9]   Research Center Seibersdorf, Institute of Radiation Protec-
      tion, Seibersdorf, Austria (H. Sorantin)

[10] University of Salzburg, Division of Biophysics, Salzburg,
      Austria (F. Steinhausler)

[11] University of the Saarland, Institute for Biophysics, Hamburg,
      FRG (K. H. Folkerts)

[12] National Institute of Radiation Hygiene, Copenhagen, Denmark
      (K. Ulbak)

[13] Technical University of Denmark, Laboratory for Applied
      Physics I, Lyngby, Denmark (N. Jonassen)

[14]   Central Research and Design Institute for Silicate Industry, Budapest, Hungary (M. Gallyas)

[15]   Imperial College, Blackett Laboratory, London, UK (I. Siotis)

[16]   National Institute of Radiation Hygiene, Osteras, Norway (E. Stranden)

[17]   University College Dublin, Physics Department, Dublin, Ireland (J. P. McLaughlin)

[18]   National Radiological Protection Board, Chilton, Didcot, UK (A. D. Wrixon)

[19]   Commission of the European Communities, Radiation Protection Research Program, Brussels, Belgium (J. Sinnaeve)

[20]   OECD - Nuclear Energy Agency, Paris, France (O. Ilari)

[21]   National Institute of Radiation Protection, Stockholm, Sweden (G. A. Swedjemark)

[22]   Gesellschaft f.Strahlen- und Umweltforschung, Institute f.Strahlen schutz, Neuherberg, Germany F.R.

[23]   Institut fur Radiumforschung und Kernphysik, Vienna, Austria (H. Friedmann)

[24]   ENEA-PAS, CRE Casaccia, Rome, Italy (G. Sciocchetti)

[25]   University of Erlangen, Institute f.Radiobiologie, Erlangen, Germany F.R.   (H. Pfister)

[26]   Eidgenossisches Institut f.Reaktorforschung, Health Physics Division, Wurenlingen, Switzerland (H. Brunner)

[27]   Bundesgesundheitsamt, Institut f.Strahlenhygiene, Neuherberg, Germany F.R. (A. Wicke)

[28]   Rijksuniversiteit Groningen, Kernfysisch Versneller Institut, Groningen, The Netherlands (F. Wolfs)

[29]   Finnish Centre for Radiation and Nuclear Safety, Helsinki, Finland (O. Castren)

[30]   Technical Research Centre of Finland, Occupational Safety Engineering Laboratory, Tampere, Finland (M. Lehtimaki)

[31]   Swedish National Testing Institute, Division of Building Physics, Boras, Sweden (O. Hildingson)

Part Two

Microorganisms

8.

Part Two: Overview

Philip R. Morey

National Institute for Occupational Safety and Health
944 Chestnut Ridge Rd., Morgantown, WV   26505-2888

Microorganisms may cause allergic respiratory reactions as well as respiratory infections. The extent of health problems caused by microorganisms in the indoor environment is difficult to estimate. This is partly due to the broad array of microorganisms that evoke human response, and also to the broad range of environments in residential, commercial, and public buildings where exposure occurs. Despite uncertainty about the magnitude of the health effects caused by exposure, the impact of some microorganisms is appreciable.

Outbreaks of hypersensitivity pneumonitis caused by microorganisms in both large and small office buildings are well documented. Remedial measures have ranged from simply cleaning a component of the heating, ventilation and air-conditioning (HVAC) system to total replacement of the system together with all building furnishings. Cases of hypersensitivity pneumonitis in residences are also well documented. The most common causes of these cases are contaminated forced air heating systems, contaminated humidifiers, or flooding disasters.

Asthma may be caused by pollens, mites, animal dander, fungi, and insect eminations. While the indoor environment offers some protection from pollen, the growth and proliferation of other allergens especially molds and mites are encouraged by moisture and high relative humidity. Only a few studies have been made on the parameters that are associated with changing levels of fungi and mites in the indoor environment.

One of the more prominent examples of infections caused by microorganisms in buildings would be the occurrence of Legionnaires disease. For the most part Legionnaires disease has been associated with the infiltration into the building environment of aerosols containing Legionella pneumophila from external sources such as cooling towers. Pathogenic microorganisms may also be disseminated into the intramural environment by use of hot tubs (dermatitis caused by Pseudomonas spp.), whirlpools, cold mist vaporizers, and nebulizers.

133

The national concern for energy conservation has, especially
in large buildings, brought about an emphasis on air recirculation
at the expense of the introduction of outside air into the indoor
environment. The increased use of recirculated air in buildings
where susceptible individuals are present 24 hours per day (e.g.,
hospitals, retirement homes) can become a serious problem if
microorganisms are not removed by cleaning or preventive
maintenance.

Harriet Burge describes sources of airborne microorganisms in
the indoor residential environment and methods of collecting
viable and nonviable microbial aerosols. Suction samplers for
viable microorganisms and for nonviable spores are among the most
useful instruments for collecting microbial aerosols. However,
because of their diversity and the differing growth requirements
of collected organisms, the full spectrum of microbial aerosols
within any given volume of air cannot be collected by any single
sampler. Appliances such as humidifiers, cool-mist vaporizers,
and air conditioners are sources of airborne microorganisms in the
residential environment. Spores that have settled in the indoor
environment are easily reaerosolized by human activities such as
bedmaking, vacuuming, and cleaning of contaminated appliances.
Controlling indoor microbial aerosols involves elimination of
sources as well as the substrates on which microorganisms grow.
Keeping interior surfaces dry and maintaining relative humidity
below 50% will prevent microbial growth. Restricted use of
humidifiers and nebulizers and the regular thorough cleaning of
necessary appliances will also discourage microbial growth.

Peter Kozak describes illnesses caused by exposure to fungi
growing within the indoor residential environment. Depending upon
host susceptibility, particle size, frequency of exposure, and
viability or nonviability, endogenous mold-spore exposure may
cause illnesses such as allergic rhinitis, asthma, and bronchopulmonary aspergillosis. Homes with heavy shading of the roof,
considerable outdoor organic debris or poor landscape control, and
inadequate interior housekeeping may have higher levels of indoor
mold spores than homes with less shading or better maintenance.
Electrostatic filtration of air in residential forced air heating
and cooling systems can significantly reduce indoor spore levels.
Interpretation of the results of volumetric sampling in indoor
residential air is made difficult by a multiplicity of factors
including: (a) some spores or spore components are hydrophilic
whereas others are hydrophobic, (b) a single large diameter spore
(e.g., Alternaria spp.) will deliver much more antigen mass to the
airways than a small diameter spore (e.g., Penicillium spp.), and
(c) some spores (e.g., Aspergillus fumigatus) may colonize the
airways of susceptible individuals resulting in excessive antigen
production and life-threatening disease.

The importance of airborne transmission of viable and
nonviable aerosols in the hospital environment is controversial.

Ruth Kundsin describes instances where transmission of viable
aerosols has led to hospital acquired infection. Airborne counts
of microorganisms are proportional to incidence of respiratory
tract infections in high-risk patient areas. In some operating
rooms, the upper concentration limit for particles containing
viable bacteria has been set at 175 per m^3. Sedimentation and
impaction of viable and nonviable particles into open wounds are
of considerable importance with regard to hospital acquired
infection. Environmental monitoring for microbial aerosols is
needed in some critical care areas in hospitals. More attention
needs to be given to the protection of both patients and hospital
personnel from microbial aerosols especially with the likelihood
of diminished contaminant control as the operation of hospital
HVAC systems is modified in response to energy conservation
measures. Techniques (e.g., filtration, ultraviolet disinfection,
and industrial clean room technology) are currently available to
prevent the air within a hospital from becoming the medium of
disease transmission.

James Feeley points out that respiratory infection accounts
for about 50 to 60% of all community acquired illness. Although
most of these infections are of a viral etiology, bacterial
diseases do occur and have caused substantial problems in settings
such as hospitals, day-care centers, and schools. Legionella
pneumophila has caused outbreaks of Legionnaires disease and
Pontiac fever in hospitals, office buildings, and hotels. Some of
these outbreaks were shown to be due to transmission of the
bacteria by means of HVAC systems. Other airborne bacteria such
as Neisseria meningitidis and Haemophilus influenzae are
responsible for illness outbreaks in day-care centers and schools,
but do not appear to be transmitted by recirculated, conditioned
air. Control of viable aerosols is facilitated by fastidious
maintenance of each of the following: (a) reservoirs such as
cooling towers and filters where microorganisms accumulate, (b)
amplifiers such as cooling towers and humidifiers where
microorganisms proliferate, and (c) disseminators such as HVAC
systems and their outdoor air intakes through which microorganisms
may be transmitted to the indoor environment.

Air is the common medium in the indoor environment. Transmission of viable and nonviable microbial aerosols that cause
disease in the indoor environment is not acceptable. The 1981
National Academy of Sciences (NAS) report on Indoor Pollutants
stated that allergens and pathogens were a significant but
unquantified risk factor to both general and certain susceptible
populations. In the time since the publication of the NAS report
there has been little effort to study systematically the nature or
the impact of indoor allergens and pathogens. Nevertheless, given
the continuing trend toward tightening the indoor environment of
residences, schools, hospitals, and public buildings, research is
needed both to determine if levels of allergens and pathogens are
increasing and if there is an associated health risk. In the

interim, awareness is needed both that proper maintenance is essential for potential reservoirs, amplifiers, and disseminators of microorganisms and that microorganisms disseminated from man himself must be controlled. Large volumes of inexpensive air are no longer available to dilute indoor microbial aerosols as was the case up to the middle of the past decade. The public, as well as professional groups including architects, and building engineers and managers, should be aware that energy conservation measures that result in warm, moist indoor environments may increase human health risk. Awareness is also needed that certain pathogenic microorganisms such as <u>Legionella pneumophila</u> can enter the indoor environment in subtle ways such as from building vents located close to cooling towers, and also from building hot water services.

Many control techniques have been advocated in diminishing the impact of microbial aerosols in the indoor environment, but the scientific documentation of the effectiveness of these techniques is limited. Preventive maintenance or housekeeping together with protection from floods and moisture incursion will reduce the chance of microbial contamination in the indoor environment. Elimination of the use of water-spray systems in office building HVAC systems, restriction in the use of humidifiers, careful use of respiratory therapy equipment, proper location of cooling towers with respect to building vents, and the utilization of available filtration-air disinfection devices have been advocated as control measures for specific indoor environments.

Not enough data is available to accurately estimate or make guidelines for exposure to airborne microorganisms in the indoor environment. Nevertheless, an important research issue is whether it will be possible to determine upper limits of exposure for environments housing susceptible populations (facilities used to care for immunosuppressed individuals, operating rooms, day care centers and retirement homes)and subsequently for other populations (offices, residences). This task is made complex by variables such as the enormous diversity of viable microorganisms, the differing amounts of antigen mass in inhaled particles, sensitization or other host factors, and the absence of standard sampling protocols for airborne allergens and pathogens.

In order to quantify accurately risk due to exposure to viable and nonviable microbial aerosols, standard sampling protocols for viruses, bacteria, fungi, protozoans, and mites are needed. Protocols should describe what type of sampling is needed for specific illnesses and how sampling results should be interpreted. Specific sampling techniques relying on immunologic mechanisms are needed for problem organisms (e.g., <u>Legionella</u> spp., <u>Aspergillus fumigatus</u>). Groups such as the American Society of Heating, Refrigerating, and Air-conditioning Engineers, American Institute of Architects, American Conference of Governmental Industrial Hygienists, American Industrial Hygiene Association, American Hospital Association, and American Society of Microbiology should develop recommendations for sampling protocols and for control of microbial contamination within the indoor environment of concern to each professional association. Standardized microbial sampling protocols should be utilized concurrently with medical and epidemiological studies of occupants so as to increase our ability to understand sampling results for problem buildings and residences.

9.

Indoor Sources for Airborne Microbes

Harriet A. Burge

University of Michigan Medical School
Ann Arbor, Michigan

Indoor biological pollution, while of intimate concern to all who inhabit interiors, is only beginning to receive the attention afforded outdoor or even indoor chemical pollution [1]. This apparent lack of interest is tied to the difficulties of sampling biological aerosols and their variable health effects. The airborne bioflora is inherently complex and variable to a point that defies quantification. The air in a single room in a "clean" house may contain hundreds of different kinds of biological particles and technology does not exist to quantify all of them. As many as four distinct sampling modalities are required if those particles that are measurable are to be accurately assessed. Health effects of biological aerosols are basically different from those of their chemical counterparts. The majority of bioaerosols are nonpathogenic and cause disease only in sensitized people. Even pathogenic microorganisms are usually able to infect only susceptible hosts. In addition, levels of either saprophytes or pathogens required to produce disease differ with each particle type and are unknown for most.

Given that the field is still in its infancy, I would like to briefly summarize what is known about sources, sampling methods and control of a few especially common bioaerosol types.

SOURCES OF INDOOR BIOAEROSOLS

The Outdoor Environment

The majority of biological particles found in interior situations come from the outdoor environment [2]. Pollen is almost entirely from outdoor sources and can be used to assess structure penetration by particles between 15 and 50 μm. The majority of fungus spores encountered indoors are derived either directly (by penetration) or indirectly (by penetration and subsequent growth on surfaces) from outdoor air. Most fungi are primary decay organisms and are abundant on dead or dying plant and animal materials. Many are plant pathogens and can be present abundantly in air wherever their host plants grow or are grown. A few fungi are human pathogens. These are, however, primarily saprophytic decay organisms that can opportunistically cause human disease and almost always primarily exist in outdoor environments. During the growing season (or, in fact, whenever outdoor temperatures are above freezing) outdoor fungus spore levels exceed those indoors, especially for such spore types as basidiospores, ascospores, and spores of the plant pathogens, which only rarely are produced on interior substrates. Only in cases of serious contamination do indoor spore levels exceed outdoor levels.

Many bacteria, including some that cause human disease, thrive in outdoor reservoirs. The most infamous of these is Legionella, which is basically soil borne, and is probably introduced into cooling towers and other enrichment environments during excavations for roads or buildings. A wide variety of other bacteria, most of which are not pathogenic for people, are abundant on outdoor substrates and may penetrate and possibly grow on interior surfaces. Actinomycetes (filamentous bacteria) fall into this latter class.

Other biological particles that are occasionally common in outdoor air and that can contribute to indoor aerosols include algae, insects, and arachnids. Algae can be abundant outdoors near ponds or lakes and become airborne and enter interiors through wind action. Outdoor insect infestations (a prime example being gypsy moths) can result in airborne fragments small enough to penetrate even closed buildings. Other insects (e.g., cockroaches) and spiders, while primarily outdoor creatures, can abundantly colonize interiors and produce body fragments, excrement, and web material that can accumulate indoors.

Indoor Contamination

Although most indoor contamination results ultimately from outdoor sources, interior situations can become heavily contaminated with biological entities to the point of presenting severe health hazards. Indoor buildup of bioaerosols results from two general processes: material being shed and accumulating indoors, and actual growth on interior substrates.

Materials that are shed and accumulate indoors include particulates such as human and animal skin scales (dander), bacteria and dermatophytes shed from animate surfaces, and insect and arachnid fragments. Bacterial endotoxins and (theoretically) fungal mycotoxins can accumulate indoors from microorganisms growing on interior surfaces. Adverse health effects from endotoxin accumulating in commercial environments have been reported [3]. Mycotoxins have not been measured in air, but have been accused of implication in cases of leukemia following exposure to <u>Aspergillus</u> spores.

Most severe indoor biological pollution problems result from growth of the offending organism on surfaces within structures. Virtually <u>any</u> substrate that includes both a carbon source (no matter how esoteric) and water will support the growth of some microorganism. Human (or animal) skin scales, for example, support not only a lively mite population, but mesophilic bacteria and fungi such as <u>Aspergillus</u> <u>amstelodami</u> and <u>Wallemia</u> <u>sebi</u>, both of which can withstand relative dryness. Cellulose or lignin based materials, when damp, provide ideal enrichment media for a wide variety of bacteria and especially the common saprophytic fungi which have, in fact, evolved to decay just those substrates in nature. Leather, especially when loaded with skin oils from use, can support massive fungal growth. Plastics and nylon are long-chain polysaccharides, similar in structure to cellulose, and, while more difficult to digest than cellulosic materials, are readily utilized by some of the more specialized microorganisms. Likewise soap, grease, and other hydrocarbon films on surfaces can support a variety of microorganisms.

An absolute requirement for all of these kinds of interior contamination is a more or less consistent source of moisture. In some cases, high relative humidity is sufficient. Relative humidity between 25% and 70% directly affects fungal spore levels, probably by increasing growth on surfaces absorbing water. Above 70% is apparently optimal and further increases in relative humidity do not appear to affect airborne spore levels. Skin scales (as well as many fabrics, leather and wood materials) absorb sufficient water from the air to support fungal growth. Unless relative humidity is very high, (>75%) most surfaces (finished wood, paper, paint, tile) do not become dangerously wet. However, a very small leak in a roof or water pipe is sufficient to support abundant fungal growth, as do cold surfaces such as window panes or cooling coils with adequate condensation.

Unfortunately, a variety of modern appliances provide standing water reservoirs which, when not absolutely clean, provide ideal enrichment situations for microorganisms. Humidifiers, including vaporizers and water spray conditioners [4,5], evaporative coolers, self-defrosting refrigerators, and flush toilets, all have water reservoirs with the potential for contamination and, in fact, most have built-in mechanisms for airborne dispersal. In these appliances, standing, usually contaminated, water is in contact with a moving air stream which can pick up small particulates (bacteria, antigens) and spray them into room air. Vaporizers and spray coolers that aerosolize the water itself are especially dangerous. Air conditioners have cold surfaces often bearing an abundant and constant supply of water condensed from the air. When contaminated, microorganisms are readily blown from these surfaces into room air.

Any of these water-related air treatment units that are combined
with a building-wide forced air system can create problems that are
often difficult to solve without, at worst, removing the offending
units, or, at best, maintaining constant (and expensive) vigilance.

FACTORS AFFECTING INDOOR MICROBIAL LEVELS

Indoor surface contamination by bacteria, fungi, insects, arachnids or other biological particles is dangerous for the most part
only when the particles become airborne and are inhaled. While some
microorganisms produce volatile and/or soluble substances, most must
become airborne in particulate form. Several factors can cause
aerosolization of surface microorganisms. Air is almost never
still, and even the most delicate of air currents can cause dry
fungus spores to become airborne. Such fungi as Aspergillus and
Penicillium produce their spores on stems (conidiophores) that project above the surface growth to take advantage of this passive type
of dispersal. Even in still air, some fungi can forcibly discharge
their spores. Mushrooms and shelf fungi are particularly adapted
for spore discharge in still air. Mushroom fruiting bodies
(Merulius lacrymans) produced on rotting timbers in war damaged
British buildings have been shown to produce millions of (allergenic) spores even in still basements, which are readily spread by
air currents throughout the building.

Air currents produced by convection from radiant heat and, of
course, by air mechanically circulated by forced air systems are
more than adequate to spread dust (including entrained biological
particles) as well as mobilizing surface growth; and, of course,
growth within forced air systems is subjected to even greater wind
velocities.

Many of the appliances mentioned above can produce microbial
aerosols by their very operation. Thus humidifiers are designed to
put aerosols into the air; air conditioners work by air circulation.
Less obvious are the dangers posed by refrigerators and clothes
dryers. Self-defrosting refrigerators have an "evaporation" pan
where defrost water collects and is evaporated when warm air from
the refrigeration heat exchangers passes by. If these systems function perfectly there is little danger since water doesn't accumulate
to any extent. In older, less efficient units, however, significant
accumulation can occur with resultant microbial growth and the
potential for aerosolization. Clothes dryers present a different
problem. Warm humid air passes through dryer exhaust pipes often
laden with minute (and highly nutritious) lint particles which build
up on pipe surfaces. Thermotolerant and thermophilic organisms can
thrive in this situation, and produce abundant spores which enter
the air stream during dryer operation. As long as the dryer is
externally vented, this contamination source is relatively unimportant. However, the energy crunch has prompted many to try to

recover the heat energy from clothes dryers by venting them into the house, often with nylon stockings as the only final filter. You gain not only heat but humidity this way. Unfortunately, nylon stockings do not block bacteria or actinomycete spores and interior levels of these during improperly vented dryer operation can reach thousands per cubic meter of air. Additional appliances that can increase air biopollution are commercial dishwashers and flush toilets. Dishwashing equipment in hospitals poses particular risks especially for kitchen workers, as disease-causing organisms can reach potentially infective levels when such units are improperly sealed. Flush toilets while surely aerosolizing bacteria, do not pose great disease risk unless heavily contaminated and poorly maintained, since one is generally exposed mainly to one's own microbial flora.

Practically any human or pet activity can increase airborne microbial loads. Especially effective are vacuuming, sweeping, dusting, scrubbing contaminated surfaces, bed making, etc. None of these activities should be undertaken by or in the presence of sensitive or susceptible individuals.

SAMPLING FOR INDOOR MICROBIAL POLLUTION

There is no single method of choice for sampling airborne microbial particles (6). Each method has strong and weak points, and in general more than one is necessary to accurately assess most situations. There are three major sampling modalities:

Viable Particle Sampling

Viable particle sampling is historically the oldest and, in fact, the very existence of airborne microorganisms was demonstrated in this way. The most widely used method for viable sampling is the "settle" plate where a dish of culture medium is set out uncovered for a period of time and viable spores that happen to land on the medium are encouraged to grow. Attractive and easy as this sounds, it is virtually useless, especially where small spores (e.g. Aspergillus) or bacteria are of interest. Chances of impingement on a surface in absolutely still air (a purely theoretical condition in any normal environment) is directly related to a particle's mass. Large particles fall fastest. No matter how long you expose a petri dish (unless you allow every particle to fall out), large particles will be over-represented. When air is moving, the error is compounded and made more complex since aerodynamic size becomes more important. Small diameter particles tend to remain in a turning air stream and are less likely to be impacted than larger particles. Also, of course, gravity methods can never be quantitative. Therefore, accurate site to site and time to time comparisons cannot be made.

A variety of volumetric cultural sampling devices are commercially available. Most draw air through a defined orifice via a vacuum pump, accelerating the air to the point that most particles impact. It is important with suction samplers to remember that for the units to sample a wide range of particle sizes at stated efficiency, they must be wind oriented so that incident airflow is parallel to and going in the same direction as the suction flow. Air speed is also important although rarely controllable. At very low wind speeds (i.e. in still air) small particles may be diverted into the sampler at a greater rate than expected. At high wind speeds the smaller particles tend to slip around the orifice and are underestimated. In practical terms, air movement within buildings is usually in a range where, unless you sample next to an air return or fan, these errors are not significant. Viable sampling, even under ideal sampling conditions, always underestimates actual microbial levels. Of course, only living particles are collected and for hypersensitivity conditions, viability is probably not important. In addition, only particles that will grow under the given culture parameters (culture medium, oxygen tension, light/dark, temperature) will be recovered. Many fungi produce volatile and/or soluble substances that inhibit growth of neighboring colonies. In fact, percent recovery of airborne spores with an Andersen sampler varies inversely with actual spore levels. At levels exceeding $1000/m^3$ the underestimate exceeds 90% for _Cladosporium_ (even with shortened sampling times), a fungus that grows readily under most cultural conditions. Therefore, choice of culture parameters (culture medium, etc.) must be carefully made, and the decision to use a viable method carefully considered. Good reasons include the need to measure living, potentially infective organisms, and the need to sample organisms that cannot as yet be visually or immunologically recognized.

Particulate Sampling with Visual Assessment

Particulate sampling with visual assessment is obviously limited to recognizable particles. This is an enormously restrictive condition when dealing with anything other than pollen or complex fungus spores. Assuming one can accept this restriction, there are two methodologies from which to choose: impaction or suction. Impaction samplers (e.g., the rotorod) spin narrow sticky surfaces through the air, which entrap the particles. Efficiency is related to diameter of trapping surface, speed of rotation, and, most important, particle size. Commercially available units are very efficient for particles larger than 15 to 20 microns. This includes, in fact, most pollen and a good percentage of recognizable fungus spores (but not _Cladosporium_, the most abundant spore type). _Aspergillus_ spores are not efficiently collected but also are not identifiable as such visually.

Suction traps efficient over a wide particle size range are available and are governed by the same principles discussed above

for viable suction samplers. These samplers, while more expensive
than rotorods, can produce time discriminated samples over a 7-day
period which can be permanently mounted and saved for years. This
is the method of choice where identifiable particles are of concern.
Actually, even spores of _Aspergillus_ and _Penicillium_, which are not
identifiable to specific taxon, can be recognized as fungus spores
and counted where more accurate identification is not required. The
Burkard Manufacturing Company (Rickmansworth, England) offers not
only a wind oriented 7-day outdoor sampler of this type, but an
indoor, lighter weight model that operates for shorter periods of
time.

Immunological Sampling

The third major sampling modality for bioaerosols--suction sampling
with immunological detection--is new and still under development [3,
6]. It uses the same principles as viable and particulate suction
samplers but requires neither viability nor visual discrimination.
Samples are drawn from large volumes of air and either impinged on a
filter, dissolved or suspended in a liquid, or frozen from the air
on the walls of cooled containers. Samples are then used to inhibit
assays for antigen-specific antibodies. So far, this method has
been used to examine, for example, _Alternaria_, _Aspergillus_ and thermophilic actinomycete antigens, cockroach components in house dust,
airborne animal dander and ragweed pollen antigens, and offers great
promise for a wide variety of other antigens. Obviously, however,
you must decide in advance which antigen to use in each assay (you
have to know what you are looking for before you sample), you must
have antibody that recognizes each antigen, and you should have some
feeling for possible cross-reactivity between the antigen of
interest and other possible airborne particulates. Given advances
in monoclonal antibody technology, the antibody requirement may not
be limiting in the future but does add time and expense to the
method. The technique is certainly useful for detection of commonly
occurring contamination such as cockroach, ragweed, or _Aspergillus_
where at least some cross-reactivity work has been done.

     One relatively crude shortcut method that can be used to test
whether or not samples prepared for immunological testing contain
substances causing hypersensitivity disease in specific people is to
use double diffusion in agar gel and look for precipitates resulting
when precipitating antibody comes into contact with appropriate
antigen. This is a test with a high percentage of both false positive and false negative results and is strongly dependent on both
antigen and antibody concentrations. It has been successfully used
to connect specific symptoms to airborne building contamination,
especially with humidifier or air treatment slimes where specific
casual organisms could not be pinpointed.

CONTROL MEASURES

     Assuming you have in fact discovered a contamination problem in
a building, or even if your observations and tests have all been
negative but you are intuitively convinced that biological pollution
is at fault, the next steps are to attempt immediate remedial measures followed by more permanent efforts of control. Remedial measures require that the actual source of the problem has been identified and can be attacked directly. For example, humidifiers can be
removed, leaks can be fixed, pets can be banished, and surfaces can
be temporarily disinfected. If the heating system for a 30-story
building is contaminated, the problem is more severe, but not impossible. You can take the system apart and disinfect it [8] being
sure to modify whatever condition caused the problem, or, more
drastically, you can remove and replace the system. Often, however,
specific identifiable sources will not be found and control measures
must be more general.

     Measures involving actual air cleaning fall into two
categories: reducing outdoor influx of contaminants, and removing
contaminants from indoor air. Closing doors and windows, while
effective in blocking particle penetration also reduces fresh air
exchange and must be supplemented by a carefully filtered air
source. Air conditioning with filters down wind of the cooling
coils is highly effective although penetration of submicronic aerosols into air-conditioned buildings has yet to be assessed. As
emphasized earlier, the system must contain no standing water or
continuously damp surfaces. Electrostatic precipitators are used
both to prevent particle influx and to clean recirculated indoor
air. Unfortunately, they do neither as well as commercial claims
indicate and probably add little if anything to a good, filtered air
conditioning system. Precipitators work only when absolutely clean,
and unless the air they are treating is already nearly particle free
(in which case you don't need the precipitator) they stay clean for
periods measured in hours rather than days, weeks, or months. A
variety of centrally installable and, even more common, console type
air cleaners are available that utilize HEPA-type filtration often
combined with charcoal. Filters in these units are not actually
HEPA filters: fans required to move air through HEPA filters would
be prohibitively large and expensive. The filters will, under ideal
conditions, remove most particles down to less than 1 micron from
some quantity of air. If there is no continuing source of contamination in a room and an air cleaner is run long enough, the air will
become cleaner. Unfortunately, conditions are rarely ideal. In
most situations where air cleaners are used, there are continuing
sources (outside air, dust, people, pets), and the cleaners strike a
balance depending on how active the source. Usually, the balance is
not on the side of really clean air. Actual efficiency of these
units measured under field conditions remains to be accurately
assessed.

As an adjunct to treating the air, source prevention is essential and maintaining relative humidity as low as possible consistent with comfort is of primary importance. Relative humidity less than 50% is necessary to keep mold levels below average except in exceptionally clean environments. De-humidifiers are necessary in most climates although central air conditioning alone is effective in some situations. Needless to say, humidification should be limited both to keep relative humidity down and because of its inherent potential for contamination. If used, a humidifier must be maintained completely free of scale and/or slime, at all times. This may require weekly or at the very least, monthly cleaning. A preferable alternative would be a steam vaporizer.

Maintaining a dust-free environment is necessary for keeping airborne microorganisms at a minimum even with top quality air filtration. This means no carpeting, no stuffed furniture, no pets. All fabric must be washable (and regularly washed). Bedding (mattresses, pillows) should be sealed in plastic and dust catchers (books, knick-knacks) should be kept to a minimum. In an office environment these stringencies are relatively easy to maintain. In the home environment they are often unacceptable (although highly sensitive people will do whatever is necessary). Often a single room (usually a bedroom) can be maintained dust free.

Antimicrobial agents are less effective in preventing contamination than the basic environmental changes discussed. They can be useful in cleaning known sources but unless basic conditions change, recontamination will surely occur. A variety of agents are available for disinfection. None are ideal, all are more or less irritating to the people who must use them. Household bleach is probably the safest, followed by phenol compounds (e.g., Lysol). These agents will not prevent humidifier contamination when added to the water, or effect long term prevention of recontamination on surfaces. Ethylene oxide can be used by professionals to fumigate moldy or contaminated items that cannot be discarded (books for example). Paraformaldehyde can also be used in closed, uninhabited situations as a fumigant, however, levels required to affect fungus spores are well above levels safe for the human respiratory tract. Ultraviolet light, while effective for bacterial decontamination, does not affect most fungal spores.

To briefly summarize, prevention of outdoor air influx, maintenance of low relative humidity, and elimination of dampness or standing water of any kind will go far to prevent serious indoor air biopollution problems.

REFERENCES

[1] Spengler, J. D. "Indoor Air Pollution: A Public Health Per-
     spective," Science 221 (4605):9-17 (1983).

[2] Solomon, W. R. and H. A. Burge. "Allergens and Pathogens," in
     Indoor Air Quality, P. J. Walsh and C. Dudney, Eds. (Boca
     Raton, FL: CRC Press Inc., 1983).

[3] Rylander, R. and P. Haglind. "Airborne Endotoxins and Humidif-
     ier Disease," Clin. Allergy 14:109-112 (1984).

[4] Reed, C. E., Swanson, M. C., Lopez, M., Ford, A. M., Major,
     J., Witmer, W. B., and T. B. Valdes. "Measurement of IgG Anti-
     body and Airborne Antigen to Control and Industrial Outbreak
     of Hypersensitivity Pneumonitis," J. Occup. Med. 25(3):207-
     210 (1983).

[5] Ager, B. P. and J. A. Tickner. "The Control of Microbiological
     Hazards Associated with Airconditioning and Ventilation Sys-
     tems," Ann. Occup. Hyg. 27(4):341-358 (1983).

[6] Solomon, W. R. "Sampling Techniques for Airborne Fungi," in
     Mould Allergy, Y. Al-Doory and J. F. Damson, Eds., (Philadel-
     phia, PA, Lea and Febiger, 1984).

[7] Habenicht, H. A., Burge, H. A., Muilenberg, M. L., and W. R.
     Solomon. "Allergen Carriage by Atmospheric Aerosol II.
     Ragweed Pollen Determinants in Submicronic Atmospheric Frac-
     tions," J. Allergy Clin. Immunol. 74(1):64-67 (1984).

[8] Berstein, R. S., Sorenson, W. G., Garabrant, D., Beaux, C.,
     and R. D. Treitman. "Exposure to Respirable, Airborne
     Penicillium from a Contaminated Ventilation System: Clinical,
     Environmental and Epidemiological Aspects," Am. Ind. Hyg. J.
     44(3):161-169 (1983).

10.

Endogenous Mold Exposure: Environmental Risk to Atopic and Nonatopic Patients

Peter P. Kozak, Jr. and Janet Gallup

University of California at Los Angeles
Orange, California 92668

Leo H. Cummins and Sherwin A. Gillman

University of California – Irvine
Irvine, California

INTRODUCTION

As allergists we have had a particular interest in mold sensitivity. As clinicians we have also had the opportunity to evaluate our patients in the real world and to begin to relate environmental factors to quantitative mold spore exposure. Our patient population has been very cooperative, allowing us to perform extensive studies in their homes and, on occasion, allowing us to photograph interesting problems.

The overview of endogenous mold exposure is in part based on research carried out in our allergy practice and also reflects conditions encountered in Southern California. Because of the great differences in climate, home construction, use of various heating, cooling, humidifying, and filtering devices, the data may not be applicable to all areas of the United States. This work has been accomplished as several small projects over the past 7 to 8 years, and, in part, has been published or presented at various medical meetings. We are grateful to several organizations that have supported this research effort, including the American Lung Association of California, Honeywell Corporation, and the Los Angeles Chapter of the Allergy and Asthma Foundation.

Many patients and health professionals equate allergy care to treatment with immunotherapy. The allergist actually has several additional modalities that are used to reduce and control allergic symptoms. These include the use of diet, environmental controls, and symptomatic medications. Recent advances in immunology and protein chemistry have raised serious questions regarding the composition of commercial mold extracts and their reliability for skin testing and immunotherapy [1-3]. This has significantly restricted the use of mold immunotherapy by many allergists and increased their efforts in areas of environmental avoidance.

Like many allergists, we initially resorted to gravity-culture plate collection in homes of patients suspected of having mold allergy in an attempt to define the mold spores responsible for their problems. These early studies were not very fruitful, and the information derived was rarely helpful. We now obtain a more detailed history of the home, with particular emphasis on prior water damage. This information is then used to plan the strategy for the survey and to maximize the yield of useful data. The information form presently being used is shown as Table 1.

We are presently evaluating the home environment using two different volumetric sampling techniques in addition to Scotch-tape imprints of suspected mold-contamination items. The sampling methods used have been described previously [4]. A volumetric Andersen sampling is obtained from one outdoor location and several indoor areas. At least one Andersen and one rotorod study are obtained from the area of the home suspected of having an endogenous mold problem. We have found the Scotch-tape imprint technique to be especially helpful, and we use it for rapid identification of mold colonies growing on organic materials. The presence of mycelial structures with attached spores confirms that the mold is actively growing on the item in question.

FACTORS OF IMPORTANCE IN DETERMINING DISEASE MANIFESTATION

Host Factors

Environmental mold exposure has been the primary concern of the allergist, pulmonologist, and epidemiologist. Health care specialists must reassess patient mold exposure and determine what, if any, effect it could have on their patients. Certainly patients who are being treated with immunosuppresive measures should be protected from opportunistic infections, including those due to fungi. In addition, avoiding or minimizing exposure of patients with cystic fibrosis to Aspergillus might reduce some of their pulmonary complications and improve the quality of their lives. Similar avoidance might also reduce the risk of mold-sensitive patients in developing allergic bronchopulmonary aspergillosis.

Table 1. ENVIRONMENTAL SURVEY INFORMATION FORM

DATE_____PHONE_____

NAME_____ADDRESS_____

DWELLING TYPE_____SQUARE FEET_____AGE_____

CONSTRUCTION OF HOME: Concrete slab_____Hardwood floor_____

PERCENT CARPETED_____SCORE DUST CONTROL COMPLIANCE_____
NUMBER OF OCCUPANTS_____CHILDREN_____AGES_____
NUMBER SMOKERS IN HOME_____AMOUNT SMOKED_____

OUTDOOR SURROUNDINGS: Cultivated fields_____Number of trees_____

                       Landscaping_____Shade level_____

AVAILABLE MATERIAL FOR SAPROPHYTES (Organic debris)_____
OUTDOOR SOIL DRAINAGE_____
OTHER COMMENTS_____

TYPE OF HEATING: Forced air - Gravity - Space Heater - Electric,
                  On or Off During Survey

FILTERING SYSTEM - HEPA - Electrostatic - Other
                     Portable or Central / On or Off During Survey

AIR CONDITIONING - Portable or Central / On or Off During Survey

HUMIDIFICATION OR DEHUMIDIFICATION DEVICE - If present, describe
                  type/frequency of use/cleaning, etc.
On or Off During Survey

INDOOR PETS_____WHERE BEDDED_____HOUSE PLANTS (number/location)

ANY STANDING WATER_____CONDENSATION_____

WICKER/STRAW ITEMS (especially important if unfinished)
    Check for history of water damage_____

HAS THERE BEEN ANY WATER DISASTER IN THE HISTORY OF THE HOME? Y N
(e.g., water softener-washer overflow, roof leak, crack in ,
foundation etc.) - Describe in detail_____

ANY VISIBLE MOLD - If so, obtain Scotch-tape imprint and record

WEATHER AT TIME OF SURVEY_____
ANY UNUSUAL OUTDOOR ACTIVITY PRIOR TO OR DURING SURVEY_____

PARTICLE COUNTS (when appropriate) Indoor_____Outdoor_____
COMMENTS_____
       _____

Mold Spore Factors

Many physicians consider exposure to environmental molds to be occurring at a constant rate. In actuality, exposure to most mold spores is sporadic, reflecting the impact of outdoor activity, vegetation, and weather conditions, in addition to a variety of indoor factors. Based on the results of skin testing to spore-rich extracts, the relative frequency of isolation, and the concentration of exposure we have observed in Southern California, we suspect that colonization of the airway may be of clinical importance. If colonization of the airway is very transient, sensitivity may not occur. On the the other hand, if there is prolonged or repeated colonization, reaginic sensitivity could be established. Chronic colonization of the airway by mold spores could theoretically lead to chronic perennial allergic rhinitis and asthma.

Molds are generally considered to be an important cause of asthma and allergic rhinitis. Their contact with the respiratory mucous membranes is thought to be relatively brief, and their clearance from these areas is believed to be similar to clearance mechanisms for dust and pollen. For some mold spores, this in fact may be accurate. For other genera, especially for those capable of sustained growth at 37°C, prolonged contact with the respiratory tract may be possible through colonization of the airway. Mold spores are very complex packets of chemicals that contain a wide variety of enzymes and toxins. Inhaled mold materials could act through a variety of mechanisms, including complement activation [5] or through interaction or activation of other mediators of inflammation. Kauffmann and de Vries [6] have studied Aspergillus antigens and have reported the presence of enzymes with trypsin activity for Aspergillus fumigatus precipitants 13 and 18, with additional chymotryptic activity for precipitant 18. Mold genera, which because of their size, aerodynamic characteristics, and viability at 37°C are likely to colonize the airway, include various species of Aspergillus, Candida, Scedosporium, Scopulariopsis, Geotrichium, and Paecilomyces.

INTERPRETATION OF DATA DERIVED FROM ENVIRONMENTAL MOLD SURVEYS

Much of our effort has been directed toward defining "normal mold spore exposure," indoor and outdoor factors of importance in determining indoor mold levels, and clarifying indoor conditions that significantly alter endogenous mold exposure. To properly interpret our data, you have to understand how they were generated. Virtually all data are derived from mold surveys performed in homes of our patients. This represents a very select group of middle- to upper-class families residing in Southern California. Surveys were performed only during periods of stable weather; surveys were not scheduled during periods of wind or rain. Patients were asked to avoid changes in their usual pattern of cleaning and housekeeping.

Most patients, however, had been instructed in general dust control
and were to some degree attempting to comply with these instruc-
tions.   Homes in Southern California rarely have basements.   Central
humidifiers and swamp coolers are infrequently present.   Many of our
patients had installed central electrostatic filtering units, but,
because of electrical cost, many were using these units only intermittently.   A comparison of data from different sections of the
United States [7-9] indicates significant seasonal and regional
differences.   Additional quantitative mold surveys are required in
various sections of the country to further define local exposure and
to evaluate factors that may influence patient exposure.   Caution is
recommended when using data derived from other sections of the country to formulate treatment strategies for patients living under
vastly different conditions.

   Mold spore counts have limited value in assessing the importance of a particular genera in causing disease.   For atopic
diseases, the total count is not all that is important.   Factors
such as solubility of surface-spore antigens, antigenic mass, ability of the allergens to stimulate IgE production, antigenic cross
reactivity with other genera, and the aerodynamic characteristics of
the spores are crucial in determining the importance of a particular
mold genera.   As previously noted, viability may be important for a
select group of molds.   The relative concentration and aerodynamic
characteristics of aerosolized dust containing mold could be important in producing pulmonary changes by complement activation or
activation of other mediators of inflammation.

PRESENTATION OF DATA

   In our original study, we attempted to evaluate a number of
indoor and outdoor factors that could affect the level of indoor
mold spores.   Although the first study was not as comprehensive as
our present surveys, it does supply a data base for comparison.

   In the initial study [9], we evaluated 68 homes.   Because of
limited resources, we obtained only one indoor Andersen sample for
study of viable mold spores.   These homes were surveyed during
stable weather to minimize the effects of weather conditions; no
attempt was made to alter the usual activities of a family or to
change their life style.   The Andersen study was performed in either
the living room or family room.   The mean viable spore count was
437.7 spores/m^3, with a range of 36 to 5984.   The 39 taxa identified are listed in Table 2.   In retrospect, there were several homes
in the initial group that had endogenous mold problems.
Cladosporium was identified in all 68 homes, with a mean concentration of 437.7 spores/m^3.   Other commonly isolated genera included
Penicillium (91.2%), Alternaria (87%), Epicoccum (52.9%),
Aspergillus species (48.5%), and Drechslera (38.2%).   Aspergillus

Table 2.   Isolation, Frequency, and Concentration of Viable Molds
Identified in a Survey of 68 Homes in Southern California
(Reproduced with the permission of <u>Annals of Allergy</u>)

Mold Genera	Percent of Homes in Which Genera Isolated	Range of Spores/m^3	Mean of Spores/m^3
Cladosporium	100	12–4637	437.7
Penicillium species	91.2	0–4737	168.9
Nonsporulating mycelia[a]	89.7	0–494	44.3
Alternaria	87.0	0–282	30.7
Streptomyces	58.8	0–212	28.1
Epicoccum	52.9	0–153	9.6
Aspergillus species	48.5	0–306	15.0
Aureobasidium	44.1	0–294	8.0
Drechslera (Helminthosporium)	38.2	0–94	6.9
Cephalosporium	36.7	0–59	5.3
Acremonium	35.3	0–188	3.6
Fusarium	25.0	0–47	4.5
Botrytis	23.5	0–54	2.9
Aspergillus niger	19.1	0–59	2.9
Rhizopus	13.2	0–24	1.4
Rhodotorula	11.8	0–29	1.5
Beauvera	10.3	0–12	0.7
Chaetomium	8.8	0–47	1.2
Unknown	8.8	0–34	1.2
Scopulariopsis	8.8	0–25	0.9
Mucor	7.4	0–41	1.4
Curvularia	7.4	0–12	1.1
Rhinocladiella	4.4	0–12	0.5
Verticillium	4.4	0–12	0.4
Plenozythia	4.4	0–6	0.3
Pithomyces	2.9	0–25	0.4
Zygosporium	2.9	0–18	0.4
Paecilomyces	2.9	0–12	0.3
Stachybotrys	2.9	0–12	0.3
Aspergillus fumigatus	2.9	0–5	0.2
Nigrospora	2.9	0–5	0.1
Stysanus	2.9	0–6	0.1
Leptosphaerulina	1.5	0–18	0.3
Botryosporium	1.5	0–6	0.1
Trichoderma	1.5	0–12	0.2
Chrysosporium	1.5	0–6	0.1
Phoma	1.5	0–6	0.1
Sporobolomyces	1.5	0–6	0.1
Trichothecium	1.5	0–6	0.1
Ulocladium	1.5	0–5	0.1
Yeast	1.5	0–5	0.1
Geotrichum	1.5	0–3	0.04

[a]Subcultures of nonsporulating mycelia from one home (grown
on Moyer's multiple media)   subsequently produced <u>Torula herbarum</u>
colonies.

niger was identified in 19.1% of the homes, and A. fumigatus in only 2.9%.

Outdoor factors that appeared to effect the indoor mold spore level (Tables 3-5) included marked shade, marked levels of organic debris near the home, and natural or basically uncared for property. Outdoor levels of shade had the greatest effect on the mean indoor viable mold spore levels (2228 spores/m^3). Statistically significant higher indoor mold counts were also noted when there were marked levels of organic debris near the home; a mean viable spore level of 1312/m^3 was noted in homes with a high level of organic debris compared with levels of 461 and 401, respectively, for homes with low and moderate levels of organic debris. Homes without formal landscaping, or where there was no attempt at maintaining the landscaping, had a mean indoor spore count of 1047.

Table 3. Effect of Shade Level Near the Home
on Indoor Mold Spore Isolation
(Reproduced with the Permision of Annals of Allergy)

	Shade[a]		
	Minimal	Moderate	Marked
Mean	408	421	2228
Minimum	36	36	1109
Maximum	3828	2004	5984
Homes/category	40	22	6

[a]Mold levels reported in spores/m^3 (p < 0.0003).

Table 4. Effect of Organic Debris Level Outdoors
on Indoor Mold Spore Isolation
(Reproduced with the Permission of Annals of Allergy)

	Level of Debris[a]		
	Low	Moderate	Marked
Mean	461	401	1312
Minimum	77	36	130
Maximum	3828	2004	5984
Homes/category	27	34	7

[a]Mold levels reported in spores/m^3 (p < 0.02).

Table 5. Effect of Landscaping/Maintenance
on Indoor Mold Spore Isolation
(Reproduced with the Permission of <u>Annals of Allergy</u>)

	Average	Lush	Natural
Mean	427	424	1047
Minimum	77	36	130
Maximum	1923	3828	5984
Homes/category	29	30	9

Mold levels reported in spores/m^3 (p < 0.046).

Two indoor characteristics were associated with statistically
lower levels of spore isolation. The greatest effect was noted with
central electrostatic filtration. As shown in Table 6, the mean
mold spore count was 687/m^3 in homes without electrostatic filtration. If a filter was present but used only intermittently, the
mean level dropped to 344/m^3; homes where the electrostatic filter
was operated continuously had mean spore levels of 155/m^3. A very
significant effect was also observed in homes where there was good
compliance with dust control. Homes in which compliance was the
least (level 5 and 6) had the highest spore levels (822/m^3), whereas
homes with the best compliance (level 10) had the least amount of
mold spores (292/m^3).

Table 6. Comparison of Viable Mold Spore Levels[a] in Homes
with Central Electrostatic Filtration (CEF) vs those Without
(Reproduced with the Permission of <u>Annals of Allergy</u>)

	No CEF	Intermittent CEF	Continuous CEF
Mean	687	344	155
Minimum	106	125	36
Maximum	5984	1038	755
Homes/category	40	8	13

[a]Mold levels reported in spores/m^3 (p < 0.00005).

Solomon [7] assessed the indoor prevalence of viable mold
spores in Michigan (using Andersen sampling) and noted significant
differences between mold isolates recovered during the winter and
spring-summer. During the frost-free period, indoor viable spore
levels were approximately 25% of the levels outdoors and basically
reflected a similar type of genera. During the winter, when snow
cover was generally present, the outdoor viable mold level never

exceeded 230 isolates/m^3. Indoor levels ranged from 20 to over
14,000 isolates/m^3, with 18% of the homes having levels in excess of
1000. Penicillium, Aspergillus, Oospora, and Sporothrix represented
the most frequent genera isolated.

In a second study, Solomon [8] reported more extensive studies
involving 150 single-family homes in Michigan that were evaluated
during the winter when snow cover was generally present. Although
the outdoor viable spore count never exceeded 230/m^3, indoor levels
ranged from 10 to over 20,000 isolates/m^3. The dominant genera
encountered indoors were Penicillium, Aspergillus, Cladosporium,
Rhodotorula, and nonpigmented yeast. In some homes, however, other
genera, including Cephalosporium, Sporobolomyces, Verticillium, and
Sporothrix, were frequently isolated. Solomon [8] associated higher
levels of mold isolates with the relative humidity present at the
time of the study. In homes without central humidification devices,
the relative humidity at the time of the survey did not exceed 31%.
The viable mold levels in these homes never exceeded 370
isolates/m^3, with mean levels of only 43/m^3. Fourteen of the homes
had central electrostatic filtration. When compared with similar
homes having the same relative humidity at the time of the survey,
Solomon noted a 16% reduction for mold isolates in homes with central electrostatic filtration. Of the 21 electrostatically filtered
homes studied, 8 had viable mold isolates exceeding 500/m^3, with 3
having levels in excess of 1200.

Between 1976 and 1979, we studied 32 homes surveyed before and,
on two occasions, after installation of a central electrostatic
filter [10]. Andersen sampling was performed outdoors and in two
indoor areas. Location A was either the living room or family room,
and location B was an adult bedroom. The same site was consistently
sampled on all occasions. The occupants were asked to keep the windows and doors closed, and the forced-air fan was run continuously
for at least 24 h prior to the study. Repeat studies were performed
under similar conditions on two additional occasions after installation of a central electrostatic filter and with the unit and fan
operating continuously. Twenty-five different genera were isolated
from the living/family room area, with twenty-six genera isolated
from the bedroom. Twenty-one genera were isolated in both areas,
with a pattern similar to the outdoors. Results of the prefilter
installation surveys are summarized in Tables 7 and 8. The mean
viable spore count for the living/family room area was 660/m^3, with
a range of 11 to 3708. Summary results of these studies are shown
in Table 9. Because of the wide variability of the data, we used
the logarithm of the spore count to achieve a normal distribution
for statistical analysis. During the control period, the median
indoor viable spore level in the living/family room area was 31.4%
of the outdoor level; the bedroom was 30.2%. A reduction to 16.3%
(p <0.04) and 15.5% (p <0.005) of the outdoor levels occurred in the
living/family room and bedroom, respectively, when the homes were
evaluated after installation of the central electrostatic filter.

Table 7.   Summary of 32 Homes Surveyed Prior to Installation of a
Central Electrostatic Filter (Outdoor Area)

Mold Genera	Frequency of Isolation (%)	Mean[a] for Total Isolation	Range of Spores/m^3
Total molds/m^3		1283	212–3884
Acremonium	31	134	35–776
Alternaria	81	143	35–424
Aspergillus fumigatus	6	35	35
Aspergillus niger	13	62	35–141
Aspergillus flavus			
Aspergillus species	31	85	35–282
Aureobasidium	25	35	35
Beauveria	13	115	35–353
Botryosporium			
Botrytis	19	41	35–71
Cephalosporium	19	47	35–71
Chaetomium			
Cladosporium	97	648	35–2436
Curvularia	3	35	35
Drechslera	16	70	35–141
Epicoccum	25	57	35–106
Fusarium	16	49	35–106
Geotrichum	3	212	212
Mucor			
Nigrospora			
Nonsporulating mycelia	86	141	11–671
Paecilomyces			
Phoma	6	71	35–106
Penicillium	75	236	35–1978
Pithomyces	3	35	35
Plenozythia	3	35	35
Rhinocladiella	13	35	35
Rhizopus			
Rhodotorula	3	35	35
Sporobolomyces	3	35	35
Stachybotrys	3	35	35
Streptomyces	38	115	353
Stemphyllium			
Scop. species	3	35	35
Ulocladium	3	35	35
Unknown	13	132	35–247
Zygosporium	3	288	288
Others			

[a]Determined as mean for only those studies in which genera
was isolated.

Table 8. Summary of 32 Homes Surveyed Prior to Installation of a Central Electrostatic Filter (Living/Family Room Area)

Mold Genera	Frequency of Isolation (%)	Mean[a] for Total Homes	Range of Spores/m^3
Total molds/m^3		660	11-3708
Acremonium	31	39	11-83
Alternaria	75	58	11-353
Aspergillus fumigatus	3	12	12
Aspergillus niger	9	31	11-59
Aspergillus flavus			
Aspergillus species	31	36	12-106
Aureobasidium	19	24	12-35
Beauveria	6	18	12-24
Botryosporium			
Botrytis	16	16	11-35
Cephalosporium	16	26	11-83
Chaetomium			
Cladosporium	94	293	23-1306
Curvularia	3	24	24
Drechslera	34	24	12-71
Epicoccum	38	29	11-106
Fusarium	9	12	12
Geotrichum			
Mucor	6	24	12-35
Nigrospora	3	12	12
Nonsporulating mycelia	75	62	12-223
Paecilomyces	6	29	24-35
Phoma	6	36	24-47
Penicillium	86	227	12-2942
Pithomyces	6	18	12-24
Plenozythia	3	12	12
Rhinocladiella	6	18	12-24
Rhizopus	3	11	11
Rhodotorula	6	12	12
Sporobolomyces	13	27	12-47
Stachybotrys			
Stemphyllium			
Streptomyces	59	31	11-129
Ulocladium			
Unknown	9	59	12-129

[a]Determined as mean for only those homes in which genera was isolated.

Table 9.   Summary of Andersen Sampling of 32 Homes Before and After
Installation of a Central Electrostatic Filter

	Living/Family Room	Bedroom	Outdoors
Prior to Installation			
Median	341	317	1,112
Mean	660	565	1,283
Minimum	11	35	212
Maximum	3,708	3,542	3,884
First Postinstallation study			
Median	157	93	706
Mean	192	197	1459
Minimum	11	11	71
Maximum	1,246	1,141	20,833
Second Postinstallation study			
Median	122	122	742
Mean	160	173	1,167
Minimum	0	0	71
Maximum	445	717	6,991

Two factors appear to account for the lower levels of mold isolates in homes with electrostatic filtration. The first is a barrier effect of having the home "closed" and reducing the influx of
outdoor mold spores into the indoor environment. A second effect
appears to be caused by the filter itself.

Most recently, we have begun an analysis of another 186 homes
surveyed over the past 4 years. Eighty of these homes had a rotorod
study performed indoors, in addition to the routine Andersen sampling. Of the 80 homes, 63 had a history of interior water damage
and were strongly suspected of having an endogenous mold problem.
The characteristics of homes with endogenous mold problems were studied. Of the 63 homes, 83% had a forced-air heating system; 19% had
central electrostatic filtration, with 58% of the units operating
continuously. Eighty-nine percent of the homes either did not have
air conditioning, or the unit was not operating at the time of the
study. Fifty-one percent of the mold-problem homes had only one
problem. Two mold problems were identified in 36.4% of the homes,
with three problems noted in 12.6% of the mold-problem homes. The
most likely area to have a problem was the bathroom (31%), followed
by the living room (18.4%), family room/den (16.5%), and adult bedrooms (11.7%). The material most likely to be damaged was the
jute-backed carpeting/baseboard (53.4%), followed by wicker/straw
baskets (17.5%) and walls, ceilings, and window frames (13.6%).

The cause of the problem was chronic water spills in 35% of the homes, followed by recurrent water leaks (in 20.4% of homes) from plumbing fixtures. Approximately 10% of the problems were attributed to each of the following: a one-time isolated water disaster, a roof leak, influx of outdoor water into the structure, or a construction/structural defect. Either the structure itself or its furnishings were wet for longer than 14 d in approximately 90% of the homes studied.

Excluding homes with electrostatic filtration or outdoor factors (or both) that could increase environmental mold (e.g., shade), we compared viable mold spore levels for adult bedrooms and family rooms/dens selected from the 80 homes. Twenty-nine adult bedrooms from mold-problem homes are compared with eleven adult bedrooms from control homes without a history of a mold problem (Table 10). The mean viable spore count for problem homes was 4200 spores/m^3 compared with 834 spores/m^3 in the nonproblem homes. Increases in mean spore levels were noted for specific genera, including Penicillium (mean 2405 vs 108), Aspergillus species (mean 813 vs 12), and Cladosporium (mean 816 vs 496). An even more dramatic difference was noted when family rooms/dens from 26 mold-problem homes were compared with family rooms/dens from 14 nonproblem homes (Table 11). The mean viable spore count was 7595 in the problem homes and only 597 in the control family rooms/dens. The same three genera were again noted to be elevated. The mean level for Penicillium was 5512 in the problem homes compared with only 219 in nonproblem homes. Increases were noted for the Aspergillus species (mean of 591/m^3 in problem homes compared with 16/m^3 in the controls) and for Cladosporium (mean of 759 in problem homes vs 253 in nonproblem homes).

The effects of central electrostatic filtration on the indoor mold spore level was evaluated for the 63 mold-problem homes. The results of Andersen sampling are shown in Table 12. The mean viable spore levels for the 51 homes without electrostatic filtration was 4075/m^3. Homes with intermittently operating units had mean levels of 1052 spores/m^3, whereas homes with units used continuously had mean levels of 567. Decreases in mold isolates involved mainly Alternaria, Aspergillus species, Chaetomium, Cladosporium, and Penicillium. A similar analysis of rotorod surveys (Table 13) from the same 63 homes also indicated a decrease in the total mold spore level for Alternaria, Chaetomium, and Cladosporium. The mean rotorod spore count for the 51 unfiltered homes was 3426, with mean levels of 599 in homes with intermittently operated units, and 341 when the filter was used continuously. Most of the hyaline spores detected by rotorod sampling were of the genera Aspergillus and Penicillium. A total of 2154 spores were identified in the nonfiltered homes vs 95 and 57, respectively, in the intermittently filtered and continuously filtered homes. The high level of mycelial fragments, Stachybotrys, and Torula herbarum could not be detected with Andersen sampling.

Table 10.   Andersen Study of Adult Bedrooms in Mold-Problem
Homes Compared with those in Nonmold-Problem Homes
(most frequently isolated fungi, spores/m^3)

Mold Genera	Nonmold Problem		Mold Problem	
	No. Homes	Mean	No. Homes	Mean
Total molds/m^3	11	834	29	4200
Alternaria	8	55	21	45
Aspergillus species	1	12	15	813
Beauveria	2	18	6	28
Cladosporium	11	496	28	816
Fusarium	1	35	5	17
Mucor			3	43
Nonsporulating mycelia	9	55	22	72
Penicillium	10	108	29	2405
Rhinocladiella	1	35	4	59
Streptomyces	9	50	16	138
Ulocladium			4	15
Unknown			2	18

Table 11.   Andersen Study of Family Room/Den Areas in
Mold-Problem Homes Compared with those in Nonmold-Problem Homes
(most frequently isolated fungi, spores/m^3)

Mold Genera	Nonmold Problem		Mold Problem	
	No. Homes	Mean	No. Homes	Mean
Total molds/m^3	14	597	26	7595
Alternaria	7	47	19	65
Aspergillus species	3	16	15	591
Beauveria	1	12	3	43
Cladosporium	14	253	25	759
Fusarium	1	18	2	30
Mucor	1	12	6	34
Nonsporulating mycelia	11	32	21	138
Penicillium	14	219	26	5512
Rhinocladiella	1	12		
Streptomyces	5	42	17	71
Ulocladium	2	24		
Unknown	1	24	5	158

Table 12. Andersen Study of 63 Mold-Problem Homes
with and without Central Electrostatic Filtration
(most frequently isolated fungi, spores/m^3)

Mold Genera	No CEF[a]		Intermittent CEF		Continuous CEF	
	No. Homes	Mean	No. Homes	Mean	No. Homes	Mean
Total molds/m^3	51	4075	5	1052	7	567
Alternaria	38	64	3	39	5	21
Aspergillus species	35	686	4	59	5	50
Chaetomium	12	56	1	17	1	12
Cladosporium	50	847	5	561	7	217
Nonsporulating mycelia	44	89	5	86	2	18
Penicillium	51	1913	5	92	6	83
Stachbotrys	6	153			1	635
Streptomyces	32	101	4	21	2	206
Ulocladium	5	779	1	12		

[a]Central electrostatic filtration.

Table 13. Rotorod Study of 63 Mold-Problem Homes
with and without Central Electrostatic Filtration
(most frequently isolated fungi, spores/m^3)

Mold Genera	No CEF[a]		Intermittent CEF		Continuous CEF	
	No. Homes	Mean	No. Homes	Mean	No. Homes	Mean
Total molds/m^3	51	3426	5	599	7	341
Alternaria	49	130	3	76	6	70
Chaetomium	14	97	1	19	2	50
Cladosporium	49	956	5	163	7	86
Hyaline spore	26	2154	2	95	1	57
Mycelial fragment	47	170	4	71	6	46
Stachbotrys	10	290			1	76
Torula herbarum	11	2487	2	29	1	8
Ulocladium	10	41	2	48	1	34

[a]Central electrostatic filtration.

A summary of rotorod results comparing the 63 mold-problem homes with the 17 normal controls is presented in Table 14. To remove the effect of electrostatic filtration, homes with these units were excluded from the analysis. A number of potential aeroallergens clearly originate outdoors, and their presence indoors is not being fully appreciated. These include Algae, Ascospores, Basidiospores, Myxomycetes, Rusts, and Smuts. The frequency of isolation and the concentrations present indoors make these materials suspect as potential causes of environmental allergens. Increases in mean isolates for several genera were noted, including Cladosporium, Hyaline spores, Stachybotrys, and Torula herbarum. Based on our observations and the data presented, we feel that concurrent sampling should be performed using both a viable and nonviable quantitative device.

Table 14.   Rotorod Study Comparing 14 Control (Nonproblem)
Homes with 47 Homes with Endogenous Mold Problems[a]

Mold Genera	Nonmold Problem			Mold Problem		
	(%/Home)	Mean	Range	(%/Home)	Mean	Range
Total molds/m^3		1,150	99-4,966		3,641	87-40,192
Algae	35.7	26	15-57	8.5	241	18-833
Alternaria	85.7	165	8-568	95.7	133	18-682
Ascospores	14.3	65	15-114	19.1	36	9-114
Aureobasidium	64.3	50	15-133	46.8	64	9-227
Basidiospores	57.1	53	15-152	63.8	53	10-212
Botrytis	14.3	38	38	6.4	29	19-38
Chaetomium	21.4	70	38-95	27.7	97	16-379
Cladosporium	92.8	448	46-1,515	95.7	1,016	10-22,539
Curvularia	7.1	114	114	10.6	25	8-46
Dreschslera	35.7	49	19-76	42.5	36	9-114
Epicoccum	42.9	110	15-227	50.0	79	8-341
Fusarium	7.1	8	8	2.1	76	76
Hyaline spores	57.1	78	19-227	48.9	2,418	19-32,956
Mycelia fragment	92.9	267	15-2,008	91.5	173	10-1,023
Myxomycetes	21.4	22	8-38	29.8	91	15-379
Pithomyces	28.6	21	8-38	8.5	43	19-76
Rust	42.9	60	23-152	78.7	71	8-265
Smuts	64.3	98	19-417	68.1	99	8-492
Stachbotrys	7.1	19	19	19.1	309	19-1,723
Stemphyllium	7.1	19	19	6.4	24	15-38
Torula herbarum	7.1	19	19	23.4	2,487	16-25,872
Ulocladium				19.1	45	9-114
Unknown brown	7.1	190	190	25.5	140	17-890
Others	7.1	38	38	38.3	120	15-1,515

[a]Homes with electrostatic filtration or outdoor characteristics that could increase endogenous mold excluded from analysis

Most of the information regarding prevalence and concentration of indoor mold exposure comes from studies performed in a stable setting with no impact from activity of the inhabitants or weather conditions. On several occasions we have had the opportunity to study structures where the conditions or findings were unique. Examples of several surveys are given in the addendum to this chapter along with a brief history to highlight the studies.

SUMMARY

We are constantly being exposed to molds in our environment. Indoor concentrations of mold spores can be attributed to several outdoor factors including shade, level of organic debris, and landscape maintenance. Indoor mold exposure in Southern California is relatively stable in homes without endogenous mold problems. Exposure to mold spores indoors can be reduced by basic dust control and the use of central electrostatic filtration.

Endogenous mold problems generally occur after prolonged or repeated water damage to a variety of organic materials including unfinished wood, jute-backed carpet, wallboard, window frames, wallpaper, books, leather goods, and wicker and straw baskets.

To fully evaluate a home or other structure, several studies are recommended:

1. at least one outdoor sampling for viable mold to determine the variety and concentration of spores present;

2. multiple indoor studies for viable mold spores, with particular emphasis in areas that historically have had repeated water damage; and

3. at least one indoor rotorod survey (or similar volumetric nonviable evaluation) to microscopically identify mold spores present in the area suspected of having a mold problem.

With the present emphasis on energy-efficient structures, we can anticipate some reduction in the influx of outdoor mold spores. If repeat or chronic interior water damage occurs in the presence of organic material, a significant increase in endogenous mold levels would be expected. Additional problems will undoubtedly occur that are secondary to contamination of humidification-control and filtering devices. Extensive studies are required to better define indoor mold exposure and to monitor changes that occur as we manipulate the indoor environment. Our goal should be to evolve an indoor environment that would afford maximum health to all inhabitants at a reasonable energy cost.

ADDENDUM

The results of environmental surveys illustrate several points that have not been fully discussed. Virtually all of our data have been obtained during periods of stable weather. The effects of the Santa Ana winds on one of our homes is shown in Table 15. The outdoor viable spore count was 43,946/m^3 compared with our usually expected outdoor count of 1000 to 1500. A total of 3344 mold isolates/m^3 were recorded for the living room, and 10,961 were identified in the adult bedroom. Many genera were increased, most of which was attributed to Cladosporium, Fusarium, and Penicillium.

The effects of outdoor activity on mold exposure have not been extensively studied. Patients repeatedly observe increased wheezing with yard work and grass cutting. If windows and doors to the home are open at the time of these activities, one would certainly expect some increase in the indoor mold flora. Two examples of outdoor activity on the outdoor mold spore levels are shown in Tables 16 and

Table 15. Andersen Study of the Effect of Santa Ana
Winds on Indoor Mold Exposure

Mold Genera	Living Room	Bedroom	Outdoor
Total molds/m^3	3,344	10,961[a]	43,946[a]
Alternaria	182	374	214
Aspergillus niger		32	54
Aureobasidium	65	62	214
Cladosporium	2,017	8,403	32,965
Curvularia	13		
Dreschslera	79		
Epicoccum		32	535
Fusarium	13	62	7,492
Nonsporulating mycelia	117	32	277
Penicillium	598	1,200	1,766
Streptomyces	104	500	321
Miscellaneous	156	264	108

[a]Presence of slower-growing genera probably obscured
by more-rapidly growing genera.

Table 16. Andersen Study of the Effects of
Disturbing Compost on Outdoor Mold Spore Level

Mold Genera	Before	After
Total molds/m^3	493	1,231,612[a]
Aspergillus fumigatus		676,822
Aureobasidium	35	
Cladosporium	388	206,775
Epicoccum		336,716
Penicillium	35	11,299
Streptomyces	35	

[a]Presence of slower-growing genera probably
obscured by more-rapidly growing genera.

17. Patient K.R. has allergic bronchopulmonary aspergillosis. He
had a major problem with outdoor exposure to Aspergillus fumigatus
that was detected only after a second Andersen sampling was performed after compost had been disturbed. Outdoor Andersen sampling
of the M.L. home occurred prior to the arrival of gardeners. A
second study performed while the yard work was being done offers an
interesting contrast.

Patients, especially children, are exposed to multiple environments in the course of a day. Results of surveys made in a mold-
damaged playhouse and in a normal school are presented in Tables 18
and 19, respectively.

Table 17. Andersen Study of the Effect
of Yard Work on Outdoor Mold Spore Level

Mold Genera	Before	During
Total molds/m^3	142	21,026[a]
Alternaria		566
Aspergillus species		743
Aureobasidium		71
Botrytis		35
Cephalosporium		106
Cladosporium	71	13,877
Drechslera		248
Epicoccum		35
Fusarium		106
Nonsporulating mycelia		2,124
Penicillium	71	2,372
Streptomyces		743

[a]Presence of slower-growing genera probably
obscured by more-rapidly growing genera.

Table 18. Andersen Study of the Effect of a Water-Damaged
Outdoor Playhouse Compared to Child's Bedroom

Mold Genera	Bedroom	Playhouse
Total molds/m^3	2,520	28,206[a]
Alternaria	106	671
Aspergillus fumigatus		247
Aspergillus species		1059
Aureobasidium	12	106
Cladosporium	1,377	20,509
Fusarium	12	71
Nonsporulating mycelia	24	318
Penicillium	388	4,377
Rhizopus	12	71
Streptomyces	106	706
Miscellaneous	483	71

[a]Presence of slower-growing genera probably obscured
by more-rapidly growing genera.

Table 19.   Andersen Study of Multiple Classrooms
in Nursery School

| | Classroom | | | | | |
Mold Genera	1	2	3	4	5	Outdoors
Total molds/m^3	530	414	353	402	330	4626
Acremonium	12					
Alternaria	24	24	35	24	24	282
Aspergillus niger		24		12		
Aspergillus species			12	12		35
Aureobasidium		24			12	530
Botrytis			12			
Cephalosporium		12				71
Cladosporium	200	224	188	200	129	3249
Drechslera					12	71
Epicoccum	12					35
Fusarium	12		24			106
Nonsporulating mycelia		24	35	24		141
Penicillium	235	82	47	47	141	71
Rhizopus				12	12	
Streptomyces	35			71		
Ulocladium						35

Adult exposures can also vary, depending on hobbies and work exposure. Table 20 shows results of a survey performed in a home and a detached greenhouse. Results of a survey made in a water-

Table 20.   Andersen Study – Comparison of Mold Exposure Outdoors
with Living Room and Detached Greenhouse

Mold Genera	Living Room	Greenhouse	Outdoors
Total molds/m^3	2,638	36,112[a]	2,012
Alternaria	12		247
Aspergillus species	129		35
Aureobasidium			71
Botrytis	24		
Chaetomium		71	
Cladosporium	1,495	22,239	1,165
Drechslera	24		
Mucor	12		
Nonsporulating mycelia	141		141
Penicillium	777	13,767	353
Rhizopus		35	
Stemphyllium	24		

[a]Presence of slower-growing genera probably obscured by more-rapidly growing genera.

damaged business office are presented in Table 21. Additional examples of significant mold exposure can be found in an earlier publication [11].

Table 21.   Andersen Study – Water–Damaged Office
and Adjacent Waiting Room

Mold Genera	Office	Waiting Room	Outdoors
Total molds/m^3	5649	389	424
Alternaria	247		71
Aspergillus species	1412	177	
Cladosporium	2684	71	177
Drechslera		35	
Epicoccum	35		35
Nonsporulating mycelia	141		71
Penicillium	1130	106	35
Streptomyces			35

REFERENCES

[1]   Yunginger, J.W., R.T. Jones, and G. J. Gleich. "Studies of
      Alternaria Allergens.II. Measurement of the Relative Potency
      of Commercial Alternaria Extracts by the Direct RAST and by
      RAST Inhibition," J. Allergy Clin. Immunol. 58:405 (1976).

[2]   Aukrust, L., and K. Aas. "The Diagnosis of Immediate-type
      Allergy to Cladosporium herbarium," Allergy 33:24-29 (1978).

[3]   Aas, K., et al. "Immediate-type Hypersensitivity to Common
      Molds. Comparison of Different Diagnostic Materials," Allergy
      35:443 (1980).

[4]   Kozak, P. P., J. Gallup, L. H. Cummins, and S. A. Gillman.
      "Currently Available Methods for Home Mold Surveys. I.
      Description of Techniques," Ann. Allergy 45:85-89 (1980).

[5]   Marx, J. J., Jr., and D. K. Flaherty. "Activation of the Com-
      plement Sequence by Extracts of Bacteria and Fungi Associated
      with Hypersensivity Pneumonitis," J. Allergy Clin. Immunol.
      57:328 (1976).

[6]   Kauffmann, H. F., and K. de Vries. "Antibodies Against
      Aspergillus fumigatus. II. Identification and Quantification
      by Means of Cross Immunoelectrophoresis," Int. Arch. Allergy
      Appl. Immunol. 62:265-275 (1980).

[7]   Solomon, W. R. "Assessing Fungus Prevalence in Domestic Inte-
      riors," J. Allergy Clin. Immunol. 56:235-242 (1975).

[8]   Solomon, W. R. "A Volumetric Study of Winter Fungus Prevalence
      in the Air of Midwestern Homes," J. Allergy Clin. Immunol.
      57:46-55 (1976).

[9]   Kozak, P. P., J. Gallup, L. H. Cummins, and S. A. Gillman.
      "Factors of Importance in Determining the Prevalence of Indoor
      Molds," Ann. Allergy 43:88-94 (1979).

[10] Kozak, P. P., J. Gallup, L. H. Cummins, and S. A. Gillman.
      "Effect of Central Electrostatic Filtration on Indoor Viable
      Mold Spore and Dust Levels," presented at the annual meeting
      of the Academy of Allergy in Atlanta, Feb. 19, 1980.

[11] Kozak, P. P., J. Gallup, L. H. Cummins, and S. A. Gillman.
      "Currently Available Methods for Home Mold Surveys. II. Exam-
      ples of Problem Homes Surveyed," Ann. Allergy 45:167-176
      (1980).

11.

Hygienic Significance of Microorganisms in the Hospital Environment

Ruth B. Kundsin

Brigham and Women's Hospital, Harvard Medical School, Boston, Ma.

Urinary tract infection, surgical wound infection, and pneumonia rank highest in the hospital's rate of nosocomial infections. The role the hospital environment plays in each can be scientifically determined. Each hospital is unique. Each hospital has different standards of cleanliness, different habits associated with patient care, and different microorganisms in the environment. The philosophic ideology of patient care also varies greatly. Scientific validation is necessary, however, to support the ideology. Constant reevaluation of practices needs to be done particularly with the introduction of new equipment for patient care.

Microorganisms can be found in the environment associated with droplet nuclei, droplets, or dust. Their state of suspension determines the role they play in the transmission of human disease, and simultaneously defines their final disposition as well as the methods most appropriate for their destruction. Droplet nuclei are the smallest particles. They are the bacterial or viral residues from the evaporation of larger particles expelled by coughing, sneezing, talking. These residues are airborne for long periods and are carried by air currents until inhaled or vented. Their control is more elusive than that of dust and droplets, the large particles, which settle out rapidly onto horizontal surfaces and can be eliminated by germicides. Microorganisms in both airborne droplet nuclei and accumulations of dust and droplets on surfaces can be destroyed by properly designed exposure to ultraviolet irradiation.

Table I from Wells (1), shows a comparison of characteristics of dust, droplets, and droplet nuclei as to source, mode of suspension, particle diameter, and settling velocity. These characteristics determine the

pattern and type of infections that result. For
example, the seasonal patterns of spread of measles and
chickenpox are compatible with the transmission by
airborne droplet nuclei. In the winter when people
congregate indoors, the spread of these infections
peaks. Those peaks and valleys disappear as air-
conditioning during the summer becomes more readily
available and air is recirculated while fresh air
intake is reduced to a minimum in the interest of energy
conservation. With air-conditioning, a winter "high
peak" environment is maintained all year round.

Table I

Comparison of Dust, Droplets, and Droplet Nuclei

	Dust	Droplets	Droplet nuclei
Sources	solid matter, cellulose, skin squames etc.	Fluids from nose and throat	Solid residues of evaporated droplets
Production	Attrition	Atomization of fluids	Evaporation of droplets
Mode of suspension	Air wafted	Projected into air by sneezing, coughing, etc.	Caught in air
Particle diameter	10-100 microns	>100 microns	2-10 microns
Settling velocity	1 ft./min.to 1 ft./sec.	>1 ft./sec.	<1 ft./min.

     The measurement techniques for particles are
dependent on their aerodynamics. The large-sized
particles, dust and droplets can be collected as these
settle on standard petri dishes. The fallout can be
quantified by timing exposure and dividing the total
count obtained following incubation by the number of
fifteen minute increments. The number and type of
microorganisms recovered describe the fallout per square
foot per minute.

Another parameter, the accumulation of microorganisms on surfaces can also be used to evaluate housekeeping procedures as well as microorganisms shed by patients and personnel. Accumulation can be detected by the use of a Rodac plate. This is a shallow plastic plate, two and one eighth inches in diameter, filled with an appropriate nutrient medium. It is firmly pressed against a surface to pick up particles. Following incubation, the quantity and types of microorganisms that have accumulated can be precisely defined.

Airborne droplet nuclei can only be detected by the use of volumetric air samplers. There are a number of satisfactory ones available such as the Andersen sampler and the slit sampler. These samplers detect the microorganisms in transit. Organisms per cubic foot or cubic meter can be determined, quantified and identified. An even more sophisticated determination can be made by dividing fallout per square foot per minute by the organisms per cubic foot obtained by a volumetric sampler. This value gives the settling velocity in feet per minute and can be helpful in describing particle sizes in different environments. All the original work has been done in the British system using feet and inches, and it is clumsy to translate these observations into the metric system.

The size of the particle determines the category of hazard the patient is exposed to. Droplet nuclei penetrate to the lung. Therefore aerosols produced by humidifiers and respiratory therapy equipment, as well as droplet nuclei expelled by infected patients, can be implicated in the 8%-33% of nosocomially acquired pneumonias. Airborne transmission of viruses in hospitals has been documented (2). Clearly transmission of chicken pox in a children's hospital can be by droplet nuclei.

Dust and droplets, because of their high settling velocity, fall out on horizontal surfaces. Since the exposed tissue during surgery is just such a horizontal surface, the number of bacteria falling into the wound is a factor of the length of exposure and the level of aerial contamination. When inhaled, dust particles are trapped in the nose and throat because of their size. This serves to create carriers. Because of inhalation of ambient microorganisms, occupants of any environment carry organisms characteristic of that environment. This has been repeatedly demonstrated in the hospital. Personnel who work in the cleanest hospital area, the operating room, have the lowest rate of colonization with <u>Staphylococcus aureus</u>. Personnel in open wards, however, have the highest rate of colonization.

Table II, taken from our own data, demonstrates that operating room nurses who spend their entire day in the cleanest hospital area, the operating room, have the lowest rate of <u>S. Aureus</u> colonization. The orderlies, on the other hand, who spend their day in the open wards placing patients on stretchers for transport to the operating room, have the highest rate of colonization.

Table II

Staphylococcus aureus

Carriers Among Operationg Room Personnel

O.R. nurses	21%
Surgeons	33%
Anesthetists	57%
Orderlies	71%

Patients and personnel can, therefore, be victims as well as sources of environmental microorganisms: victims when they are colonized by hospital flora, sources when they shed these microorganisms.

The airborne component of surgical wound infections depends on two factors, the number of shedding carriers in the operating room and the length of the procedure. A benign fallout of two organisms per square foot per minute becomes a thousand microorganisms per square foot of exposed tissue over an eight hour operation.

In a study done at the Peter Bent Brigham hospital, 2% of the total of 8% of postoperative wound infections were found to be related to carriers present in the operating room at the time of surgery. Moreover, a carrier in the periphery of the room was proven to be implicated in two wound infections (3).

Airborne particulates augment invasiveness. Fibres from textile or disposable clothing or drapes, talc from gloves have all been found associated with postoperative wound infections (4,5).

Cruse and Foord (6) in a prospective study of
23,649 surgical wounds, found a steady increase in the
rate of postoperative wound infections with increased
length of surgery.   Their data are shown in Table III.

Table   III

Duration of Operation and Clean-Wound Infection Rate

Duration of operation (hours)	Total No.	Number Infected (%)
0-1	12,238	182 (1.4%)
1-2	4,051	114 (2.8%)
2-3	584	26 (4.4%)

These investigators reported a direct and
consistent relation between length of operating time and
infection rate.   The clean rate was found to roughly
double every hour. In a collaborative study on the
incidence of infection the same observation was made
(7).   These data are shown in Table IV.

Table IV

Duration of Operation and Infection Rate

Duration of Operation (minutes)	Number	Number of Infections(%)
0-59	4395	229 (5.2%)
60-119	5671	363 (6.4%)
120-179	2806	253 (9.0%)
180-239	1295	129 (10.0%)
240-299	651	71 (10.9%)
300-359	337	52 (15.4%)
360 or more	267	47 (17.6%)

The increase in infections over the length of time
of surgery can only be attributed to the fallout in the
environment of the operating room.   In the data from
Cruse the average increment was 1.5% increase per hour.
In the collaborative study the average increment was a
2.5% increase per hour.   It is apparent from these data
that environmental contributions to infection differ in
the hospitals studied.   Because the increments are
constant over time in the two reports,   the inescapable
conclusion is that environmental fallout is predictably
different in the two reports, yet consistent over time
for each.

Cruse and Foord found that their infection rate was influenced as follows:

1. Increased by longer preoperative stay
2. Decreased by a hexachlorophene detergent shower before operation
3. Decreased by keeping shaving to a minimum
4. Decreased by prevention of contamination at the time of surgery
5. Decreased by careful surgical technique
6. Decreased by expediting surgery
7. Decreased by particular care in the treatment of the elderly, the obese, the malnourished, and diabetic
8. Increased by the weary surgeon
9 Decreased by distribution of wound infection statistics to all surgeons

Of the characteristics listed as influencing infection rates, preoperative stay and length of surgery are an obvious reflection of the patients' exposure to the microbiology of the environment.

Kelsen and Mcguckin (8) have reported additional evidence of the role of environmental microbiology in nosocomial pneumonia by describing a significant relationship between the respiratory attack rate and airborne microbial counts. Figure 1 demonstrates this relationship.

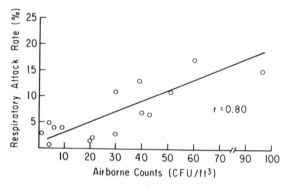

Figure 1. Relationship between airborne bacterial counts and respiratory tract attack rates. Correlation coefficient (r = 0.80) determined by least-squares regression analysis (p < 0.05).

SUMMARY AND CONCLUSIONS

The microbiology of the hospital environment and the microbiology of hospital acquired infections are inseparable. Health personnel have been reluctant to accept this fact. Unfortunately, finding a solution depends on acceptance of the problem.

The Centers for Disease Control have trivialized the microbiological environment of patients. They have officially discouraged all environmental monitoring. Monitoring is now done by the patients who become ill by inhaling the airborne microorganisms and by wallowing in a contaminated environment. An example is the history of Legionnaire's disease. No one cleaned air conditioners. They were installed and forgotten. When people became ill, epidemiology implicated the cooling towers, the air conditioners, the shower heads, and even the hot water from the tap. Patients were the experimental animals. While many deplore the use of animals in research, it would seem that it is the use of humans as animals that should be deplored.

One hospital reported that the <u>Pseudomonas</u> <u>sp.</u> in the chilling units in a hospital's air-conditioning system was also isolated from clinical specimens. Antibiograms suggested that patient strains and strains in the water system were identical, even though the air-conditioning system was a closed system (9). The Centers for Disease Control would not have approved of doing the microbiology of the chilling units.

What is the solution?   Recognition and acknowledgment of the problem is obviously the first step. Methods developed for housekeeping should be just as scientific as the procedures used in medical practice. The clinicians and hospital administrators should take an active part in establishing the procedures and reenforcing the concept that housekeeping is an integral part of patient care.

What proportion of the three major types of nosocomial infections can be eliminated by appropriate environmental practices?

Urinary tract infections are a response to indwelling catheters. Desautels (10) has stated that the causes of urinary infections are clearly demonstrable on careful analysis and, are avoidable to a great degree by the use of proper techniques.

The environmental component may be significant, because the techniques that Desautels has used so successfully depend on a lavish use of germicide: irrigation of the distal urethra with aqueous benzalkonium chloride (1:750), washing about the catheter and meatus or urinary sinus with liberal amounts of aqueous benzalkonium chloride, daily in the male patient, at least twice daily in the female patient, use of an alcohol sponge wrapped around the junction of the connecting tube and the catheter when the tube is separated from the catheter. All these procedures are designed to intercept contamination of the catheter from outside. With these practices longterm sterility has been maintained in patients with indwelling catheters.

Air-borne microorganisms have been reported associated with respiratory attack rates. Fallout into the wound has been implicated in the studies cited. What actual evidence is there that interception of these viable particles in transit will lower the attack rates?

Kelsen and Mcguckin reported that following their observations that airborne microorganisms were related to respiratory tract infection, air filtering was instituted. As a result, a reduction in airborne bacteria and a simultaneous reduction in the incidence of both respiratory and nonrespiratory tract infections were observed. These investigators suggest that surveillance of the airborne environment may prove to be a useful epidemiologic tool in the study and control of nosocomial respiratory tract infections in high risk, patient—care areas.

Ultraviolet radiation and a reduction of deep wound infection following hip and knee arthroplasty has been reported. The term "refined clean surgery" is usually applicable to joint replacement procedures. The joints rarely have been the site of previous surgery or infection. Because endogenous contamination is unlikely, contamination must come from exogenous sources. Dr. Lowell who does these procedures at the Brigham and Women's Hospital was anxious to develop a protocol for lowering the rate of deep—wound infection. After consultation with Dr. Deryl Hart and Dr. J. Leonard Goldner of Duke University, with Dr. Carl Walter of the Harvard Medical School and with support from the Surgical Bacteriology Laboratory, ultraviolet radiation was installed in the operating rooms where joint replacement surgery was done. Records have been kept of all patients undergoing total hip and knee arthroplasty since June 1970 until the present. All patients have been followed for at least 6 months. The following table is a compilation of the results for hip and knee surgery (11).

Table V

Infection Following Hip and Knee Arthroplasty With and Without Ultraviolet Irradiation

Pre-UV radiation

Primary Operations

Category	Total	Infections (%)
Hips	519	11 (2.1%)
Knees	63	6 (9.52%)

With UV radiation

Primary Operations

Category	Total	Infections (%)
Hips	1516	6 (0.4%)
Knees	1424	4 (0.28%)

Hips $P = .0002$
Knees $P = < .0001$

In summarizing the experience with ultraviolet radiation for securing these remarkably low levels of infection, the report concludes that the lamps have been simple to install, easy to use, productive of minor inconvenience, effective, relatively inexpensive and with few complications. Thus by intercepting the fallout of environmental microorganisms, the rate of infection was significantly lowered.

Thus far, only contributions to nosocomial infections arising out of indoor microbial contamination have been considered. The quality of outdoor makeup air also warrants consideration. For example, demolition and construction work was implicated in the deaths of 10 patients due to aspergillosis in a bone marrow transplant unit. What was particularly noteworthy in this outbreak was that environmental studies revealed the average air exchanges per hour varied among the six

rooms in the unit. The aspergillosis attack rate was
the greatest in rooms with high air exchange rates,
indicting an outside source of aspergillus spores.

The only resolution for the problem of nosocomial
infections is to learn from the experience of industry.
Hospitals should mimic industry. Industry has been
compelled to monitor and eliminate airborne particulates
where precision processes are used. Clean rooms of
different categories have been minutely defined and
refined. Surgery despite complexity and implantation
of prostheses is often done with complete abandon as to
the viable and nonviable particulates circulating in the
operating suite.

There are no standards for viable or nonviable
particulates in the operating room, or in any hospital
area. There are no federal or state regulations
regarding air filtration systems. Safeguards urgently
need to be developed to protect the immuno-suppressed
patient, the routine patient, as well as the personnel
who occupy the hospital environment.

References

1. Wells, W.F. Airborne contagion and air hygiene
Harvard University Press, Cambridge, Ma. 1955.

2. Leclair, J.M., J.A. Zaia, M.J. Levin, R.G. Congdon,
D.A. Goldmann Airborne transmission of chickenpox in a
hospital N.Eng.J. Med. 302:450-453 1980

3. Walter, C.W., R.B. Kundsin, M.M. Brubaker The
incidence of airborne wound infection during operation
J.Am.. Med. Assoc.186:908-913 1963

4. Tinker, M.A., I. Teicher, D. Burdman Cellulose
granulomas and their relationship to intestinal
obstruction Am. .J.Surg.133:134-139 1977

5. Janoff, K., R. Wayne, B. Huntwork, H. Kelley, R.
Alberty Foreign body reactions secondary to cellulose
lint fibers Am.J.Surg. 147:598-600 1984

6. Cruse, P.J.E. and R. Foord A five-year prospective
study of 23,649 surgical wounds Arch. Surg. 107:206-210
1973

7. Altemeier, W.A., J.F. Burke, B.A.Pruitt, Jr., W.R.

Sandusky, Eds. Manual on Control of Infection in Surgical Patients   J.B. Lippincott, Philadelphia-Toronto   1976

8. Kelsen S.G.and M. McGuckin   The role of airborne bacteria in the contamination of fine particle nebulizers and the development of nosocomial pneumonia Ann. N.Y. Acad. Sci.353:218-229   1980

9. du Moulin, G.C.,G.O. Doyle, J. McKay, J. Hedley-Whyte Bacterial Fouling of a hospital closed-loop cooling system by Pseudomonas sp. J. Clin. Microbiol. 13:1060-1065 1981

10. Desautels, R.E.   The causes of catheter-induced urinary infections and their prevention   J. Urol. 101:757-760 1969

11. Lowell, J.D., R.B.Kundsin, C.M.Schwartz, D.Pozin Ultraviolet radiation and reduction of deep wound infection following hip and knee arthroplasty   Ann. N.Y. Acad. Sci.353:285-293   1980

12.

Impact of Indoor Air Pathogens on Human Health

James C. Feeley

Center for Infectious Diseases, Centers for Disease Control,
Atlanta, GA

INTRODUCTION

Respiratory infections have been estimated to cause 50% to 60%
of all community acquired illnesses. The economic impact of this
both from the standpoint of health care cost and loss of
productivity in the work-place is in the millions of dollars.

Although most of these infections are of viral etiology,
bacterial diseases such as tuberculosis, streptococcal pneumonia,
and meningococcal meningitis do occur and have caused substantial
problems in hospitals, hotels, day care centers, nursing homes,
mental institutions, and schools. Most of these diseases are
spread via person-to-person. This is by either close contact such
as kissing, hugging, and touching or by "droplets" or "droplet
nuclei" present in the air. Droplets emanate from the mouth or the
nose during talking, singing, coughing, or sneezing. Those of
large size usually drop to the floor within 1 meter. Smaller
"droplets" can dry and form "droplet nuclei" which can remain
suspended in the air for long periods of time. They are the only
ones that should be considered truly airborne [1]. Particles less
than 5 microns in size can easily penetrate to the alveoli of the
human lung [2]. If they are viable, virulent, and in sufficient
number, infection can result. Since these particles can be
circulated by air-handling systems, they should be the concern of
all who are interested in air quality and its effect on human
health.

NEW RESPIRATORY DISEASE AGENTS

A recently recognized genus of bacteria, Legionella [3], has
been identified as one of the major causes of respiratory illness
worldwide. It is estimated that between 1 and 13% of all
pneumonias seen in hospitals in the United States, Canada, the
United Kingdom, and Germany are caused by this group of bacteria
called legionellae [4]. The disease they cause, legionellosis,

183

should be of special interest to the attendees of this symposium on
air quality. Instead of originating from an ill person or a
healthy carrier, they are present in the environment in most
surface waters [5]. Its transmission, in some circumstances, has
been linked directly to the quality of air being provided by
air-conditioning systems. For this reason, the factors that
influence the transmission of this new group of organisms in
air-conditioning systems and the intervention methods used to
prevent their transmission will be discussed by reviewing an
epidemic of Pontiac fever [6].

PONTIAC FEVER

This disease is one of the 2 clinical forms of legionellosis. It
is a febrile illness that has a high attack rate, approximately 90
percent, short incubation time, 2 to 3 days, and usually resolves
without hospitalization. In contrast Legionnaires' disease, the
other form of legionellosis is a very severe multisystemic
illness. Although it has a lower attack rate, 2 to 3 percent, and
a longer incubation time, 4 to 10 days, it causes a severe
pneumonia that requires hospitalization. Even with administration
of proper antimicrobials, 2 to 3 percent of the cases are fatal.

OUTBREAK OF PONTIAC FEVER

    The epidemic started Monday evening, July 1, 1968, among
employees of the Oakland County Health Department in Pontiac,
Michigan. By Tuesday evening 67 persons were ill with malaise,
myalgia, high fever and headache. By Wednesday evening an
additional 22 employees and 15 visitors had become similarly ill.

Epidemiologic Investigation

    Three CDC investigators arrived at the health department on
Saturday, July 7, and worked in it while the air-conditioning
system was turned off for the weekend. All 3 became ill late
Tuesday night after they continued to work in the building on
Monday and Tuesday while it was being air-conditioned. When 3
additional CDC investigators developed illness 36 hours after they
had started to work in the air-conditioned building, it was
hypothesized that the air-conditioning system might be the vehicle
of transmission. Consequently the building was closed to the
public on July 12. Altogether 144 individuals became ill, 95
percent of the employees and 29 percent of the visitors. Most
recovered within 2-5 days without treatment.

Engineering Flaws

    Since the air-conditioning system of the building was
epidemiologically implicated as the vehicle of transmission, it was
thoroughly inspected. Investigation revealed that it consisted of
2 separate air-handling systems and that defects in them allowed

water from the evaporative condenser to enter the general air circulation ducts. Two possible sites for contamination of the conditioned air were discovered. One on the roof where the exhaust from one system was being vented less than 2 meters from the air intake of the second system. This improper design allowed contaminated air to be sucked back into the building. The second was cracks in the adjacent ducts of the exhaust and intake systems that allowed contaminated water condensate to travel through them and contaminate the incoming air.

Laboratory studies

To prove that the air in the health department's building was contaminated with some type of agent, air samples were collected. After no etiologic agent was detected by conventional air sampling procedures [7], sentinel guinea pigs were placed in the building where illness occurred. For the purpose of a negative control, additional guinea pigs were placed in a neighboring building where illness had not occurred. Guinea pigs exposed to the health department's conditioned air developed nodules in their lungs; the control guinea pigs did not. To prove that the condensate water contained the agent, a series of aerosolization studies were performed in the laboratories of CDC in Atlanta. Additional guinea pigs were subjected to aerosols of the condensate water that had been boiled or filtered or untreated. Only guinea pigs that were exposed to aerosols of untreated water developed lung nodules. This suggested that the agent was living and that it was the approximate size of a bacterium. All attempts to isolate the agent during the time of the investigation failed. Only after the outbreak of Legionnaires' disease in Philadelphia [8] and the development of special laboratory procedures and reagents [9,10] was the agent, Legionella pneumophila, isolated [11].

Transmission Factors and Control

Although the primary intervention measure used to stop the outbreak consisted of simply not turning on the air-conditioning system when the building was reopened, it was highly effective. No new cases of either Pontiac fever or Legionnaires' disease have occurred in this building after the air-conditioning system was completely cleaned, dismantled, and replaced with a properly designed one. The reason for the high effectiveness of this control measure was that it eliminated 2 of the 3 factors or links in the transmission chain of the organism, namely the amplifier and the disseminator. This was accomplished by preventing condensate water (the amplifier) to collect and support the growth of legionellae and by renovating the air-handling system (the disseminator). Although the reservoir for the organism in the outbreak was not specifically identified, the organism was probably present in most of the nearby surface waters. For this reason it would have been impossible to eliminate this transmission factor.

CONCLUSION

The findings of the Pontiac, Michigan investigation provided the first well documented evidence of airborne transmission of a new disease, legionellosis. It clearly showed that air-conditioning systems must be properly designed and maintained. Subsequent to the Michigan epidemic, several outbreaks of airborne legionellosis have occurred that have confirmed this.

REFERENCES

1. Langmuir, A. D. "Epidemiology Of Airborne Infection," Bact. Reviews 25(3):173-181 (1960).

2. Hatch, T. E. "Distribution and Deposition of Inhaled Particles in Respiratory Tract," Bact. Reviews 25(3):237-240 (1960).

3. Brenner, D. J., J. C. Feeley, and R. E. Weaver, "Legionellaceae," in Bergey's Manual of Systematic Bacteriology, N. R. Krieg and J. G. Holt, Eds. (Baltimore: Williams and Wilkins, 1984), pp. 279-288.

4. Broome, C. V. "Current Issues in Epidemiology of Legionellosis, 1983," in Legionella, Proceedings of the 2nd International Symposium. C. Thornsberry, A. Balows, J. C. Feeley, and W. Jakubowski, Eds. (Washington, DC: American Society for Microbiology, 1984), pp. 205-209.

5. Fliermans, C. B., W. B. Cherry, L. H. Orrison, S. J. Smith, D. L. Tison, and D. H. Pope. "Ecological distribution of Legionella pneumophila," Appl. Environ. Microbiol. 41:9-16 (1981).

6. Glick, T. H., M. B. Gregg, B. Berman, G. Mallison, W. W. Rhodes, and I. Kassanoff. "Pontiac fever: an epidemic of unknown etiology in a health department. 1. Clinical and epidemiologic aspects," Am. J. Epidemiol. 107:149-160 (1978).

7. "Sampling Microbiological Aerosols," Public Health Monograph No. 60 (Washington, DC: Government Printing Office, 1959).

8. Fraser, D. W., T. R. (sic) Tsai, W. Orenstein, W. E. Parkin, H. J. Beechman, R. G. Sharrar, J. Harris, G. S. Mallison, S. M. Martin, J. E. McDade, C. C. Shepard, P. S. Brachman, and the Field Investigation Team. "Legionnaires' disease: description of an epidemic of pneumonia," N. Engl. J. Med. 297:1189-1197.

9. McDade, J. E., C. C. Shepard, D. W. Fraser, T. R. (sic) Tsai, M. A. Redus, W. R. Dowdle, and the Laboratory Investigation Team. "Legionnaires' disease. Isolation of a bacterium and

demonstration of its role in other respiratory disease."
N. Engl. J. Med. 297:1197-1203 (1977).

10. Feeley, J. C., G. W. Gorman, R. E. Weaver, D. C. Mackel and
     H. W. Smith. "Primary isolation media for the Legionnaires'
     disease bacterium." J. Clin. Microbiol. 8:325-329 (1978).

11. Kaufmann, A. K., J. E. McDade, C. M. Patton, J. V. Bennett,
     P. Skaliy, J. C. Feeley, D. C. Anderson, M. E. Potter, V. F.
     Newhouse, M. B. Gregg and P. S. Brachman. "Pontiac fever:
     Isolation of the Etiologic Agent (Legionella pneumophila) and
     Demonstration of Its Mode of Transmission," Am. J. Epidemiol.
     114:337-347 (1981).

Part Three

Passive Cigarette Smoke

13.

Part Three: Overview

Ira B. Tager

Harvard Medical School and Beth Israel Hospital
Boston, MA

Much attention currently is being paid to the possible health consequences of involuntary exposure to environmental tobacco smoke (ETS). A number of different types of studies (e.g. exposure studies in specially constructed chambers, epidemiologic studies of acute and chronic health effects) have suggested that adverse health outcomes which range from nuisance symptoms related to mucosal irritation to an increased risk of lung cancer can be observed in non-smokers passively (involuntarily) exposed to ETS. As a consequence of the complexities of the nature of the exposure and the multiplicity of potential outcomes, the interpretation of the existing health data and the conduct of future studies in this area require a more precise understanding of a number of issues: 1) the constituents of sidestream and mainstream smoke that are relevant to health effects related to ETS and the relationship of the concentrations of these chemicals in the environment to concentration of constituents of ETS that are easily measured 2) the optimal chemical markers of exposure to ETS, 3) the optimal combination of measurement tools (environmental monitoring, standardized questionnaires, metabolites of tobacco constituents) required to determine dose accurately, 4) the interactions of the physical properties of the smoke aerosol with the physiologic and anatomic characteristics of the respiratory system that determine deposition of ETS in the lungs; 4) the minimal requirements for epidemiologic studies of health effects. This section reviews these areas and attempts, thereby, to define current research needs and to assess the current state of knowledge.

Dr. Melvin First explores the issue of the constituents of sidestream and mainstream tobacco smoke and markers to identify them in public environments. He points out that ETS is a mixture of both sidestream smoke (SSS) and exhaled mainstream smoke, the exact proportions of which are poorly defined. Based on an estimate that, on average, 50% of mainstream smokers (MSS) is expelled, Dr. First suggests that the composition of ETS can be estimated by the addition of 50% of mainstream smoke to sidestream smoke. He notes that from the point of view of emission flux per unit time, MSS assumes considerable importance. He further notes that the expelled MSS is not precisely the same as inhaled smoke and presents data on the changes that occur. Measurements of particle sizes for sidestream smoke indicate that 90% of particles are under 0.4 ± 0.1 μm vs a median size of 0.42 μm for MSS but tend to shrink due to evaporation. This size range puts the particles at the minimum point of most particle size–lung retention curves. Dr. First notes that many substances that have been used to quantify ETS have other sources in the environment. He points to the specificity of nicotine as a marker for ETS and the fact that the development of filter tip cigarettes has altered its concentration in SSS very little. He concludes with the note that studies to establish a firm linkage between exposure to ETS and retained dose are central to further understanding of the likely health effects of exposure to ETS in public environments. Additional research recommendations also are provided by him.

The difficult issues related to the experimental designs needed for the measurement of exposure to ETS are discussed by Dr. Jan Stolwijk. He notes that our current capacity to conduct studies on the health effects of ETS is limited to some extent by our inability to characterize and assign exposure to ETS with sufficient reliability for a large number of people in either retrospective or prospective epidemiologic studies. Given the chemical and physical complexities of ETS, proxy measurements need to be identified and validated in laboratory and field settings. Depending on the goals of a particular study, markers for personal and environmental exposure are required. Dr. Stolwijk suggests that self-reported questionnaire responses, supplemented by measurements of selected components (or their metabolites) of ETS in biological fluids which, in turn, are backed up by a clear understanding of the parameters of exposure are likely to be the optimal exposure characterization scheme for population based studies of health effects. Studies aimed at the determination of exposure contribution and design of exposure reduction measures will require characterization based on space parameters such as source description, ventilation patterns, density of occupation, etc. These later studies will require physical and chemical monitoring techniques and have little use for questionnaires or biological monitoring. Data are presented to illustrate the features and limitations each of these design elements and combinations thereof.

Central to any understanding of the biological potential of exposure to ETS is a characterization of the factors that

influence exposure-dose relationships for the constituents of
ETS. Dr. Joseph Brain in his section addresses this issue.
Particle size, breathing pattern, individual anatomical differences and pre-existing disease are identified as factors which
influence the deposition of ETS in the lungs. Dr. Brain discusses the ICRP model of particle deposition in the respiratory
tract which provides an estimate that 30-40% of particles in the
size range of these in cigarette smoke will deposit in the
alveoli and 5-10% in the tracheobronchial tree, although recent
data suggest that the alveolar fraction may be too high by a
factor of 2. He also discusses data on aerosol deposition based
on studies of airway cast models and data concerning factors
that affect particulate retention in the lung. He concludes
with a detailed list of research recommendations for studies in
this area.

Among the health effects attributed to involuntary exposure to
ETS, none has raised more concern than that related to lung
cancer risk. Dr. Jonathan Samet discusses in detail the data
which relate to this problem. A brief review of methodologic
issues relevant to epidemiologic studies in this area is provided. Dr. Samet identifies 9 studies which have addressed this
issue. Amongst 3 studies showing an excess risk of lung cancer
in non-smokers (usually spouses) exposed to ETS, relative risks
in the range of 1.5-3.0 have been observed, including a trend
toward dose-effect relationships. Dr. Samet notes that the
study by Japanese investigators provides the strongest evidence
in this regard and that criticisms leveled at the study have
been addressed satisfactorily by the investigators without alterations in the findings. Amongst studies failing to find an
excess risk of lung cancer, problems with small numbers, variable
definitions of exposure, and reductions in effect due to misclassification are sited as problems. Dr. Samet also notes
that, among the studies whose point estimates fail to show an
excess risk of lung cancer in those exposed to ETS, the confidence
intervals (usually 95% intervals) around these estimates are
compatible with studies whose point estimates have suggested an
effect. He concludes that, while the association between passive
exposure to ETS and lung cancer is not firmly established, sufficient data are at hand for preventive and regulatory steps.
Dr. Millicent Higgins concludes this section with a review of
the data related to the relationship of passive exposure to ETS
and non-malignant cardiopulmonary disease. She notes that
despite some inadequacies of study design and exposure characterization, the evidence linking passive exposure to ETS to
respiratory illness in children and to reduced lung function
(especially in children) is quite strong. Data which relate to
this issue is reviewed in some detail by Dr. Higgins. In summarizing the presentations of this section, it appears reasonable
to conclude the following: 1) The major unresolved problem for

research in this area of passive exposure to ETS involves better characterization of the features and determinants of environmental contamination and individual dose. Methods are available to address these problems, and further research clearly is needed. This conclusion should not be construed to mean that all possible chemical-biological-health effect pathways need to be characterized. Characterization of markers for classes of chemicals and categories of exposure will be sufficient in this regard. 2) Health effects definitely can be attributed to passive exposure to ETS. What remains to be addressed is a more precise characterization of the magnitude of these effects, in terms of the range of the diseases to be encountered and the number of people expected to be effected. This is especially true in the area of chronic disease outcomes. Again, methods are available which, when coupled with better definitions of exposure, can provide the data needed for these more precise estimates. Sufficient concern about these health effects justifies a continued research effort. While impetus for research in these areas derives from the immediacy of concern for the public well-being, it also derives in equal part from the scientific need for a better understanding of the interaction of complex low-dose pollution exposures and human health. The issue of regulation and control must be separated from these later basic scientific needs.

14.

Constituents of Sidestream and Mainstream Tobacco Smoke and Markers to Quantify Exposure to Them

Melvin W. First

Harvard School of Public Health, Boston, Massachusetts

INTRODUCTION

"Tobacco smoke in the environment and its possible effects on non-smokers is a subject which has been widely discussed over many years (1)." So begins a report from the most recent international workshop on ETS (environmental tobacco smoke). It is noteworthy, perhaps, that "discussion" was the characteristic first commented upon, rather than celebrating the research achievements that had taken place since 1974, when a similar workshop was conducted. In truth, since 1974 little has been accomplished with the difficult task of measuring ETS and determining its health effects. Most of the studies that have been published on these matters have been seriously flawed, though loudly hailed by partisans.

A rational delineation of the effects of ETS on human health and comfort are totally dependent on the development and sound implementation of methods for identifying and quantifying the tobacco smoke component in indoor air. In the absence of quantitative measurement methods it is not surprizing that little progress has been made in defining human effects, nor is this situation likely to change for the better until agreement has been reached on a definition of ETS and on standard methods for its measurement. Therefore, it is appropriate that a discussion of the constituents of sidestream tobacco smoke and markers to quantify exposure begin this segment of the program. That my assignment was a little broader, is reflected in the title, "Constituents of Mainstream and Sidestream Tobacco Smoke and Markers to Quantify Exposure to Them." I perceive that there is a need to include a consideration of mainstream tobacco smoke even when the topic is ETS for two reasons: 1) most of what we know about tobacco smoke composition, its measurement, and its effect on experimental animals and humans is derived from studies of mainstream tobacco smoke 2) ETS is a mixture of sidestream smoke and exhaled mainstream smoke, the exact proportions being but one of the many fundamental facts about ETS that are presently obscure.

195

PHYSICAL AND CHEMICAL CHARACTERISTICS OF ENVIRONMENTAL TOBACCO SMOKE

Although the smoke fraction entering the atmosphere is usually referred to as "sidestream", the word has been more precisely defined in the Dictionary of Tobacco Terminology: "In a closed smoking system (for analytical purposes), sidestream is the smoke that does not issue from the mouth end of a cigarette but rather from the burning end, through the paper, etc. In a free smoking situation, it is all of the smoke issuing from any part of a cigarette except that which is drawn through the mouth end during puffing. In free smoking, sidestream may issue from the mouth end during static burning (2)." ETS differs from sidestream smoke by the addition of residual mainstream smoke exhaled by smokers.

The results of recent measurements of the fraction of inhaled smoke that is deposited by male and female confirmed smokers are shown in Table 1 (3). The wide variations that occur from subject to subject reflect basic differences in personal smoking habits. The extremes are represented by those who expell most of the smoke after taking a brief, shallow inhalation, and by those who inhale deeply, retain the smoke for a perceptible period, and then exhale very little residual smoke. Although one might think of the shallow inhalers and rapid exhalers as merely social smokers, it should be remembered that nicotine is readily absorbed through mucous membranes of the mouth and throat and these people, too, may be confirmed smokers. Although the sample was small (11 subjects), the measurement system was designed to be accurate and to interpose no interference whatsoever to each person's normal smoking regimen. Therefore, the average fraction expelled, 50%, is believed to be a useful number for estimating this contribution to ETS.

As a first approximation, the influence of exhaled mainstream smoke on the composition of ETS can be estimated by adding 50% of mainstream smoke to 100% of sidestream smoke. Table 2 shows a comparison between mainstream and sidestream smoke for some constituents of major interest, based on the output from a single cigarette. The first item of interest is that although the weight of tobacco consumed during inhalation is only slightly less than the amount burned between puffs, the time differential between puffing and smoldering is marked, being a ratio of 1 to 27. This means that the 50% of mainstream smoke emitted to the room air is discharged in a spaced series of brief concentrated puffs whereas the sidestream smoke products are released at a much slower, but steadier, rate.

A second consideration is the nature of the expelled mainstream smoke. Table 2 shows that the total output of many smoke constituents differs significantly between mainstream and sidestream smoke, but when considered from the standpoint of emission flux per unit time, the expelled fraction of mainstream smoke assumes greater importance. For example, particle numbers emitted per second are 2.5 x 10^{10} for expelled mainstream smoke but only 6.5 x 10^9 for side-

Table 1. Cigarette smoke deposition in confirmed smokers (from reference 3).

| Group | Subjects | Measurements | Deposition % | | |
			Range	Mean	(S.D.)
All	11	22	22-75	47	(13)
Males	6	9	33-75	57	(13)
Females	5	13	22-49	40	(18)

Table 2. Comparison of mainstream and sidestream smoke constituents (mg/cig) (from reference 4).

Compound	Mainstream	Sidestream	Ratio Sidestream/ Mainstream
Tobacco burnt	347 (20 sec)	411 (550 sec)	1.2 (27)
No. particles produced	10^{12}	3.5×10^{12}	3.5
Tar	20.8 10.2^{*}	44.1 34.5^{*}	2.1 3.4^{*}
Nicotine	0.92 0.46^{*}	1.69 1.27^{*}	1.8 2.8^{*}
Benzo(a)pyrene	3.5×10^{-5}	13.5×10^{-5}	3.7
Pyrene	13×10^{-5}	39×10^{-5}	3.0
Phenols	0.228	0.603	2.6
Ammonia	0.16	7.4	46
Nitrogen Oxides	0.014	0.051	3.6
Carbon Monoxide	19	88	4.7

* Filter cigarette

stream smoke. When similar comparisons can be made for the other smoke components shown in Table 2, it becomes clear that although the total amount of expelled tobacco products is smaller, the concentration of many of the components in the two sources of environmental tobacco smoke may not differ greatly.

On the other hand, it is unlikely that the expelled smoke is precisely the same as the inhaled smoke in all its characteristics other than a 50% reduction in concentration. From physical and chemical considerations alone, we surmise that some gas phase components may be retained more completely than others and that the particulate phase may undergo a size increase from absorption of water vapor and by coagulation, thereby changing the characteristics of expelled mainstream smoke from those shown in Table 2. Indeed, absorption experiments conducted with mainstream tobacco smoke passed through an airway simulator of variable length, lined with a physiological saline-soaked membrane, show clearly that very water soluble components, such as acetonitrile and acetone, are totally absorbed during passage, whereas less soluble components, such as acrolein, are only partially reduced in concentration, and insoluble components such as acetonitrile and crotonaldehyde are expelled unchanged (4).

There has been much interest in the size characteristics of the particulate phase, that contains all of the tar component and most of the nicotine, because lung retention of ETS particles is initimately related to particle size. Measurements of the particle size of sidestream tobacco smoke with a ten-stage piezo-electric cascade impactor showed that 90% of the particles were under 0.4 ± 0.1 µm (5). This may be compared with size measurements of mainstream tobacco smoke made with an aerosol centrifuge that showed a median size of 0.42 µm (6). Considering the effect on experimental agreement of using different measurement methods in different laboratories, there appears to be reasonable agreement between the size characteristics of mainstream and sidestream smoke. When one considers that water makes up about 16% by weight of the total particulate matter and that organic liquid components with an appreciable vapor pressure make up much of the remainder, it is reasonable to expect that airborne tobacco smoke particles will tend to shrink somewhat in size as they lose vapor to the surrounding air (as long as the air remains unsaturated in the vaporizing components). From this, it follows that the longer smoke particles remain suspended in air, the smaller they are likely to become by evaporation. At the same time, the smaller the smoke particles become, the longer they will remain air-suspended. For example, particle number measurements made indoors by one group of investigators (6) showed only a 50% reduction over a 90 minute undisturbed holding period. Another group of investigators (7) showed only a 65% decrease in the nicotine content of indoor air over 45 minutes under similar holding conditions. Within the level of accuracy that lung retention measurements can be made, there is no reason to believe that retention of inhaled mainstream and sidestream smoke particles differs significantly when it is recognized that the sizes of both are close to, and on either side of, the minimum point of most particle size-lung retention curves (8).

MEASUREMENT OF ENVIRONMENTAL TOBACCO SMOKE

It is frequently stated by tobacco chemists that more than 2,300 (often, "more than 3,000" is the number cited) individual compounds have already been positively identified in tobacco smoke, and that an end to the identification of new compounds is nowhere in sight. Most of the identified compounds are present in what would have been termed "trace amounts" just a few years ago, but remarkable developments in analytical instrumentation have made the detection and measurement of even the least of these compounds border on the routine. This means that measurements of ETS are limited only by the availability of instruments and manpower - or by the presence of interfering substances. The latter restraint is likely to be the more intractable because the overwhelming majority of tobacco smoke chemicals also originate from sources other than tobacco smoke, often in larger amounts. Examples abound and I will discuss several having special importance for measurement of ETS.

A compilation of published literature on the measurement of environmental smoke appeared two years ago in the Journal of the Air Pollution Control Association (9). The substances that were measured as indicators of environmental smoke included: acrolein (3 publications), aromatic hydrocarbons (6 publications), carbon monoxide, the most frequently measured tobacco smoke product, (17 publications), nicotine (4 publications), nitrogen oxides (3 publications), nitrosamines (2 publications), airborne particulate matter (10 publications), and a catchall category called "residuals" that includes phenols, aldehydes, SO_2, and sulfates (4 publications). It will be clear from this listing that, in general, most of those attempting to measure environmental smoke use simple methods and have a strong preference for direct-reading instruments that require little or no skill for their operation. Doubtless, the wide availability of small, battery-operated direct-reading instruments accounts for the popularity of measuring CO and suspended particulate matter as surrogates for tobacco smoke. However, these substances are widely dispersed in the biosphere from sources other than tobacco combustion and both occur indoors from internal sources as well as by infiltration from the outside in concentrations of about the same general magnitude whether smokers are present or absent. The same comment applies equally to measurements of acrolein, aromatic hydrocarbons, nitrogen oxides, nitrosamines, phenols, aldehydes, sulfur dioxide and sulfate. The usual practice is to measure indoor and outdoor concentrations of the selected tracer (but not always, or even usually, simultaneously) and to ascribe any excess found indoors to airborne tobacco products. A variant of this experimental protocol is to make repeat measurements with smokers present and absent. By far the major defect associated with using the above-named gases, vapors, and particles as surrogates for environmental tobacco smoke is the likely presence of sources other than tobacco that produce concentrations sufficiently high to confound any attempt to distinguish the amount contributed solely from tobacco smoke. A few examples will serve to illustrate the point.

Restaurants, cafes, and bars (favored locations for measuring environmental smoke) habitually use gas for cooking and for heating water to be used for washing eating and cooking utensils. Often the gas-using facilities are located inside public rooms, e.g., restaurant grills and bar glass washing stations, and make a major contribution to the room CO concentration. Many restaurants, cafes, and bars are located on large, busy thoroughfares, frequently at intersections of two or more such roads, and infiltration of CO and particulate matter from traffic flow is another major contributor to indoor air contamination. Unless such a restaurant or bar is populated wall-to-wall with smokers, or has no ventilation system whatsoever, the CO contribution from tobacco smoke is unlikely to be greater than the uncertainty inherent in the analytical methods employed.

The portable, direct-reading electrochemical instruments that are widely used for CO measurements have a serious defect for indoor measurements in restaurants and bars in that they respond to ethyl alcohol vapors much more strongly than they do to CO (10). Therefore, special filters must be used at the entrance to the device to react with all ethyl alcohol and similar compounds that may be present in small concentration, or spurous CO measurements will result, especially as the instrument cannot be relied upon to read closer than 1-2 ppm CO even under the best of circumstances.

Assuming that the fractions of smokers and nonsmokers remains about the same, the more people present, the larger the numbers of smokers, but also, the greater the amount of internal dust generating activity. If one were to measure dust content in the air and attempt to relate this with the observed number of people smoking, there should be a reasonably good correlation simply because a greater number of smokers means there will be more people and, hence, more dust generating activity, aside from the contribution from smoking. This is not to say that tobacco smoking does not make its own contribution to indoor airborne particulate matter but rather to point out that in public places other sources of airborne particulate matter are likely to be of sufficient magnitude to obscure the measurement of the increment from smoking. This comment is especially pertinent for the TSI portable airborne particle sampling and analysis instrument that has been used for tobacco smoke measurements because the instrument was not designed for this type of service and lacks the sensitivity and precision needed to sense small incremental concentrations attributable to tobacco smoking.

Similar comments apply to all the other tobacco smoke surrogates that have been mentioned, except for nicotine. For all practical purposes, nicotine is a unique compound of the dried leaves of

Nicotiana tabacum and N. rustica. Aside from some commercial usage
of the extracted nicotine sulfate as an agricultural insecticide,
there is no other source than smoking tobacco. This means that in
occupied spaces away from actively growing agricultural areas, whatever nicotine is detected in the air comes from smoking tobacco and
from no other source. This makes nicotine the ideal surrogate for
tobacco smoke inasmuch as all nicotine found in the air represents
only tobacco smoke in practically all indoor locations of interest.

With the exception of water, nicotine is the largest single
component of the particulate phase of tobacco smoke; nicotine concentration is unaffected by the moisture content of the smoke; and
sensitive gas chromatography analytical methods are available for
measurement of nicotine concentrations (7). Airborne nicotine can
be sampled with a gas-tight syringe and analyzed by a GC method
capable of detecting picograms of nicotine in a 25-50 mL grab sample
(7). Alternatively, integrated samples of airborne nicotine can be
taken at a rate of 2 Lpm over 10-30 min. by passing the air through
a Cambridge or similar glass fiber filter that has been treated with
0.5 mL of 0.5 molal potassium bisulfate to suppress the volatility
of captured nicotine.

Although nicotine is specific for tobacco smoke, and can be
measured reliably in low concentration in the environment, it is
difficult to relate measured nicotine concentrations in the air to
other constituents of tobacco smoke, such as nitrosomines, because
of differences in the composition of commercial cigarettes and the
unknown effect of gas phase reactions on each of the thousands of
compounds in tobacco smoke. It has been estimated that currently
there are more than 200 cigarette choices available in the United
States and, in at least one major market, the eleven most popular
choices represent only 40% of total cigarette sales. Therefore,
use of an average nicotine content for all cigarettes makes much
more sense when sampling air in an enclosed arena containg 50,000
people than when sampling in a room of usual size containing fewer
than ten smokers. This is a complicating factor associated with the
use of nicotine as a measure of environmental smoke, but the use of
nicotine is likely to result in less uncertainty than the use of
other surrogates that are not unique for tobacco smoke.

The problem of accounting for the average nicotine content of
an unknown mix of cigarettes is probably less troublesome then
first appears. Reference to Table 2 shows that although the nicotine and tar content of mainstream smoke from filter and non-filter
cigarettes differs markedly, the nicotine and tar content of sidestream smoke is more nearly the same for filter and non-filter
cigarettes, indicating that the filter is the principal mechanism
for reducing tar and nicotine in mainstream smoke. In fact, there
tends to be a degree of uniformity of nicotine content from
cigarette to cigarette in sidestream smoke that does not exist for
most components of mainstream smoke. To the degree that sidestream
smoke is the major source of ETS; the less will be the uncertainty

in using the environmental nicotine concentration to estimate other
important constituents of ETS. Determination of the relationship
between the simple measurement of nicotine in air and an indication
of what the exposure might have been to other important constituents
of ETS is an important linkage to all environmental measurements.

DISCUSSION AND RECOMMENDATIONS

Although only metabolite studies (e.g., nicotine and cotinine
in blood, urine, and saliva) made in conjunction with human response
studies can provide a useful relationship between retained dose and
human effects, there is an urgent need to establish a firm linkage
between exposure and retained dose because it is simpler, less
costly, and far less objectionable to potential donors to measure
their environment rather than to extract biological fluids from them.
Personal sampling is more important than area monitoring for a
reliable characterization of exposures to ETS and should be encouraged.

Research needs for better characterization of ETS and for a
more accurate delineation of the significance of the measured
quantities to retained dose were cited in the report of the 1983
workship on ETS (1). Research needs included: 1) determination of
the composition of sidestream smoke from current market cigarettes,
2) relationship of ETS composition and concentration to analytical
smoke measurements of the same components in the sidestream smoke,
3) the effect of aging on ETS composition and concentration, 4)
development of chemical dosimeters and biochemical markers for easy
and reliable measurements of long term, average exposures to ETS
products, 5) development of standard methods for sampling and
analyzing the sidestream smoke components and ETS as consensus
standards within ASTM, ISO, or a similar standards-writing organization.

This is surely a large agenda. Nevertheless, we must, in my
opinion, proceed in this direction if we are ever to be able to
apply respectable science to the investigation of the effects of
ETS on human populations.

REFERENCES

1. Rylander, R. "Introduction," in ETS - Environmental Tobacco
    Smoke. Report from a workshop on Effects and Exposure Levels,
    R. Rylander, Y. Peterson, and M. -C. Snella, Eds. (Switzerland,
    University of Geneva, 1983) p.7.

2. DeBardeleben, M.Z. (Ed). Dictionary of Tobacco Terminology,
    Philip Morris Inc. Technical Information Facility, P.O. Box
    26583, Richmond, VA 1980.

3. Hinds, W., M.W. First, G.L. Huber, and J.W. Shea "A Method for
    Measuring Respiratory Deposition of Cigarette Smoke During
    Smoking," Am. Ind. Hyg. Ass. J. 44:113-118 (1983).

4. First, M.W. "Environmental Tobacco Smoke Measurements: Retro-
    spect and Prospect," in ETS - Environmental Tobacco Smoke.
    Report from a workshop on Effects and Exposure Levels,
    R. Rylander, Y. Peterson, and M.-C. Snella, Eds. (Switzerland,
    University of Geneva, 1983) pp. 9-16.

5. Girman, J.R., M.G. Apte, G.W. Traynor, J.R. Allen, and C.D.
    Hollowell "Pollutant Emission Rates from Indoor Combustion
    Appliances and Sidestream Cigarette Smoke" Prepared for the U.S.
    Dept. of Energy under Contract DE-AC03-76SF00098, Lawrence
    Berkeley Laboratory, Univ. of California, 1982.

6. Hinds, W.C. "Size Characteristics of Cigarette Smoke," Amer.
    Ind. Hyg. Assoc. J., 39:48-54 (1978).

7. Grubner, O., M.W. First, and G.L.Huber "Gas Chromatographic
    Determination of Nicotine in Gases and Liquids with Suppression
    of Absorption Effects," Analyt. Chem. 52:1755-1758 (1980).

8. Lippmann, M. "Size-Selective Health Hazard Sampling," Air
    Sampling Instruments, Ed. 6, Lioy, P.J. and Lioy, M.J.Y., Eds.,
    Am. Conf. Governmental Industrial Hygienists, Cincinnati, OH
    1983, Fig. H-6, p. H-8.

9. Sterling, T.D., H. Dimoch, and D. Kobayashi "Indoor Byproduct
    Levels of Tobacco Smoke: A Critical Review of the Literature,"
    J. Air Pollut. Control Ass. 32:250-259 (1982).

10. First, M.W., and W.C. Hinds "Ambient Tobacco Smoke Measurement,"
    Amer. Ind. Hyg. Assoc. J., 37:655-656 (1976).

15.

Experimental Considerations in the Measurement of Exposures to Sidestream Cigarette Smoke

Jan A.J. Stolwijk, Brian P. Leaderer and Marianne Berwick

John B. Pierce Foundation Laboratory
and
Department of Epidemiology and Public Health
Yale University School of Medicine, New Haven CT 06510

INTRODUCTION

There is increasing evidence that exposure to sidestream tobacco smoke has an adverse effect on human health and comfort. These adverse effects may take the form of acute effects such as irritation of mucous membranes in the eyes or upper airways (1,2,3), reduction in effectiveness of the host defense mechanisms against airway infections (4) or of delayed effects such as an increased risk of lung cancer when a non-smoker is chronically exposed to sidestream tobacco smoke for long periods of time (5).

The quantitative level of exposure to sidestream tobacco smoke is difficult to establish even under the best of controlled circumstances. The problems of the extreme complexity in the composition of the smoke, the poorly known efficiency of deposition in the respiratory system and the extreme variability in the concentration of smoke in the atmosphere all conspire to make estimates or measurements of sidestream exposure very difficult. It is difficult to make such measurements on a current basis, for prospective studies, and it is clearly even more difficult to make effective estimates of such exposures retrospectively for a period of years. Nevertheless, epidemiologic studies of the effect of sidestream tobacco smoke exposure on lung cancer risk, or on cardiovascular or pulmonary disease have to be based on such estimates. The dangers of misclassification in such epidemiologic studies are well known (6,7), and if such misclassification occurs it will tend to quickly reduce the power of such a study to determine whether or not an association or an effect exist.

To a considerable extent our ability to conduct effective studies of the adverse impact of sidestream tobacco smoke is limited by our ability to characterize and assign the exposure for a large number of people in either retrospective or prospective studies, with a known level of reliability. It is important to note that all of these people in the micro-environments in which they move around are simultaneously exposed to a number of other pollutants which may contribute to the same or other adverse health outcomes. If such other pollutants are not measured or estimated at the same time they will have the effect of obscuring or exaggerating the effect of the sidestream tobacco smoke.

Any study of the effect of sidestream tobacco smoke should, therefore, not only measure or estimate the effective exposure to tobacco smoke, but also determine the presence of other contaminants and their sources, or the study population should be selected for the absence of such sources.

METHODOLOGIES FOR MEASUREMENT AND ESTIMATION OF SIDESTREAM TOBACCO SMOKE EXPOSURES IN THE FIELD

There is no universally accepted and standardized method to characterize the personal exposure of an individual, or the exposure experienced by an individual in a given indoor space, to sidestream tobacco smoke. Given the chemical and physical complexity of tobacco smoke, and the further complexities inherent in deposition of constituents of tobacco smoke in the respiratory system, exposure assessments will be based on some acceptable proxy measurements which can be demonstrated to be justified and effective.

Such proxy measurements should be developed and validated in closely controlled laboratory settings, and subsequently tested and validated in field settings, before being committed to a full field study. Exposure estimates can take two forms: for the most optimal personal exposure estimate it is clearly of most importance to obtain an estimate of the exposure for a particular person over a given period of time. In studies in which such exposures are to be linked to personal health outcomes, this personal exposure is the most important. In studies where at least one of the objectives is the assessment of the contribution made to total personal exposure by various specific indoor environments it is clearly important to assess the exposure to environmental tobacco smoke contributed by a particular environment, as influenced by the characteristics of the sources, the ventilation and infiltration, and any deposition or inactivation which takes place in that indoor environment. For regulations or recommendations which are to reduce the total population exposure as effectively as possible it is necessary to understand the contributions made to total population exposure by different types of indoor environment.

Estimates or measurements of exposure to environmental tobacco smoke can be made by one or more of the following methodologies:

1. reports of the presence or absence of tobacco smoke in a

given environment, obtained from questionnaire responses; such estimates may be made more detailed if the number of cigarettes smoked, and the ventilation characteristics of the space are known

2. measurements obtained in the space, or by equipment carried by an individual, which monitor one of the characteristic values associated with tobacco smoke, such as respirable particulates, optical scattering, carbon monoxide

3. measurements may be made of the level of one of the compounds associated with tobacco smoke in expired air, the blood or the urine of the study population; this biological monitoring has the potential capacity of reflecting the personal dose of individuals or whole study populations without having to make environmental measurements, or without having to rely on questionnaire responses.

Each of these major avenues has advantages and disadvantages associated with it, and it would appear that a judicious combination of all of them is necessary for an optimal solution. Population based studies of the adverse health impact of sidestream tobacco smoke require exposure information on a large number of people occupying a large number of spaces over a period of weeks to years. Self — reported questionnaire responses, supplemented whenever possible by well validated and appropriate biological monitoring, in turn backed up by a clear understanding of the parameters of exposure, are likely to be the optimum exposure characterization scheme for population-based health effect studies.

On the other hand, studies aimed at determining the exposure contribution and at design of exposure reduction measures will require characterization based on the space characteristics, such as source description, ventilation characteristics, density of occupation, and on careful monitoring of representative spaces with average and known use patterns. Such studies will use physical and chemical monitoring techniques, and have little or no use for questionnaire or biological monitoring techniques.

EXPOSURE CHARACTERIZATION FROM QUESTIONNAIRE RESPONSES

Most studies which have evaluated the validity of questionnaire data have concentrated on the ability of questionnaire data to discriminate smokers from non-smokers. This type of study can rely on simultaneous physiological or biochemical measurements of tracers such as CO or thiocyanate (8,9). Although the congruence between reported tobacco exposure and biochemical levels is reported to be quite good, most of these types of questions have been directed at active smoking. Attempts to provide estimates of sidestream tobacco smoke exposures are much less numerous and the study quoted in (10) indicates that the correlation between biochemical markers and reported exposure to sidestream smoke is likely to be somewhat weaker. This is to be expected since the power of the biochemical markers will be diminished with the actual level of uptake which will be less in passive exposures, and in addition the judgement about the severity of the environmental exposure is much more difficult for the respondent than a report on recent active smoking. Severity of environmental tobacco smoke exposure is a subjective judgement which requires recall of other people's activities in a number of spaces which the respondent has occupied or visited in the recent past. The difficulties which occur even in relatively short recall situations will obviously be much more serious when very long periods of recall are required, such as occurs in studies of the possible carcinogenic action of sidestream tobacco smoke (5) or in studies of the long term effect of such exposures on respiratory health (11,12). Determinations of the environmental tobacco smoke exposure (and the total absence of primary exposure) in retrospective studies, by asking the survivors, also deserve very careful scrutiny, and confirmation by independent sources whenever feasible.

It is likely that questionnaire responses will always be an important source of sidestream tobacco smoke exposure estimates in any population based study of the adverse health effects of such exposures. It is clear that there will be a need for independent affirmation and confirmation of such estimates by as many independent routes as possible. Confirmation by direct measurements of concentrations in the air of indoor spaces is possible in only a limited number of spaces and for limited amounts of time. Personal exposure assessment through biological monitoring provides another important avenue of determining the sensitivity and specificity of questionnaire data.

Since exposure determination will be one of the most serious problems in population based studies of the health effects of environmental tobacco smoke, considerable thought should be given to methods of selecting subjects which have relatively easily defined exposures with an absence of simultaneous exposures to alternative sources which might produce similar adverse health effects.

EXPOSURE   CHARACTERIZATION   THROUGH   PHYSICAL   AND   CHEMICAL MEASUREMENTS

Direct measurement of one or more of the specific constituents of sidestream tobacco smoke is clearly the most desirable assessment of exposure at a given site, for a given span of time covered in the measurement. It is equally clear that this is not a very realistic objective, given the number of sites, and the length of time which would need to be covered in a study involving a large number of subjects over a period of months or years. If a realistic characterization of exposure can not be based on direct measurements alone, such measurements must be at the base of any other strategy which is adopted for exposure characterization. Tobacco smoke contains a very large number of contaminants and many of these have

biological significance. Carbon monoxide has been followed as an
indicator of sidestream tobacco smoke concentration. Under
precisely controlled conditions as occur in exposure chamber
research the incremental CO concentration can be used as an
indicator of tobacco smoke exposure (1,2), but CO sources other than
tobacco smoking are very likely to interfere under other
circumstances.   Cain   and   Leaderer   (13,14,15)   studied   the
concentration of CO and respirable particulates associated with
different levels of smoking occupancy and different ventilation
rates in a well controlled environment and found that CO
concentrations ranged from 2.0 to 20.0 ppm in a chamber in which
between 4 and 24 cigarettes per hour were smoked at effective
ventilation rates of 20 to 150 liters/second.   Under the same
conditions the total particulate mass ranged from 100 to 1600
micrograms per cubic meter of air.   When these exposures are
compared with short term ambient air quality standards of 9 ppm of
CO and 250 micrograms/cubic meter of total suspended particulates
it would appear that the particulate exposure is the more serious
hazard of the two. CO measurements are commonly made with an electrochemical method which is simple, lightweight in its implementation
and relatively inexpensive. It is subject to interference by other
substances, specifically ethanol and nitrogen oxides. Other instruments using non-dispersive infrared aborption as a measuring technique are much bulkier, more costly but less subject to interference.

The suspended particulates associated with tobacco smoke can be
measured gravimetrically by collecting the particles on a filter
through which known amounts of air have passed.   Other methods
deposit the electrically charged smoke particles on to a quartz
crystal which changes its resonant frequency as a result of the mass
loading from the particles in a given amount of air. Tobacco smoke
particulates are of a size which very effectively scatters visible
light.   By measuring back-scattered light from a constant light
source it is possible to obtain a reading which is proportional to
the instantaneous concentration of suspended particulates in the
air.

Characterization of sidestream tobacco smoke exposures of
individuals, and of the contributions made to such exposures by
different environments would be greatly facilitated by the
availability of a simple, reliable and inexpensive passive monitor
of cumulative exposure to respirable particulates; such a device,
properly validated and calibrated would open up new opportunities in
the field of health effects research on environmental tobacco
smoke.

Careful   and   detailed   measurements   of   tobacco   smoke
concentrations and of the various constituents of tobacco smoke are
essential for the understanding of sidestream tobacco smoke exposure
in different environments. Validation of biological and biochemical

markers of exposures requires controlled exposures of volunteers to precisely known concentrations of tobacco smoke for relatively long periods of time in suitably designed exposure chambers.

EXPOSURE CHARACTERIZATION THROUGH BIOLOGICAL MONITORING

Tobacco smoking produces a large number of contaminants which are released into the atmosphere surrounding smokers. From this large number a much smaller number has been evaluated for their suitability as a biochemical marker of a subject's exposure to sidestream tobacco smoke. Specific examples of such "tracer" contaminants are carbon monoxide, nitrogen dioxide, thiocyanate, and nicotine. Ideally a tracer contaminant should be highly specific to tobacco smoke, and not be introduced into the environment or into the human body from other sources. The ideal tracer should also have a suitable biological half-life, measurements should be simple and inexpensive, and samples should be easy to obtain with a minimum of inconvenience to the subject.

Jarvis and Russell (10) critically evaluated carbon monoxide, nicotine and its metabolic derivative cotinine, and thiocyanate in a study which also obtained the usual questionnaire responses. As was to be expected, measurements of CO concentration in expired breath or of the carboxyhemoglobin levels in the blood did not correlate well with the self-reported level of exposure to sidestream tobacco smoke. At the levels of CO in indoor spaces resulting from tobacco smoking (usually less than 10 ppm) there will be serious interference from CO due to other sources such as automobile traffic which produces similar levels of CO in the outdoor atmosphere and somewhat lower levels indoors. The half-life of carboxyhemoglobin is of the order of several hours and has been reported to be as high as 11 hours (16), so that carboxyhemoglobin levels or expired breath CO levels provide information about exposures over a period of several hours prior to the taking of the sample. It seems likely, however, that the need to simultaneously describe the non-tobacco related CO background concentrations will limit the usefulness of CO as a marker for sidestream tobacco smoke exposure.

Thiocyanate levels in non-smokers in plasma, urine or saliva have been reported to be unrelated to self reported exposure to sidestream tobacco in subjects in whom simultaneous determinations of urinary cotinine levels were confirmatory of the self reported exposures (10). Thiocyanate is therefore not to be expected to be a likely candidate as a biochemical marker for sidestream tobacco exposure.

Nicotine is a highly specific chemical in that its presence in the atmosphere is due to tobacco. Nicotine and its principal metabolite cotinine are found in plasma, in the urine and in saliva. Although in most cases the route of uptake will be through inhalation of main stream or sidestream tobacco smoke, some individuals will take in nicotine through ingestion or sniffing. Jarvis and Russell found that urinary cotinine in their study produced the most correlation with reported exposure to sidestream tobacco smoke, while levels of cotinine in the plasma and saliva were less well correlated. Nicotine levels also were less well connected to self reported exposure (10). Urinary cotinine varied from 1.5 to 12 ng/ml depending on exposure. Cigarette smokers by comparison have levels of the order of 200 to 500 ng/ml of urinary cotinine. Matsukura et al (17) report good agreement between the level of cotinine per mg of creatinine in the urine, and the number of cigarettes smoked by others in the subjects' indoor environment. Cigarette smokers had urinary levels of· 8.5 micrograms of cotinine per milligram of creatinine, as compared with levels of 0.5 to 1.5 micrograms of cotinine per milligram of creatinine for non-smokers, depending on the smoking activity around them.

DISCUSSION

In contemplating and designing population based studies of the possible health effects of sidestream tobacco smoke it is clear that exposure characterization is an important, if not the most important factor limiting the power of such studies. A number of such studies are likely to produce results suggesting the absence of adverse health effects. In the case of relatively weak effects such an absence of significant association can easily be the result of even relatively low frequencies of misclassification of exposure. All of this suggests that health effects studies of sidestream tobacco smoke should devote a great deal of attention to the design of the exposure characterization scheme. The scheme would probably use a combination of questionnaire data, physical and chemical measurements in the environment, as well as biological or biochemical monitoring. In addition, thought might be given to schemes which would select the study population for the ease with which its exposure may be characterized, and for the absence of other confounding conditions.

REFERENCES

1. Weber, A., T. Fischer, and E. Grandjean. Objektive und subjektive physiologische Wirkungen des Passivrauchens. Int. Arch. Occup. Environ. Hlth 37:277-288 (1976)

2. Weber, A., T. Fischer, and E. Grandjean. Passive smoking in experimental and field conditions. Environ. Res 20:205-216 (1979)

3. Muramatsu, T., A. Weber, S. Muramatsu and F. Ackermann. Experimental study on irritations and annoyance due to passive smoking. Int. Arch. Occup. Environ. Hlth 51:305-317 (1983)

4. Holt, P.G., and D. Keast. Environmentally induced changes in immunological function: Acute and Chronic Effects of Inhalation of

Tobacco Smoke and other atmospheric Contaminations in man and experimental animals. Bact. Rev. 41:205-216 (1977)

5. Trichopoulos D., A. Kalandidi, L. Sparros. Lung cancer and passive smoking: conclusion of the Greek study. The Lancet, September 17, 1983:677-678

6. Shy, C.M., D.G. Kleinbaum and H. Morgenstern. The effect of misclassification of exposure status in epidemiological studies of air pollution health effects. Bull. N.Y. Acad. Med. 54:1155-1165 (1978)

7. Ozkaynak, H., P.B. Ryan, J.D. Spengler and N.M. Laird. Bias due to misclassification of personal exposures in epidemiological studies of indoor and outdoor air pollution. Proc. 3rd Internat. Conf. on Indoor Air quality and Climate, Stockholm (1984)

8. Vogt, T., S. Selvin, G. Widdowson and S. Hulley. Expired air carbon monoxide and serum thiocyanate as objective measures of cigarette exposure. Am.J.Publ.Hlth 67: 545-549 (1977)

9. Petitti, D., G. Friedman and W. Kahn. Accuracy of information on smoking habits provided on self administered research questionnaire. Am.J.Publ. Hlth. 71: 308-311 (1981)

10. Jarvis, M.J. and M.A.H. Russell. Measurement and estimation of smoke dosage to non-smokers from environmental tobacco smoke. Eur. J. Resp. Dis. 65:68-75 (1984) Suppl. 133

11. Ware, J.H., D.W. Dockery, A. Spiro III, F.E. Speizer and B.G. Ferris Jr. Passive Smoking, Gas Cooking, and Respiratory Health of Children Living in Six Cities. Am.Rev.Respir.Dis. 129:366-374 (1984)

12. Hasselblad, V., C.G. Humble, M.G. Graham and H.S. Anderson. Indoor Environmental Determinants of Lung Function in Children. Am.Rev. Respir. Dis. 123:479-485 (1981)

13. Cain, W.S., and B.P. Leaderer. Ventilation requirements in occupied spaces during smoking and non-smoking occupancy. Env. Int. 8: 505-514 (1982)

14. Cain, W.S., B.P. Leaderer, R. Isseroff, L.G. Berglund, R.J. Huey, E.D. Lipsitt and D. Perlman. Ventilation requirements in buildings -I. Control of occupancy odor and tobacco smoke odor. Atmosph. Env. 17:1183-1197 (1983)

15. Leaderer, B.P., W.S. Cain, R. Isseroff and L.G. Berglund. Ventilation requirements in buildings - II. Particulate matter and carbon monoxide from cigarette smoking. Atmosph. Env. 18:99-106 (1984)

16. Lynch, C.J. Half-lives of selected tobacco smoke exposure markers. Eur. J. Resp. Dis. 65:63-67 (1984) Suppl. 133

17. Matsukura, S., T. Taminato, N. Kitano, Y. Seino, H. Hamada, M. Uchihashi, H. Nakajima and Y. Hirata. Effects of environmental tobacco smoke on urinary cotinine excretion in non-smokers. N.Engl.J. Med. 311: 828-832 (1984)

16.

Biological Potential and Exposure-Dose Relationships for Constituents of Cigarette Smoke

Joseph D. Brain and Brenda E. Barry

Harvard University School of Public Health
Boston, MA

INTRODUCTION

Most of this session will focus on characterizing and quantifying possible responses to the inhalation of passive cigarette
smoke. However, it is essential to consider dose as well as
response. Typically, for smokers, dose is given in terms of
cigarettes smoked per day or cumulative pack years. For passive
smoking, exposure may be characterized in terms of smoke concentration (ug/m^3). But what do we really know about the total
integrated dose to the respiratory tract resulting from passive
smoking? What fraction is deposited and fails to exit with the
expired air? Moreover, what is the fate of the deposited smoke?
What is the anatomic distribution of the constituents of smoke that
are retained? After briefly considering the chemical composition
and toxic potential of tobacco smoke, we will describe the size and
aerodynamic properties of smoke and relate it to the fraction of
inspired smoke that deposits in the lungs. We will also consider
where the smoke deposits and describe its possible fate.

COMPOSITION OF PASSIVE SMOKE

Although highly variable, passive smoke includes many of the
same constituents as the smoke entering the active smoker's lungs.
Both particulate and gaseous phases are present. An aerosol is
defined as a suspension of solid or liquid particles in a gas (1).
In the case of cigarette smoke, the aerosol includes ambient air as
well as the gases, liquids, and solids produced during tobacco

combustion.   Thus, side-stream smoke and smoke expired by smokers
into indoor air includes a mixture of gases as well as tiny liquid
droplets and other particles.   When mainstream smoke is condensed
by cooling or by passing it through a filter, a dark brown material
is obtained, usually referred to as tar.   Over a thousand different
chemical compounds have been identified in the particulate fraction.   They include hydrocarbons, aldehydes, ketones, organic
acids, alcohols, nicotine, and phenols.   Metallic compounds,
including radioactive lead and polonium, are also present. Most of
these components are also present in passive smoke.

     The gas phase is also complex; in addition to nitrogen and
oxygen contained in the air, increased amounts of carbon dioxide
and carbon monoxide may be present.   There are also significant
amounts of cyanides, acrolein, nitrogen oxides, and ammonia.   The
precise quantitative composition of the tobacco smoke carries with
it many different factors including the type of tobacco plant
grown, the soil used to grow the plant, the method of curing the
leaves, the temperature of combustion during smoking, and the composition and physical properties of the cigarette paper and other
additives.   As the butt length decreases, many substances that have
previously condensed on the remaining tobacco are revaporized.
Generally, as butt length shortens, the smoke from the cigarette
contains an increasing concentration of these substances.   Aging
and dilution of smoke may modify both the amount and composition of
these aerosols.

BIOLOGICAL EFFECTS

     Most of the constituents in smoke are potentially toxic to
lung tissues.   Their toxicity extends from impairment of mucociliary transport, which is critical for particle clearance from the
lungs, to carcinogenic and cocarcinogenic activities (2,3).   Other
components initiate emphysema and chronic bronchitis. Repeated edi-
tions of the Surgeon General's Report on the Health Consequences of
Smoking (4) have amply documented the potential of tobacco smoke to
cause injury.   However, to understand where the numerous particulates in second-hand cigarette smoke deposit in the lungs and how
they are removed is important for determining the pathologic
effects of smoke exposure.

CHARACTERIZATION OF CIGARETTE SMOKE

     To predict the deposition patterns of any aerosol, such as
passive cigarette smoke, it is necessary to know the size, shape,
and density of the individual particles or droplets.   Describing
the distribution of particle diameters is essential.   It is convenient to describe the range of particle sizes as aerodynamic
diameters rather than in terms of optical measurements of particle
size, because the former is a better predictor of aerodynamic
behavior in filters or in the respiratory tract (1).   Aerodynamic
diameter is defined as the diameter of a sphere of unit density

that has the same settling velocity as the particle being measured. This may be expressed as a count median aerodynamic diameter (CMAD) and mass median aerodynamic diameter (MMAD).   These are, respectively, the diameters for which half of the number or mass of the particles are less than that diameter and half exceed it.

The particulates in mainstream cigarette smoke have been measured by several investigators using a variety of analytical devices; passive smoke is more poorly characterized.   Because of different apparatus and different methods of smoke generation and dilution, results vary but are reasonably consistent.   McCusker et al. (5) used a device called the single particle aerodynamic relaxation time (SPART) analyzer to size particulates from several brands of cigarettes, with and without filters.   The mass median average diameter (MMAD) for all brands averaged approximately 0.46 um; it was not markedly different when the filters were removed. Particulate concentrations per milliliter ranged from 0.3×10^9 to 3.3×10^9 depending on whether the cigarettes were rated ultralow, low, or medium in tar content.

Hinds (6) compared the particulate size distribution in cigarette smoke using an aerosol centrifuge and a cascade impactor. Although these devices are based on different physical principles, Hinds found that the results were comparable.   The MMAD values ranged from 0.37 to 0.52 um.   Variations depended primarily on the dilution of the smoke.   Keith and Derrick (7) used a specially modified centrifuge, termed a conifuge, to analyze cigarette smoke and reported MMAD and concentration values similar to Hinds (6) and McCusker et al. (5). Particulate analysis by a light scattering photometer yielded an MMAD of 0.29 um and particulate concentrations of 3×10^{10}/mL (8).

Time and concentration importantly modify tobacco smoke. Cigarette smoke aerosols contain volatile components and evaporation gradually reduces particle diameters.   It is also true that, with the extremely high particle concentrations encountered in mainstream smoke, the aerosol can agglomerate rapidly because nearby particles collide with each other and coalesce.   If smoke is cooled (reducing the vapor pressure of volatile components) and diluted in room air (reducing the probability of particle collisions), the size will become more stable.

FACTORS AFFECTING DEPOSITION OF PASSIVE SMOKE IN HUMANS

Particle size is a critical factor.   Submicrometric particles will not only deposit in small and large airways, but considerable deposition in alveoli will also occur.   Breathing pattern is also important (9). Large tidal volumes will favor alveolar deposition. Higher inspiratory flows will promote deposition at bifurcations. Breath-holding is important.   The greater the elapsed time before the next expiration, the higher the fraction deposited (collection efficiency).   Individual anatomic differences may influence the amount and distribution of deposited particles.   The cross section

of airways will influence the linear velocity of the inspired air.
Increasing alveolar size decreases alveolar deposition. Preexisting disease can also modify the deposition of smoke.

Particle size need not be constant; it may also change within
the human respiratory tract. After air containing smoke is drawn
into the mouth and upper respiratory tract it becomes humidified.
As previously discussed, particulates can change in size due to
coagulation and evaporation. They can also grow because of their
affinity for water, termed hygroscopicity (10).

The rate and pattern of breathing can also affect the total
dose of cigarette particulates deposited in the lungs. Dennis (11)
reported that exercise increased the percent deposition of two
experimentally generated aerosols in human subjects. Increased
deposition was also measured in exercising hamsters that inhaled a
radiolabelled aerosol (12). These results are most relevant to
those who breathe air containing passive smoke when their minute
ventilation is increased while working or during periods of exercise.

DEPOSITION OF CIGARETTE SMOKE PARTICULATES

The factors discussed in the previous section illustrate that
experimental measurements of the size and concentration of smoke
aerosols in indoor environments are insufficient for prediction of
the deposition patterns. Cigarette smoke is a mutable aerosol and
this factor complicates the collection of accurate and reproducible
data regarding its particulate composition. In addition, alterations in respiratory structure and respiratory rate can affect
deposition of particulates. These complexities stress the importance of actual measurement of regional deposition of cigarette
smoke particulates in human lungs. However, there is little data
published on this important area despite the prevalence of passive
smoking and concerns about its impact on human health. The majority of the available information on deposition of particulates
present in cigarette smoke is based on theoretical or physical
models of the lungs and measurements of differences between the
concentrations of aerosol in inhaled and exhaled air.

A model to predict percent deposition of particles based on
mass median aerodynamic diameter was presented by the Task Group on
Lung Dynamics (13) of the International Commission on Radiological
Protection. The respiratory tract was divided into three main
regions: nasopharynx, trachea and bronchi, and the alveolar region.
In conjunction with estimates of particulate clearance, deposition
calculations were made for these regions at three different inhalation volumes. This model suggests that 30 to 40% of the particles
within the size range present in cigarette smoke will deposit in
the alveolar region and 5-10% in the tracheobronchial region. This
model also emphasizes the impact of particle solubility on the
total integrated dose with time. Brain and Valberg (14) developed
convenient nomograms and a computer program to demonstrate how

particulate solubility and particle size significantly affect the
net amount of particulates retained in the lungs. Although the
basic outline of the model is generally correct, more recent measurements suggest that values for alveolar deposition may be too
high by a factor of approximately 2 (15). However, the extent to
which passive smoke particles are hygroscopic and increase in size
within the respiratory tract is an important and unresolved issue.

Aerosol deposition has also been studied in airway casts.
Physical models of the upper airways of human lungs have been made
by a double casting technique to study particulate deposition at
several airway generations (16). Lungs obtained at autopsy were
filled with wax or alloy. When these materials became solid, the
tissue was removed and the casts were coated with silicon rubber or
latex. The wax or alloy was then melted and removed, leaving a
cast of the original airways. Different flow rates and particulate
sizes were used to study deposition patterns. Schlesinger and
Lippman (17) reported that there was a correlation between the
deposition sites of test aerosols in their lung casts and the most
common sites of origin of bronchogenic carcinoma in humans. Both
occured preferentially at bifurcations. Martonen et al. (18) added
an oropharyngial compartment and a replica cast of the larynx to
the tracheobronchial casts in order to better simulate air flow
patterns in the upper respiratory tract. They used these models to
evaluate the amount of cigarette smoke condensate deposited in the
airways at different flow rates. More condensate was present in
areas where airways branched and especially at the bifurcation
points. Aerosol was also deposited preferentially along posterior
airway walls of the branching regions.

Hiller et al. (10) measured the deposition fraction of an
aerosol containing three different sizes of polystyrene latex
spheres in nonsmoking humans. They measured a 10% deposition for
0.6-um (MMAD) spheres, which is similar to the results of Davies et
al. (19) and Muir and Davies (20) using 0.5-um aerosols and Heyder
et al. (15) using aerosols with 0.2 to 1.0-um range. The size
ranges of these aerosols are comparable to those experimentally
measured in cigarette smoke, as previously discussed.

PARTICULATE RETENTION IN THE LUNG

The amount of particulates actually retained at different
sites in the lungs following inhalation of an aerosol such as
cigarette smoke depends on the balance between the amount which
deposits in the respiratory tract and the efficiency of the lung
clearance mechanisms in the airways and alveoli. Clearance mechanisms are a dynamic component of normal lung function and operate to
keep the lung clean and sterile. Particles depositing in the airways are entrained in the mucus layer which lines the airway passages. This layer is swept toward the mouth by the action of ciliated cells and eventually swallowed. Mucus transport is approximately 1-2 cm in the trachea, but it is slower in smaller airways.
In addition, macrophages present in the airways may phagocytose

deposited particulates and be carried towards the mouth by the mucociliary transport system. Particulates reaching the alveolar region - those which are usually less than several micrometers - are soon engulfed by alveolar macrophages. These cells gradually migrate towards the airways and exit the lung via the mucociliary escalator. Dissolution is an additional important clearance mechanism for soluble particles.

Lung disease and cigarette smoking itself can affect particulate clearance and retention in smokers' lungs. Previous studies have shown that smokers have different aerosol deposition patterns and slower clearance rates than nonsmokers (21-23). These alterations in clearance are, in part, caused by components within cigarette smoke that are ciliotoxic (3) and can impair phagocytosis by alveolar macrophages (24). Clearance mechanisms in smokers may be further compromised by lung diseases, such as emphysema and fibrosis, and by exposure to air pollutants. Oxidants in photochemical smog, such as ozone and nitrogen oxides, are toxic to ciliated cells and macrophages (25).

Measurements of long term retention of cigarette particulates in the lungs are difficult to estimate from data obtained with airway casts or from differences between inhaled and exhaled aerosol concentration, since these methods do not take into account clearance mechanisms. Unfortunately, there are little data available regarding the actual retention and sites of deposition of passive cigarette smoke particulates in either humans or animals. The most accurate method is quantification of particulate deposits in individual pieces of tissue dissected from the lung. Impossible in living animals, this is a tedious procedure in animal lungs or human material obtained at surgery or autopsy and is especially difficult for large lungs.

In addition to describing the deposition of smoke in the lungs, one can also attempt to quantify exposure by examining saliva, serum, or urine. These body fluids are much more accessible than the lung itself and thus a more practical component of epidemiologic studies of humans. Blood carboxyhemoglobin levels (COHb) and serum or salivary thiocyanate (SCN) levels have been used as a marker of exposure to passive smoke (26). However, both measures frequently lack both sensitivity and specificity. Nicotine is specific to tobacco, but it has a short half-life in plasma.

During the last five years, cotinine, the major metabolite of nicotine found in blood and urine, has become increasingly used as a marker of tobacco smoke inhalation. Cotinine has a much longer half-life in the blood (approximately 30h); moreover, it reaches higher concentrations in the blood and urine than does nicotine.

A number of investigators have observed elevated cotinine levels in humans following passive smoking (26-29). Matsukura et al. (29) explored the relationship between qualitative estimates of smoke concentration in the environment and urinary cotinine excretion in 472 nonsmokers. At both home and work, they found highly significant differences between smokers and nonsmokers and a good dose-response relationship between urinary cotinine excretion and such parameters as the number of cigarettes consumed per day by relatives or co-workers or the number of smokers in the home or workplace. Greenberg et al. (28) studied infants and showed a direct relationship between cotinine excretion by the infants and the self-reported smoking behavior of the mothers during the previous day.

Other constituents of smoke have also been detected in the urine of nonsmokers. Both chemical analysis and bioassays have been used. While using a _Salmonella_/microsome assay, Bos et al. (30) saw an increase in the urinary excretion of products that were mutagenic in their assay in nonsmoking subjects experimentally exposed to cigarette smoke. Other investigators have reported the presence of such carcinogens as benzo(a)pyrene in the urine of nonsmokers exposed to smoky environments. Such measurements may be helpful in exploring the hypothesis that passive smoking may be associated with increased cancer risk.

CONCLUSIONS

Passive smoke in indoor environments is composed of that exhaled by smokers and side-stream smoke that is produced by burning cigarettes in between inhalations. The concentration of respirable particulates in areas where there are smokers can range from 100 to 700 ug/m^3. This is up to 25 times higher than that found in nonsmoking areas (31). Assuming mean deposition values of 10 and 70% for passive and active smokers from the data presented by Hiller et al. (10,32), the deposition would be approximately 0.10-0.70 mg for a nonsmoker over an 8-h day in a room containing 100 to 700 ug/m^3 of smoke. In comparison, a smoker would deposit approximately 400 mg of tar in his/her lungs if he/she smoked two packs of cigarettes with average tar rating of 20 mg/cigarette during the same time period.

Although the amount of smoke depositing in the lungs during passive smoking is small compared to that encountered by the active smoker, it may differ in its composition and toxicity. For example, certain constituents are present in much higher concentrations in side-stream smoke as compared with mainstream smoke (33).

Another reason for continued concern and expanded research efforts is that large numbers of persons are involved. In the United States in 1979, 36.9% of men and 28.2% of women are current smokers (4). Bonham and Wilson (34) have shown that the majority of homes have at least one smoker. Thus, passive smoking by children, even in early childhood, is widespread and likely to remain so because of the increasing frequency of smoking among U.S. adolescent girls. Further study of exposure (35), dose (36), and response (37) is warranted.

RESEARCH RECOMMENDATIONS

1. Studies are needed to determine the precise fraction of the inspired smoke that is deposited in the respiratory tract. Individual differences and the effects of age, exercise, and preexisting disease should be explored.

2. Experiments in excised human lungs or appropriate animal models should be used to determine the anatomic distribution of retained smoke.

3. Since clearance is a major determinant of retention, the clearance kinetics of different components of passive cigarette smoke should be determined.

4. Using animal and human studies, the correlation between tobacco smoke retention and serum levels or urinary excretion of smoke components (or their metabolites) should be explored.

5. Thorough studies of the concentration of tobacco smoke in indoor environments and breathing patterns of subjects should be carried out so that epidemiologic studies of the respiratory effects of involuntary smoke exposure will have accurate estimates of dose.

REFERENCES

1.   Hinds WC. Aerosol Technology Properties: Behavior and Measurement of Airborne Particles. John Wiley and Sons, New York, 1982.

2.   Wynder EL, Hoffmann D. Tobacco and health: a societal challenge. N. Eng. J. Med. 300:894-903, 1979.

3.   Battista SP. Cilia toxic components in cigarette smoke. Proceedings of the Third World Conference on Smoking and Health (DHEW Publication No. NIH 76-1221), Vol. 1, Washington DC, Government Printing Office, 1976, pp. 517-534.

4.   Surgeon General. The Health Consequences of Smoking for Women. Washington D.C.: U.S. Department of Health and Human Services, 1980.

5.   McCusker K, Hiller FC, Wilson JD, Mazumder MK, Bone R. Aerodynamic sizing of tobacco smoke particulate from commercial cigarettes. Arch. Environ. Health 38:215-218, 1983.

6.   Hinds WC. Size characteristics of cigarette smoke. Am. Ind. Hyg. Assoc. J. 39:48-54, 1978.

7. Keith CH, Derrick JC. Measurement of the particle size distribution and concentration of cigarette smoke by the "conifuge". J. Colloid Sci. 15:340-356, 1960.

8. Okada T, Matsunuma K. Determination of particle size and concentration of cigarette smoke by a light scattering method. J. Colloid and Interface Sci. 48:461-469, 1974.

9. Brain JD, Valberg PA. Deposition of aerosol in the respiratory tract. Am. Rev. Respir. Dis. 120:1325-1373, 1979.

10. Hiller FC, Mazumder MK, Wilson JD, McLeod PC, Bone RC. Human respiratory tract deposition using multimodal aerosols. J. Aerosol. Sci. 13:337-343, 1982.

11. Dennis WL. The effect of breathing rate on deposition of particles in the human respiratory system. In: Inhaled Particles III, edited by WH Walton. Unwin Brothers Ltd, Surrey, England, 1971. pp. 91-103.

12. Harbison ML, Brain JD. Effects of exercise on particle deposition in Syrian golden hamsters. Am. Rev. Respir. Dis. 128:904-908, 1983.

13. Task Group on Lung Dynamics. Deposition and retention models for internal dosimetry of the human respiratory tract. Health Physics 12:173-207, 1966.

14. Brain JD, Valberg PA. Models of lung retention based in ICRP Task Group report. Arch. Environ. Health 28:1-11, 1974.

15. Heyder J. Particle transport onto human airway surfaces. Eur. J. Respir. Dis. 63 (Suppl. 119): 29-50, 1982.

16. Schlesinger RB, Lippmann M. Particle deposition in casts of the human upper tracheobronchial tree. Am. Ind. Hyg. Assoc. J. 33:237-251, 1972.

17. Schlesinger RB, Lippmann M. Selective particle deposition and bronchogenic carcinoma. Environ. Res. 15:424-431, 1978.

18. Martonen TB, Lowe JE. Cigarette smoke pattern in a human respiratory tract model. Proc. 36th Annual Conf. Eng. Med. and Biol. 25:171, 1983. (Abstract)

19. Davies CN, Heyder J, Subba Rama MC. The breathing of half-micron aerosols: I. Experimental. J. Appl. Physiol. 32:591-600, 1972.

20. Muir DCF, Davies CN. The deposition of 0.5 um diameter aerosols in the lungs of man. Ann. Occup. Hyg. 10:161-174, 1967.

21.   Albert RE, Lippmann M, Briscoe W. The characteristics of
      bronchial clearance in humans and the effect of cigarette
      smoking. Arch. Environ. Health 18:738-755, 1969.

22.   Cohen D, Arai SF, Brain JD. Smoking impairs long-term dust
      clearance from the lung. Science 204:514-517, 1979.

23.   Sanchis J, Dolovich M, Chalmers R, Newhouse MT. Regional
      distribution and lung clearance mechanisms in smokers and
      non-smokers. In: Inhaled Particles III, edited by EH Walton.
      Unwin Brothers Ltd., Surrey England, 1971.

24.   Ferin J, Urbanhoug G, Vlokova A. Influence of tobacco smoke
      on the elimination of particles from the lungs. Nature
      206:515-516, 1965.

25.   Bils RF, Christie BR. The experimental pathology of oxidant
      and air pollutant inhalation. Intl. Rev. Exptl. Path.
      21:195-293, 1980.

26.   Jarvis MJ, Russell MAH. Environmental tobacco smoke. 2.2.
      Measurement and estimation of smoke dosage to non-smokers
      from environmental tobacco smoke. Eur. J. Respir. Dis. 65
      (Suppl. 133): 68-75, 1984.

27.   Hoffman D, Brunnemann KD, Adams JD, Haley NJ. Indoor air
      pollution by tobacco smoke: model studies on the uptake by
      nonsmokers. Proceedings of Stockholm Conference on Indoor
      Air Pollution. 1984.

28.   Greenberg RA, Haley NJ, Etzel RA, Loda FA. Measuring the
      exposure of infants to tobacco smoke; nicotine and cotinine
      in urine and saliva. N. Eng. J. Med. 310:1075-1078, 1984.

29.   Matsukura S, Taminato T, Kitano N, Seino Y, Hamada H,
      Uchihashi M, Natajima H, Hirata Y. Effects of environmental
      tobacco smoke on urinary cotinine excretion in nonsmokers.
      N. Eng. J. Med. 311:828-832, 1984.

30.   Bos RP, Theuws JLG, Henderson PT. Excretion of mutagens in
      human urine after passive smoking. Cancer Letters 19:85-90,
      1983.

31.   Repace JL, Lowrey AH. Indoor air pollution, tobacco smoke,
      and public health. Science 208:464-472, 1980.

32.   Hiller FC, McCusker KT, Mazumder MK, Wilson JD, Bone RC.
      Deposition of sidestream cigarette smoke in the human
      respiratory tract. Am. Rev. Respir. Dis. 125:406-408,
      1982a.

33. Weiss ST, Tager IB, Schenker M, Speizer FE.   The health
     effects of involuntary smoking.   Am. Rev. Respir. Dis.
     128:933-942, 1983.

34. Bonham GS, Wilson RW.   Children's health in families with
     cigarette smokers.   Am. J. Public Health 71:290-293, 1981.

35. Hinds WC, First MW.   Concentrations of nicotine and tobacco
     smoke in public places.   N. Eng. J. Med.   292:844-845, 1975.

36. Feyerabend C, Higenbottam T, Russell MAH.   Nicotine concen-
     trations in urine and saliva of smokers and nonsmokers.
     Brit. Med. J.   284:1002-1004, 1982.

37. Surgeon General.   Chronic Obstructive Lung Disease.   Washing-
     ton DC: U.S. Department of Health and Human Services, 1984.

17.

Relationship Between Passive Exposure to Cigarette Smoke and Cancer

Jonathan M. Samet

Associate Professor, Department of Medicine, and the New Mexico
Tumor Registry, Cancer Center, University of New Mexico,
Albuquerque, New Mexico 87131

INTRODUCTION

Causal associations between active cigarette smoking and cancer of the lung and other sites have been long established on the basis of extensive toxicological, experimental, and epidemiological evidence. Only recently, however, has passive exposure to tobacco smoke been considered as a potential risk factor for lung cancer in nonsmokers. This putative role of passive smoking has become an emotionally charged and highly controversial subject with potentially important regulatory and economic implications. Tobacco industry arguments defending the individual's right to free choice concerning smoking would be severely damaged if passive smoking were shown to cause cancer in nonsmokers.

The prevalence of passive smoking in the United States further emphasizes the potential public health consequences of this exposure. Friedman and co-workers (1) questioned 37,881 nonsmoking members of a health maintenance organization concerning passive smoking at home and elsewhere. Overall, 63 percent reported some exposure and 34.5 percent received at least 10 hours per week. Unpublished findings from an ongoing case-control study in New Mexico show that 29 percent of nonsmoking male and 56 percent of nonsmoking female controls have lived with a cigarette smoking spouse.

Association between passive smoking and lung cancer derives biological plausibility from the chemical composition of sidestream smoke, the confirmation of exposure in nonsmokers with biological markers, and the failure to find a threshold for respiratory carcinogenesis in active smokers. Sidestream smoke contains the same toxic and tumorigenic agents as mainstream smoke; some are present in much higher concentrations because of the burning conditions under which sidestream smoke is generated (2). Investigations with markers of tobacco smoke exposure have convincingly demonstrated that passive smoking results in inhalation and absorption of sidestream smoke components (3). For example, Wald et al. (4) recently reported increased urinary cotinine levels in exposed nonsmokers and a dose-response relationship between urinary concentration and the duration of reported exposure. In Japan, Matsukura and colleagues (5) found that the presence of smokers in the home and in the workplace, and urban residence were associated with increased urinary cotinine levels. Finally, studies of active smoking have uniformly indicated excess lung cancer risks at lower levels of cigarette smoking and none have implied the presence of a threshold (2).

This paper will review the epidemiological evidence relevant to the hypothesis that passive smoking causes lung cancer. First, methodological considerations relevant to studying this association will be addressed. Second, the available epidemiological evidence will be reviewed. Finally, the existing data will be assessed against conventional criteria for determining the causality of association - the same criteria, in fact, that were used in the 1964 Surgeon General's Report for evaluating the association between lung cancer and active smoking (6).

METHODOLOGICAL ISSUES

The association between passive smoking and lung cancer has been approached with conventional hypothesis-testing designs: the case-control and cohort studies (Tables 1 and 2). Each has well characterized advantages and disadvantages (7). The results of both may be affected by misclassification of exposure and confounding by other risk factors, whereas other types of bias uniquely influence each design. The potential for information bias, introduced by the interviewer or the subject, is of particular importance in case-control studies of this hypothesis.

Misclassification of exposure refers to the incorrect categorization of actually exposed subjects as nonexposed and of nonexposed as exposed (8). When misclassification occurs randomly in relationship to the selection of a study's subjects, it reduces measures of effect towards unity; if nonrandom, it may increase or decrease effect measures.

The questionnaire measures that have been employed in investigations conducted to date may have introduced random misclassification on exposure to cigarette smoke. While gas phase components may also be important for carcinogenesis, the following discussion will primarily consider cigarette smoke particulate. In

TABLE 1

Investigations Showing Significant Effects of
Passive Smoking on Lung Cancer Risk

Study	Findings	Comment
Prospective cohort study in Japan of 91,540 nonsmoking women, 1966-1979 (19).	Age-occupation adjusted SMRs, by husbands' smoking: Non-smokers — 1.00 Ex, or 1-19/day - 1.61 ≥ 20/day — 2.08	Trend statistically significant. All histologies.
Case-control study in Greece with 40 cases, 149 controls, 1978-1980 (22).	Odds ratios by husbands' smoking: Non-smokers — 1.00 Ex-smokers — 1.8 ≤ 20/day — 2.4 ≥ 21/day — 3.4	Trend statistically significant. Histologies other than adenocarcinoma and bronchioloalveolar carcinoma.
Case-control study in the U.S.A. with 22 female and 8 males cases, 133 female and 180 male controls (25).	Odds ratios by spouse smoking: Non-smokers — 1.00 1-40 pack-years - 1.48 ≥ 41 pack-years - 3.11	Significant increase for ≥ 41 pack years. Bronchioloalveolar carcinoma excluded.

TABLE 2

Investigations Not Showing Significant Effects Of
Passive Smoking On Lung Cancer Risk

Study	Findings	Comment
Prospective cohort study in the USA of 176,139 nonsmoking females, 1960-1972 (13).	Age-adjusted SMRs, by husbands' smoking: Non-smokers - 1.00 < 20/day - 1.27 ≥ 20/day - 1.10	All histologies.
Prospective cohort study in Scotland of 8128 males and females, 1972-1982 (16)	Age-adjusted SMRs, for domestic exposure Males - 3.25 Females - 1.00	Preliminary, small numbers
Case-control study in Hong Kong of 84 female cases and 139 controls, 1976-1977 (28,29).	Crude odds ratio associated with smoking spouse of 0.75.	All histologies. two reports are inconsistent on the exposure variable.
Case-control study the U.S.A. 25 male and 53 female cases with matched controls, 1971-1980 (24).	Odds ratios for current exposure at home were 1.26 in males and 0.92 in females.	All histologies. Findings negative for spouse smoking variable as well.
Case-control study in Hong Kong with 88 nonsmoking female cases, 1981-1982 (30).	Odds ratio for combined home and workplace exposure of 1.24 ($p > 0.40$).	All histologies.

the United States, cigarette smoking is a major source of indoor respirable particulates and thus a major determinant of variation among individuals in exposure to this pollutant (9-11). Within a room, concentrations will be determined not only by the strength of sources, such as cigarette smoking, but by building characteristics and ventilation rate (9). Time-activity patterns further modify the profile of exposure (11). Thus, with regard to domestic exposure, simple descriptions of spouse smoking behavior cannot satisfactorily define gradients of exposure. They can, however, document that exposure to tobacco smoke has occurred. Similar limitations apply to questionnaire derived indices of workplace exposure. With regard to total passive exposure to tobacco smoke, variables that do not include time outside of the home will lead to misclassification. In the population studied by Friedman et al. (1), high proportions of nonsmoking males and females reported exposure outside of the home. Workplace exposure was associated with higher urinary cotinine levels in the recent report from Japan by Matsukura et al. (5). Thus, random misclassification of exposure is likely with questionnaire indices. Studies that have used such measures may be conservative since random misclassification reduces effect measures toward unity.

REVIEW OF THE EVIDENCE

Evidence concerning passive smoking and lung cancer has been sought indirectly in descriptive data and directly with case-control and cohort studies. Time-trends of lung cancer mortality in nonsmokers have been examined with the rationale that increasing passive smoking should be mirrored by increasing mortality rates. Enstrom (12) calculated lung cancer mortality rates from various nationwide sources for the period 1914-1968 and concluded that a real increase had occurred among males after 1935. In contrast, Garfinkel (13) did not identify time trends in nonsmokers in the Dorn Study of Veterans, 1954 to 1969, or in the American Cancer Society study, 1960 to 1972. In a large autopsy series, Auerbach and colleagues (14) did not find increased abnormalities in the bronchial epithelium of male nonsmokers deceased in 1970-1977 in comparison with those deceased in 1955-1960.

While this review emphasizes lung cancer, associations of passive smoking with cancers of other sites or with other diseases would strengthen the evidence concerning passive smoking and lung cancer. An investigation of all cancer deaths in females residing in Western Pennsylvania has been frequently cited as showing an adverse effect of passive smoking (15). Miller interviewed surviving relatives of 537 deceased nonsmoking women concerning the smoking habits of their husbands. A significantly increased relative risk of cancer death was found in the women who were not employed outside of their homes. The large number of potential subjects that were not interviewed and the possibility of information bias detract from this report.   Gillis et al. (16) followed 16,171 healthy Scottish individuals, ages 45 to 64 years, over at least a 6 year period.   In a preliminary report concerning 8,128 subjects, all-cause mortality was comparable in nonsmoking males with and without domestic tobacco smoke exposure, but was increased by nearly 50 percent in exposed nonsmoking women.   A case-control study of 438 cancer cases involving multiple sites and 470 controls showed increased relative risks from exposure during childhood and during adulthood (17).   In a 25-year cohort study in Amsterdam, all cause mortality in females was not affected by the husbands' smoking status (18).

More relevant is the direct hypothesis-testing evidence provided by case-control and cohort studies.   In 1981, two papers were published which reported significantly increased risks of lung cancer in nonsmoking women whose husbands smoked cigarettes (Table 1).   Hirayama (19) conducted a prospective cohort study of 91,540 nonsmoking women in Japan.   Standardized mortality ratios for lung cancer increased significantly with the amount smoked by the husbands.   The findings were unchanged with control of potentially confounding variables and with extension of follow-up from 14 years to 16 years (20).   Overall, the relative risk from passive exposure was 1.8 whereas that from active smoking was 3.8.   Hirayama has also reported a significantly elevated relative risk (2.94) in nonsmoking men with smoking wives (21).

Following its publication, this article received intensive scrutiny and correspondence in the British Medical Journal offered concerns about statistical methodology, about population selection, about uncontrolled confounding by factors such as cooking fuel exposure and socioeconomic status, and about the seemingly high relative risk.   In his responses, Hirayama satisfactorily rebuffed most of these criticisms; in particular, confounding did not appear to explain the findings though active smoking by reportedly nonsmoking women can not be excluded.   In this regard, Hirayama (20) has reported that the findings after 16 years of follow-up are consistent with effects of passive smoking on mortality from emphysema and chronic bronchitis, nasal sinus cancer, and ischemic heart disease.   Biologically, these effects seem somewhat less plausible than lung cancer and these new associations raise concern about confounding by unreported active smoking.   Hirayama has explained the level of relative risk by the low percentages of women working outside the home in Japan, low divorce rates, small room sizes, and lack of inhibition about smoking in the presence of nonsmokers (21).   No data concerning respirable particulate levels in the subjects' homes have been provided, however.

Also reported in 1981 were the results of a case-control study in Athens, Greece (22) (Table 1).   Female lung cancer cases with a diagnosis other than adenocarcinoma or bronchioloalveolar carcinoma

were identified at three large hospitals and controls were selected
at a hospital for orthopedic disorders.   All subjects were
interviewed by the same physician and their smoking status and that
of their husbands was obtained.   Single women were considered as
married to nonsmokers and changes in marital status were
considered.   The final series included 40 nonsmoking cases and 149
nonsmoking controls.   A significant trend of increasing risk with
presumed extent of passive exposure was present when either the
husbands' current or lifetime smoking habits were used for
stratification.   The findings were unchanged when the series was
expanded to 77 cases and 225 controls (23).

Less criticism has been published concerning the Greek study
than concerning Hirayama's investigation in Japan.   As discussed by
Kabat and Wynder (24), the attempt to restrict the case series to
histologies other than adenocarcinoma appears premature at present.
Further, the diagnosis of lung cancer was made without histological
or cytological confirmation in 35 percent of the cases.
Noncomparability of the case and control series must also be
considered when they are ascertained at different institutions; in
this context, Trichopoulos et al. did demonstrate comparability of
the case and control series for key demographic variables.   The
possibility of information bias must be raised because case and
controls were interviewed by a single physician who may have been
aware of the study's hypotheses.   Finally, the investigators
assessed the statistical significance of their findings with a
chi-square for trend in proportions.   The assumption that a former
smoking husband provided an exposure intermediate between that of a
nonsmoker and a current smoker was not justified by the authors.
However, the odds ratio is significantly elevated for the stratum
with the highest level of current smoking.

The results of another case-control study, published in 1983,
also demonstrated a significant association between passive smoking
and lung cancer risk (25) (Table 1).   Correa et al. obtained
information about the smoking habits of the parents and spouses of
eight male and 22 female nonsmoking lung cancer cases and of 313
controls.   Lung cancer risk increased with the spouses' lifetime
cigarette consumption.   Maternal smoking was associated with a
significantly increased odds ratio in active smokers but not in
nonsmokers.   On stratification by sex, the increase was
statistically significant only in males.

The relatively small numbers of subjects in this investigation
mandate caution in interpreting its results.   However, the overall
findings were unchanged, as reported in a recent abstract, when
these data were combined with comparable information from two other
case-control studies (26).   The overall design wa s appropriate but
information bias may affect the results of case-control studies
that rely on interview for exposure information.   The exposure
variable, cumulative cigarette consumption, differs from the

measures used by Hirayama (19) and by Trichopoulos et al. (22).   It would be useful to reanalyze these data with comparable exposure variables.

The results of two other investigations have also been interpreted as showing an increased lung cancer risk associated with passive smoking.   In Germany, Knoth et al. (27) accumulated a series of 792 lung cancer cases of which 59 were in females. Thirty-nine of these women had not smoked but 24 of the nonsmokers had lived in households with smokers.   Because the investigators did not interview a control series, they relied on census statistics to estimate the anticipated proportion of smoking spouses in the general population.   In the age group 50 to 69 years corresponding to the husbands of most patients, the census showed only 22.4 percent currently smoking.   In another recent report, Gillis et al. (16) described the results of a cohort study of 16, 171 males and females in Western Scotland (Table 2).   Exposure to tobacco smoke in the environment was characterized by four strata: nonsmoker and not domestically exposed, nonsmoker and domestically exposed, smoker and not domestically exposed, and both a smoker and domestically exposed.   Mortality rates for lung cancer and for all other cancers were calculated separately for males and females within each stratum.   Among males, six lung cancer deaths were observed in nonsmokers; in the control stratum, the annual mortality rate was 4 per 100,000 whereas in the domestically exposed nonsmokers the rate was 13 per 100,000.   For males the rates were similar in the two actively smoking groups.   In females, with a total of eight deaths from lung cancer in nonsmokers, the variation of mortality rates did not suggest an adverse effect of domestic tobacco smoke exposure.

The methodological limitations of these two studies are evident; neither formally tests for association between lung cancer risk and passive smoke exposure.   The German report did not involve a comparison series and the appropriateness of substituting census data was not addressed (27).   The authors did not formally test for association between passive smoking and lung cancer; in fact, they used their sparse data as a platform for discussing social and political aspects of passive smoking.   Interpretation of the Scottish investigation is constrained by the small number of deaths; in this regard, statistical significance testing was not performed (16).   The lack of effect of domestic tobacco smoke exposure in females is not consistent with earlier reports (Table 1) but the number of deaths is quite small at present.

The results of four other investigations suggest lesser or no effects of passive tobacco smoke exposure (Table 2).   Chan et al. (28,29) performed a case-control study in Hong Kong that included 84 nonsmoking female cases with 139 controls.   Apparently a single question was asked concerning passive smoking exposure.   In a 1979 report, the investigators stated that 40 percent of cases and 47 percent of controls (estimated odds ratio of 0.75) replied affirmatively to a question concerning exposure at home or at work (28). In a 1982 publication, similar findings were reported but the exposure variable was described as related to spouse smoking (29). The conflicting description of this investigation's exposure variable requires clarification. Noncomparability of the case and control series with regard to place of residence and lack of histological or cytological confirmation in 18 percent of cases further limit this investigation.

A more recent case-control study from Hong Kong also did not show definite effects of passive smoking. Koo et al. (30) interviewed 200 cases ascertained through Hong Kong health facilities and 200 controls, selected from the general population to match the age, socioeconomic, and geographic distribution of the cases (Table 2). With women not exposed to smoke at home or at work as the reference category, odds ratios for exposure at home and at work were not significantly increased. Nonsmoking cases had fewer hours of total estimated exposure than controls. In contrast to the case-control study in Louisiana (25), an effect of maternal smoking was not found.

The most important of the four publications, construed by many as negative, is based on the American Cancer Society's prospective cohort study (13) (Table 2). Between 1959 and 1960, 375,000 female nonsmokers were enrolled and follow-up of mortality lasted through 1972. From this cohort, Garfinkel identified 176,739 nonsmokers whose husbands had never smoked or were current smokers, presumably on enrollment. The standardized mortality ratios for the women with smoking husbands were greater than unity but not significantly. In the smoking-exposed group, there was no evidence for a dose-response relationship. A separate matched analysis, performed to more completely control confounding, provided similar results.

The American Cancer Society study should not be characterized as contradictory to the findings of Hirayama (19), Trichopoulos et al. (22), and Correa et al. (25). First, the standardized mortality ratios are above unity for the exposed groups. Second, confidence intervals for the mortality ratios in the American Cancer Society study overlap those reported by Hirayama (19). Third, while each of these investigations employed spouse smoking as the exposure variable, the comparability of dose among the four is uncertain. Repace (31) has suggested that the mortality ratio in the American Cancer Society study has been reduced by misclassification introduced by workplace exposures. His arguments lead to an adjusted mortality ratio of 1.7 for the American Cancer Society cohort. Finally, the use of death certificates to establish diagnosis in the American Cancer Society study probably introduced misclassification of disease status.

Recent and preliminary results from a nationwide case-control study also did not demonstrate increased lung cancer risk from domestic exposure to tobacco smoke (24) (Table 2). Kabat and Wynder examined the effects of currently smoking family members and of current exposure at work in 25 nonsmoking male and 53 nonsmoking female cases with equal numbers of controls. For men, the odds ratio for workplace exposure of 2.6 was significantly increased. Current domestic exposure was not significant for males or females. In a smaller subset of cases, adverse effects of spouse smoking were not identified. The authors clearly stated that their results were preliminary and that more data are needed. While the numbers are small, they are equivalent to those in the series reported by Correa et al. (25).

CONCLUSIONS

In summary, at present, only nine published investigations provide data directly relevant to the hypothesis that passive smoking is a risk factor for lung cancer. Several others offer indirect evidence. This paucity of data contrasts sharply with the literature cited in the 1964 Surgeon General's Report which characterized active cigarette smoking as a cause of lung cancer (6). That report reviewed 29 case-control and seven cohort studies. Their results uniformly and unequivocally demonstrated the association between active smoking and lung cancer. Application of carefully considered criteria for causality to the evidence led to the designation of cigarette smoking as causally related to lung cancer in men. The association was judged on its consistency, strength, specificity, temporal relationship and coherence. The report did not explicitly define "cause" but indicated that the term is generally applied to "... a significant effectual relationship between an agent and an associated disorder or disease in the host". It also acknowledged the multifactorial etiology of lung cancer and did not require a unique relationship between smoking and malignancy.

Application of these same criteria to the data for passive smoking highlights their weaknesses. With regard to consistency, the conflicts among the published investigations are immediately evident (Tables 1 and 2). However, because of potential differences in dose among the investigations, it is not certain that each has tested for a common magnitude of effect. Furthermore, given the small numbers of cases in most of the papers, the point estimates of effect are unstable and confidence limits generally overlap from one study to another. In the positive studies, the relative risk estimates have indicated relatively modest effect levels, ranging from about two to three. These values are much lower than those associated with active smoking and could more readily be the consequence of bias. In the face of small and conflicting studies, unidentified sources of bias should not be readily dismissed as an explanation for significant but modest elevations of risk. Specificity of association, that is a unique relationship between the factor and the disease, is an irrelevant and unimportant criterion for passive smoking. With regard to the temporal association of passive smoking and lung cancer, the directionality is unquestionably appropriate; exposure precedes the development of the disease. The remaining criterion is the coherence of the association. The biological plausibility of the association between passive smoking and lung cancer has been previously reviewed and this criterion appears to be met.

In conclusion, the association between passive smoking and lung cancer does not yet meet criteria applied to active smoking in the 1964 Surgeon General's Report. While confirmation of passive smoking as a risk factor for lung cancer would offer new ammunition against tobacco, the available evidence does not permit definitive judgments. In the face of difficult methodological problems, particularly that of accurately quantifying dose, unimpeachable data will be difficult to obtain.

New approaches for studying passive smoking and lung cancer are clearly needed. The problems of dose estimation seem more difficult for lung cancer than for other putative health effects of passive smoking. The relevant exposures may begin at birth and occur under a wide variety of circumstances. Historical reconstruction of exposures by questionnaire may be the only available approach for epidemiological studies. However, further validation of the questionnaire approach is needed with comparisons against biological markers and measured concentrations of tobacco smoke components. The reliability of questionnaire assessment of passive smoke exposure has not been established nor have sources of bias been evaluated. Interviews with next-of-kin may be particularly prone to information bias, almost certainly in the direction of overreporting. In fact, as the public becomes increasingly aware of and sensitized to potential effects of passive smoking, the results of case-control studies will become increasingly difficult to interpret. Unfortunately, the case-control design is the most efficient approach for investigating the relatively small number of lung cancer cases in nonsmokers. Cohort studies, which might offer better exposure data, must involve large numbers of subjects and lengthy follow-up. Investigative approaches which examine outcomes other than lung cancer might provide more immediate answers concerning passive smoking and respiratory tract carcinogenesis. For example, sputum cytology might be evaluated in nonsmokers in relation to passive tobacco smoke exposure.

While additional investigations will certainly be performed, the available data may already be satisfactory for both regulation and prevention. For regulatory purposes, the established carcinogenicity of tobacco smoke and the high prevalence of exposure should be sufficient to prompt action. For prevention, the data on active smoking should be sufficient; smoking prevention and cessation remain the best strategies for minimizing passive exposure.

ACKNOWLEDGEMENTS

   Supported in part by a grant from the National Cancer Institute CA 27187. Dr. Samet is recipient of a Research Career Development Award 5 KO 4 HLOO951. The author thanks Dr. A. Judson Wells for his helpful comments and Lee Fernando for preparing the manuscript.

REFERENCES

1.   Friedman, G. D., Petitti, D. B., and Bawol, R. D. "Prevalence and Correlates of Passive Smoking," Am. J. Public Health 73:401-405 (1983).

2.   U.S. Public Health Service. "The Health Consequences of Smoking. Cancer. A Report of the Surgeon General," (Rockville, Maryland: U.S. Department of Health and Human Services; Public Health Service, 1982).

3.   Jarvis, M. J., and Russell, M. A. H. "Measurement and Estimation of Smoke Dosage to Non-smokers from Environmental Tobacco Smoke," Eur. J. Respir. Dis. 65 (Supplement 133):68-75 (1984).

4.   Wald, N. J., Boreham, J., Bailey, A., Ritchie, C., Haddow, J. E., and Knight, G. "Urinary Cotinine as Marker of Breathing Other People's Tobacco Smoke (letter)," Lancet 1:230-231 (1984).

5.   Matsukura, S., Taminato, T., and Kitano, N., et al. "Effects of Environmental Tobacco Smoke in Urinary Cotinine Excretion in Nonsmokers. Evidence for Passive Smoking," N. Engl. J. Med., 311:828-832 (1984).

6.   U.S. Public Health Service. "Smoking and health. Report of the Advisory Committee to the Surgeon General of the Public Health Service," (Washington, DC: U.S. Department of Health, Education, and Welfare, Public Health Service, Center for Disease Control, PHS Publication No. 1103, 1964).

7.   MacMahon, B., and Pugh, T. F. "Epidemiology. Principles and Methods," (Boston: Little, Brown, and Company, 1970).

8. Kleinbaum, D. G., Kupper, L. L., and Morgenstern, H.
    "Epidemiologic Research. Principles and Quantitative
    Methods," (Belmont, California: Lifetime Learning
    Publications, 1982).

9. National Research Council. "Committee on Indoor Pollutants.
    Indoor Pollutants," (Washington, DC: National Academy Press,
    1981).

10. Spengler, J. D., Dockery, D. W., Turner, W. A., Wolfson, J.
    M., and Ferris, B. G., Jr. "Long-term Measurements of
    Respirable Sulfates and Particles Inside and Outside Homes,"
    Atmos. Environ. 15:23-30 (1981).

11. Spengler, J. D., and Soczek, M. L. "Evidence for Improved
    Ambient Air Quality and the Need for Personal Exposure
    Research," Environ. Sci. Technol. 18:268A-280A (1984).

12. Enstrom, J. E. "Rising Lung Cancer Mortality Among
    Nonsmokers," JNCI 62:755-60 (1979).

13. Garfinkel, L. "Time Trends in Lung Cancer Mortality Among
    Nonsmokers and a Note on Passive Smoking," JNCI 66:1061-1066
    (1981).

14. Auerbach, O., Hammond, E. C., and Garfinkel, L. "Changes in
    Bronchial Epithelium in Relation to Cigarette Smoking,
    1955-1960 vs. 1970-1977," N. Engl. J. Med. 300:381-386
    (1979).

15. Miller, G. H. "Cancer, Passive Smoking and Nonemployed and
    Employed Wives," West J. Med. 140:632-635 (1984).

16. Gillis, C. R., Hole, D. J., Hawthorne, V. M., and Boyle, P.
    "The Effect of Environmental Tobacco Smoke in Two Urban
    Communities in the West of Scotland," Eur. J. Resp. Dis. 65
    (Supplement No. 133):121-126 (1984).

17. Sandler, D., Wilcox, A., and Everson, R. "Cumulative Passive
    Exposure to Cigarette Smoke and Cancer Risk (abstract),"
    Am. J. Epidemiol. 120:482 (1984).

18. Vandenbroucke, J. P., Verheesen, J. H. H., DeBruin, A.,
    Mauritz, B. J., van der Heide-Wessel, C. and van der Heide,
    R. M. "Active and Passive Smoking in Married Couples: Results
    of 25 Year Followup," Br. Med. J. 288:1801-802 (1984).

19. Hirayama, T. "Non-smoking Wives of Heavy Smokers Have a
    Higher Risk of Lung Cancer: A Study From Japan," Br. Med. J.
    282:183-185 (1981).

20. Hirayama, T. "Passive Smoking and Lung Cancer," Presented at the Fifth World Congress on Smoking and Health, Winnepeg, Canada. July, 1983.

21. Hirayama, T. "Non-smoking Wives of Heavy Smokers Have a Higher Risk of Lung Cancer (letter)," Br. Med. J. 283:916-917 (1981).

22. Trichopoulos, D., Kalandidi, A., Sparros, L., and MacMahon, B. "Lung Cancer and Passive Smoking," Int. J. Cancer 27:1-4 (1981).

23. Trichopoulos, D., Kalandidi, A., and Sparros, L. "Lung Cancer and Passive Smoking: Conclusion of Greek Study (letter)," Lancet 1:677-678 (1983).

24. Kabat, G. C., and Wynder, E. L. "Lung Cancer in Nonsmokers," Cancer 53:1214-1221 (1984).

25. Correa, P., Pickle, L. W., Fontham, E., Lin, Y., and Haenszel, W. "Passive Smoking and Lung Cancer," Lancet 2:595-597 (1983).

26. Dalager, N., Pickle, L., Mason, T., and Ziegler, R. "Passive Smoking and Lung Cancer (abstract)," Am. J. Epidemiol. 120:482 (1984).

27. Knoth, A., Bohn, H., and Schmidt, F. "Passivrauchen als Lungenkrebsursache bei Nichtraucherinnen," Med. Klin. 78:66-69 (1983).

28. Chan, W. C., Colbourne, M. J., Fung, S. C., and Ho, H. C. "Bronchial Cancer in Hong Kong 1976-1977," Br. J. Cancer 39:182-192 (1979).

29. Chan, W. C., and Fung, S. C. "Lung Cancer in Non-smokers in Hong Kong," In Grundmann E, ed, Cancer Campaign, Vol 6, Cancer Epidemiology. Stuttgart: Gustav Fischer Verlag, 199-202 (1982).

30. Koo, L. C., HO, J. H-C, and Saw, D. "Is Passive Smoking an Added Risk Factor for Lung Cancer in Chinese Women?," In Press. J. Exp. Clin. Cancer Res.

31. Repace, J. L. "Consistency of Research Data on Passive Smoking and Lung Cancer (letter)," Lancet 1:506 (1984).

18.

Critical Review of the Relationship Between Passive Exposure to Cigarette Smoke and Cardiopulmonary Disease

Millicent Higgins, M.D.
University of Michigan School of Public Health

This review of the relationship between passive exposure to cigarette smoke and cardiopulmonary disease will be limited to non-malignant conditions and restricted to evidence from epidemiological and clinical studies. Several comprehensive reviews of the literature on passive smoking and respiratory diseases have been published recently (1-3) and reports from several workshops are also available (4-5). I will describe a few studies to illustrate the kind and quality of evidence that is available, and give an overview and evaluation of the results of published studies.

The cardiopulmonary conditions and diseases which have been associated with passive smoking in some studies are shown in Table 1.

Table 1. Cardiopulmonary Diseases and Conditions Associated with Passive Smoking

---

Pathophysiologic conditions
    carboxyhemoglobin increased
    heart rate increased
    blood pressure increased
    platelet function decreased
    exercise capacity reduced
    pulmonary function; acute changes
                      persistent reduction

Morbidity
    respiratory symptoms
    respiratory infections and illnesses
        acute and chronic bronchitis
        pneumonia
        asthma
        anginal pain

Mortality
    chronic bronchitis and emphysema
    ischemic heart disease
    lung cancer

They include acute responses such as: increases in carboxyhemo-globin (COHb), heart rate and blood pressure (6,10); decreases in exercise capacity, platelet sensitivity and pulmonary function (6,7,10,11); and acute manifestations of clinical illnesses such as respiratory infections, attacks of asthma and decreased duration of exercise before onset of anginal pain (3,6,12,13). Involuntary smoking has been associated with symptoms of chronic respiratory disease such as chronic cough, phlegm, wheeze and dyspnea, with diagnoses such as chronic bronchitis, emphysema, and asthma and with chronic obstructive lung disease. (3,12-15) There is much less information about passive smoking and mortality from nonmalignant cardiopulmonary diseases but a few reports have been presented (16,17).

Exposure to Environmental Tobacco Smoke.

Most epidemiological studies use crude measures of passive exposure to cigarette smoke and rely on available information about smoking habits of associates of nonsmokers. Parental smoking habits are frequently used to identify children who are, or are not exposed to environmental smoke at home. Sometimes maternal and paternal cigarette use are considered separately but often the two are not distinguished. Similarly, nonsmoking wives, or nonsmoking husbands, are considered to be passive smokers if their spouses smoke. Generally there is no information about exposure to environmental smoke from other smokers at home, at work, or in the community. Some attempts at measuring the dose of environmental tobacco smoke have been made by ascertaining the number of smokers with whom the nonsmoker is in contact, the number of cigarettes smoked per day ,by these smokers, or the duration of exposure. Sometimes the nonsmoker is asked to estimate the number of hours per week he is exposed to other peoples smoke and/or the degree of smokiness of environments to which he is exposed. (3,5,18) Very few studies have used a comprehensive set of questions to ascertain exposure to environmental tobacco smoke. Nevertheless, some recent studies have validated questionnaire information against biochemical markers (19,20).

The usefulness of biochemical markers as objective measures of exposure is limited because they reflect only recent exposure and because some of them such as carbon monoxide (CO) in expired air, COHb, and thiocyanate levels in body fluids are influenced by other inhaled or ingested substances. Although cotinine and nicotine are specific to tobacco and have longer half-lives, they also reflect exposure only in the recent past. However, they are useful for identifying those who falsely claim to be nonsmokers and they have provided objective evidence of exposure in nonsmokers who said they were exposed to other peoples smoke. Levels of urinary cotinine excretion were found to be higher in non-smokers who lived in homes or worked at sites where smokers were present and mean levels were higher in non-smokers from urban rather than rural areas. (19) Similar results have been published by several workers who measured a variety of biochemical markers. (20)

There have been some attempts to measure exposure to indoor pollutants as markers of passive smoking; usually CO and particulates are measured by area or personal sampling and rarely nicotine has been included. Unfortunately, only nicotine is specific to tobacco smoke whereas other indoor and outdoor sources contribute to levels of CO and particulates. Long-term measurement of individuals' exposure to environmental tobacco smoke is not feasible on the scale needed for epidemiological studies. However, studies to validate questionnaires against biochemical markers and pollutant levels are feasible and necessary to develop and standardize questionnaires and biochemical tests for measuring exposure.

Study Designs

In addition to prospective, retrospective and cross-sectional epidemiological studies of the health effects of exposure to environmental tobacco smoke, responses to acute, short-term, passive exposures have been ascertained under controlled experimental conditions. Both healthy and diseased subjects have been exposed to known amounts of tobacco smoke or its constituents at rest and during exercise in well ventilated and poorly ventilated conditions. The populations which have been studied include general populations, healthy volunteers and patients with chronic obstructive pulmonary disease, asthma, or coronary heart disease.

Results of Epidemiological Studies

Acute Respiratory Illnesses

The first reports suggesting an adverse effect of passive smoking indicated that acute respiratory illnesses were more frequent among children whose parents smoked than among children whose parents did not smoke. A recent review of 11 studies published between 1969 and 1984 showed that bronchitis, pneumonia, and unspecified lower respiratory illnesses occurred more frequently in infants whose parents smoked. (3) Rates of illness or of medical consultation for illness or hospitalization increased with the number of parental smokers and with the number of cigarettes they smoked per day. Relative risks for children with passive exposure to parental cigarette smoke ranged from 1.4 to 2.7. A few published studies did not find a relationship between parental smoking and respiratory illnesses in older children. (3) Nevertheless, despite short-comings of individual studies which did not always consider maternal and paternal smoking in detail, or ascertain other sources of environmental tobacco smoke, or investigate potentially important confounding factors, the consistency of these results for young children justifies the conclusion presented in the Surgeon General's 1984 Report that, "the children of smoking parents have an increased frequency of bronchitis and pneumonia early in life." The association with maternal smoking is stronger and more consistent than that with paternal smoking (3).

Chronic Respiratory Symptoms and Reduced Pulmonary Function

Information about respiratory symptoms and illnesses and measurements of pulmonary function were collected in several prospective epidemiological studies which focussed on the health effects of active smoking initially. The Tecumseh Community Health Study was one such study. It was designed to identify causes and risk factors for the major chronic diseases including coronary heart disease and chronic obstructive pulmonary disease (COPD) (34,35). Relationships between respiratory symptoms, and illnesses, ventilatory lung function, and passive smoking were investigated recently by Burchfiel in analyses of information on 3,600 young people who were examined and who had both parents examined (12,13).

Age-adjusted prevalence rates of chronic respiratory symptoms and illnesses are shown in Figures 1 and 2 for males and females grouped with reference to the cigarette smoking habits of their parents. When only one parent smoked, it was the father for 89% of the offspring and the mother for only 11%.

Rates of reporting phlegm, wheeze, and chest colds were higher and study physicians diagnoses of asthma were more frequent in males with two smoking parents than in males with one or no smoking parent, and these differences were statistically significant.

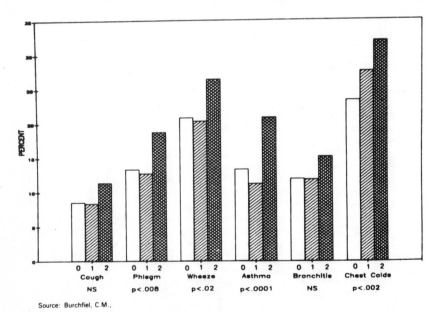

Source: Burchfiel, C.M.,

Figure 1. Age-adjusted prevalence of respiratory conditions by number of parental smokers; male, 0-19 year-old nonsmokers; Tecumseh, 1962-1965.

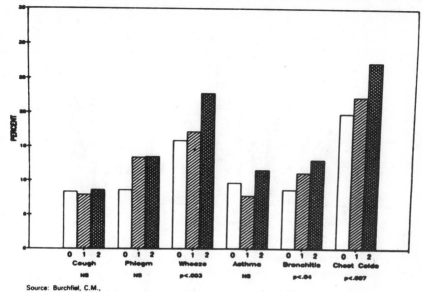

Source: Burchfiel, C.M.,

Figure 2. Age-adjusted prevalence of respiratory conditions by number of parental smokers; female, 0-19 year-old nonsmokers; Tecumseh, 1962-1965.

Similar, but non-significant differences were found for histories of cough and bronchitis.

Trends were similar in females but only those for wheeze, bronchitis, and chest colds were statistically significant.

Mean values of three measures of lung function were compared among nonsmokers aged 10-19 years. (Figure 3). In males, values were consistently lowest for those with two smoking parents. Mean values of FEV_1 and FVC were highest for offspring of two nonsmokers; the difference in mean FEV_1 was 144 ml. for sons of two smoking parents, compared with sons of two nonsmoking parents. There were virtually no differences in mean values of FEV_1 or FVC in females, but mean values of $Vmax_{50}$ were highest in those with two nonsmoking parents. Mean values of FEV_1 and FVC in males and $Vmax_{50}$ in females were significantly lower by about 5% in offspring of parents who were both current smokers compared with offspring of parents who were both nonsmokers.

In very detailed analyses of cross-sectional and longitudinal observations, Burchfiel used several different indices of passive smoking incorporating information on duration and amount of current and past smoking habits of both parents and other members of the child's household. He also controlled for potential confounding by age, sex, parental education, family size, parental respiratory

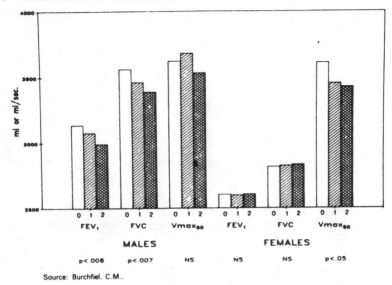

Figure 3. Mean lung function, adjusting for age and height using analysis of covariance, by number of parental smokers; 10-19 year-old nonsmokers; Tecumseh, 1962-1965.

symptoms and illnesses and for active smoking reported by the young people for themselves. The differences persisted but were generally smaller. He found that prevalence rates of respiratory symptoms and illnesses tended to be higher and lung function tended to be lower in young people if the only parental smoker was the mother rather than the father. However, the number of children with a smoking mother and a nonsmoking father was small and firm conclusions about the relative importance of maternal and paternal smoking cannot be drawn from this study. In general, differences were more frequently significant for young males than young females, but significant differences were present in both sexes. Differences were also more often significant at the younger rather than the older end of the 0-19 year age range. (12)

In their extensive longitudinal studies of about 1200 5-19 year old residents of East Boston and their families, Tager and colleagues found that rates of cigarette smoking were highest among mothers of children with lung function in the lowest fifth of the distribution, and lowest among mothers of children with the highest FEV's (21). Rates of growth in FEV_1 and FEF_{25-75} were significantly decreased among children of mothers who smoked in this study. The authors reported that the effect of maternal smoking was a decrease in expected lung growth of 3-5%.

The 1984 volume of the Health Consequences of Smoking summarizes 12 reports of pulmonary function and involuntary smoking

in children. All but two of these studies showed some association
between exposure to parental smoke and reduced lung function,
although this was not present in all comparisons within each study;
the two discrepant studies involved children in Arizona (22,23) and
in one of these, (23) non-significant trends were in the expected
direction. At a recent workshop it was suggested that differences
in results of studies might be "caused by regional and geographic
variations in levels of indoor pollution that might result from
differences in housing and lifestyle" or might be "due to
methodologic differences in data collection and/or analysis and in
the way in which potentially confounding variables have been
handled." (4)

The Health Consequences of Smoking 1984 presents an appropriate
summary of current evidence. It states, "the children of smoking
parents appear to have measurable but small differences in tests of
pulmonary function when compared with children of nonsmoking
parents. The significance of this finding to the future development of lung disease is unknown." (3)

There have been fewer studies of respiratory symptoms, diseases
or pulmonary function in adult nonsmokers exposed or not exposed to
environmental tobacco smoke. Schilling and colleagues reported no
effect of one parents smoking on the others lung function (24). In
his study of populations in Washington County, Maryland, Comstock
found no statistically significant association between impaired
pulmonary function in nonsmokers and exposure to a spouse's
cigarette smoke. (25). In contrast, White and Froeb reported that
involuntary inhalation of tobacco smoke at the worksite was
associated with reduced mean levels of FEF_{25-75} and FEF_{75-85} in
nonsmokers. In this study, passive smokers, light smokers, and
smokers who did not inhale had similar reductions in lung function;
heavy smokers had even lower values. Absolute differences in mean
values of FEV_1 were small and amounted to .16 liters in FEV_1 for
male and female nonsmokers in smoky environments compared with
nonsmokers whose work environments were not smoky. These subjects
were being evaluated for physical fitness profiles and are unlikely
to represent the general population. In addition 42 percent were
excluded from the study because of cardiac or pulmonary symptoms or
illnesses or because of exposures to occupational, industrial, or
general air pollution. (15)

More recently, Kauffmann analyzed information on 7,800 adult
residents of seven cities in France and compared results of pulmonary function tests in male and female nonsmokers classified
according to whether their spouses smoked or did not smoke.
Nonsmokers of both sexes had significantly lower mean values for
FEF_{25-75} if their spouses smoked at least 10g of tobacco a day. In
women over 40 years of age, FEV_1 was reduced to a statistically
significant extent in passive smokers and there was a dose response
relationship among nonworking women. These results were not
explained by differences in social class, educational level,
exposure to air pollution, or family size (14).

Gillis and his colleagues studied approximately 16,000 apparently healthy 45-64 year old men and women in a random sample of residents of Renfrew and Paisley, Scotland. They classified members of each household according to their own smoking habits and their partners smoking habits. In comparison with controls who were nonsmokers with nonsmoking partners, nonsmoking males with wives who smoked had a higher prevalence of persistent spit, dyspnea and hypersecretion, and nonsmoking wives with smoking husbands reported shortness of breath significantly more often. Results of pulmonary function tests were not presented in this report. (17)

Chronic Bronchitis and Emphysema

Hirayama followed-up a census-population based cohort which included 91,540 nonsmoking wives whose husbands' smoking habits were known. During the 16 year interval, 130 wives died from emphysema or chronic bronchitis. Within each age group, death rates for chronic bronchitis and emphysema were lower for wives of nonsmokers than for wives of smokers and ex-smokers, and rates for wives of heavier smokers tended to be higher than rates for wives of ex-smokers and lighter smokers combined. Hirayama calculated a standardized mortality ratio which was a weighted point estimate of the rate ratio. It was set at 1.00 for wives of nonsmokers, and found to be 1.29 for wives of ex and light smokers, and 1.60 for wives of men smoking 20 or more cigarettes per day. The rates for the most heavily exposed wives were significantly higher than those for unexposed wives. (16)

Cardiovascular Conditions

There have been fewer reports of relationships between involuntary smoking and cardiovascular conditions, despite the potential availability of data from several longitudinal epidemiological studies and clinical trials.

Gillis and colleagues compared the prevalence of cardiovascular conditions in their groups with varying experience of active and passive smoking and found no differences between controls and those with environmental tobacco smoke exposure; prevalence rates of angina and ECG abnormalities were higher in both groups of male smokers but there were virtually no differences between smoking and nonsmoking women. (17)

Hirayama calculated mortality rates for ischemic heart disease in his cohort of nonsmoking wives in Japan, among whom there were 495 deaths from ischemic heart disease. The weighted point estimates of the rate ratios were set at 1.00 for wives of nonsmokers, and found to be 1.1 for wives of light and exsmokers and 1.3 for wives of heavier smokers; these rate ratios are significantly different. The differences persisted when the husbands occupation was considered in the analysis. (16)

Results of Clinical Studies

There is very little information from epidemiological or clinical studies about involuntary smoking and the course of cardiovascular or pulmonary disease. Persons who are sensitive to tobacco smoke may experience symptoms and bronchoconstriction on exposure to tobacco smoke. In a study by Speer, higher proportions of allergic nonsmokers than of non-allergic nonsmokers reported that cough and wheezing were precipitated by tobacco smoke. (26) Some asthmatic children have been reported to improve when their parents stop smoking, but no large scale well designed randomized clinical trial has been done.

A few experimental studies have been designed to detect and quantify acute effects of passive exposure to cigarette smoke or some of its constituents under carefully controlled conditions. Ideally assignments to test and sham exposures are randomized, assessment of responses is double-blind and studies are done with variables other than the experimental interventions held constant. Some studies meeting most if not all of these requirements have been published.

Acute effects of environmental tobacco smoke on the pulmonary system were studied in healthy young adults by Pimm and colleagues (10) and by Shepherd (8) who measured pulmonary function and heart rate at rest and with exercise in subjects confined to exposure chambers; sham exposures were used as control conditions. Small decreases in flow rates were observed in the presence of tobacco smoke at rest or with exercise in both studies and in the study by Shepherd, cough and nasal discharge or stuffiness were reported by some subjects; there was no objective evidence of broncho-constriction. The slight increase in tidal volume and minute ventilation was possibly due to anxiety, in the investigators' opinion.

In a similar study of 14 asthmatics exposed for two hours to smoke from seven cigarettes burned in a closed room, or to a similar environment without environmental tobacco smoke, 36% of the asthmatics complained of wheezing and 43% complained of tightness in the presence of smoke; heart rate increased slightly, but FVC and lung volumes did not change. (9)

Dahms exposed 10 asthmatics and 10 controls to sidestream smoke for one hour and induced a rise in carboxyhemoglobin. In asthmatics FEV_1, FVC and FEF_{25-75} decreased progressively during the course of the exposure whereas the controls did not manifest such changes. There was no sham exposure in this study. (7)

Aronow measured exercise performance in 10 hypoxemic patients with COPD before and after they were exposed for one hour to either 100ppm CO or to purified air. Mean exercise time until marked dyspnea was reduced from 218 seconds to 147 seconds with CO exposure which was associated with an increase in COHb from 1.4% to

4.1% Blood pressure, heart rate, blood gases and pH did not change with exposure to CO. (27)

Although these studies suggest that small changes in pulmonary function may occur when healthy young adults are exposed to tobacco smoke, the numbers tested are small, and the results variable and far from conclusive. The two studies of asthmatics gave conflicting results. There is insufficient evidence to permit conclusions about acute effects of passive smoking on patients with asthma or chronic obstructive lung disease but it is likely that some unknown proportion of them will be adversely affected.

A considerable amount of attention has been given to studies of the acute effects of exposure to carbon monoxide on patients with coronary heart disease. Revisions to the National Ambient Air Quality Standards for CO were proposed partly because a series of studies by Aronow, Anderson and others indicated that modest elevations of COHb levels were associated with deleterious effects. Aronow reported that exercise tolerance was decreased in normal subjects exposed to CO and duration of exercise before onset of chest pain or leg pain was shortened after two hours exposure to 50ppm CO in anginal patients and patients with intermittent claudication. COHb increased from 1.03% to 2.7% which is within the range produced by passive smoking. (28-30) A study by Anderson also showed that COHb levels of about 3 percent were associated with adverse effects on the cardiovascular system. (31)

The study most directly relevant to this discussion of passive smoking involved 10 patients with clinically stable angina and angiographic evidence of narrowing of at least one major vessel. Aronow studied these patients at rest and with exercise during an experimental exposure to cigarette smoke in a well ventilated or an unventilated room. Exposure to cigarette smoke was associated with increased COHb, heart rate, systolic and diastolic blood pressure at rest and with decreased duration of exercise until the onset of anginal pain; levels of heart rate and systolic blood pressure at onset of pain were lower in the presence of tobacco smoke than under control conditions. (6)

The Aronow studies of CO exposure have been reviewed recently by a committee appointed by EPA. The Conclusion of this committee reported in the Federal Register of August 9, 1984 was "that EPA should not rely on Dr. Aronow's data due to concerns regarding the research which substantially limit the validity and usefulness of the results." However, after reviewing the scientific evidence currently available, EPA and CASAC are of the opinion that, "adverse health effects may be experienced by large numbers of sensitive individuals with COHb levels in the range of 3 to 5%. (32) New studies are under way to measure effects of exposure to CO in patients with angina.

DISCUSSION

Individuals who may be sensitive to environmental tobacco smoke include those with cardiovascular or pulmonary diseases. The prevalence of these conditions is substantial and amounts to millions of people including an estimated 5.7 million with coronary heart disease, 2 million with emphysema, 7.9 million with chronic bronchitis and 7.2 million with asthma. (33) In addition, the very young and the very old are known to be susceptible to environmental circumstances which healthy young adults can tolerate. Furthermore, adverse consequences of exposure to environmental tobacco smoke are not confined to the cardiopulmonary conditions considered in this review.

Passive exposure to environmental tobacco smoke is nearly universal. Significant and frequent exposure occurred in the home for as many as 80% of children in some populations in the U.S. (12) and over 60% of adults reported some passive smoking in a recent study by Friedman. (18) Even though cigarette smoking rates are declining, and nonsmoking areas are being provided, involuntary exposure to tobacco smoke is still frequent.

Thus, the frequency and involuntary nature of the exposure together with the presence of large numbers of sensitive people make passive smoking an important public health issue.
Associations between passive smoking and respiratory symptoms, illnesses and impaired pulmonary function have been found with a reasonable degree of consistency in children, but this is not the case in adults. For some conditions, especially cardiovascular conditions, too few studies have been done. When associations are present they are generally not strong; the increased risk to the exposed is usually less than twofold.

Although there is some evidence of a biological gradient especially for respiratory symptoms, illnesses and reduced pulmonary function in children where estimates of the dose of environmental tobacco smoke are probably more accurate, better measurements of exposure are needed to determine whether there is a dose-response relationship.

In analyzing the evidence currently available, and in collecting new information, attention should be given to getting more accurate estimates of exposure. This may be accomplished through the us⁻ of standardized validated questions or by measuring biochemical markers of exposure. Consideration should be given to similarities and differences between passive smoking and active smoking. Although there are substantial differences in the quality and quantity of smoke inhaled by the two types of smokers, the vast accumulation of knowledge concerning the health consequences of smoking lends support to the hypothesis that passive smoking causes certain pulmonary and cardiovascular conditions. However, these diseases have multiple etiologies and we should remember that it was more difficult to establish a causal relationship between smoking and coronary heart disease, than between smoking and lung cancer or COPD. Smoking and CHD relationships were especially difficult to detect in women and most adult passive smokers are women. Changing smoking patterns in men and women may result in larger numbers of passive smoking males but it will take several decades to collect epidemiological information on the health consequences of passive smoking in men. Attempts to produce disease in man are usually unethical although precipitation of reversible changes under carefully controlled conditions may be justified if information gained in this way leads to preventive or therapeutic benefits. Experimental exposure of patients with coronary heart disease, asthma, COPD and other conditions to environmental tobacco smoke could answer important questions about the risks which such patients incur when they are exposed to other people's cigarette smoke.

The possibility of testing the hypothesis that reduction of passive exposure to cigarette smoke is beneficial by doing a randomized controlled trial to prevent disease, or ameliorate its course, is attractive but probably not feasible except for well defined groups of sensitive subjects or patients since the effect is small, confounding factors are numerous and smoking cessation interventions are difficult.

The evidence linking passive smoking with acute respiratory illnesses, chronic respiratory symptoms and mild impairments of pulmonary function in children is quite strong. Available information about these conditions, coronary heart disease and other manifestations of cardiopulmonary disease in adults, is strong enough to justify further well designed studies in healthy and sensitive subjects including those with established disease. Standardized, validated measures of exposure and outcome should be used and potential confounding factors should be controlled.

SUMMARY

Non-malignant pulmonary diseases and conditions which have been associated with passive exposure to cigarette smoke in epidemiological or clinical studies include acute respiratory infections and illnesses, acute and chronic bronchitis, pneumonia, asthma and emphysema, as well as chronic respiratory symptoms, acute changes in pulmonary function and reduced lung function. In some studies, passive smoking was associated with higher rates of hospitalization for respiratory illnesses, with more frequent visits to physicians or more days of bed disability or restricted activity. However, these relationships were not found in all studies and they were sometimes present in only one sex or they were limited to persons in a certain age range. Incidence of respiratory illnesses was related to maternal but not paternal smoking habits in some studies of children.

In one study, mortality from ischemic heart disease was higher in nonsmokers whose husbands smoked than in nonsmokers whose husbands did not smoke. In a few studies, duration of exercise

before onset of anginal pain was shorter with exposure to
environmental tobacco smoke or low levels of carbon monoxide.
Increases in heart rate and blood pressure were observed as acute
reactions to environmental tobacco smoke in healthy young people
and in patients with angina.

Evaluation of these findings is difficult because evidence is
sparse and sometimes inadequate; some results are inconsistent.
Measures of passive smoking have usually been derived from
available information on smoking habits of parents, spouses, or
other associates of nonsmokers. Potential confounding factors such
as age, socio-economic status, family size, and other environmental
exposures have not always been ascertained or controlled adequately
and the possibility that some subjects smoke themselves cannot be
ruled out.

Nevertheless, the evidence linking passive smoking to
respiratory symptoms and illnesses and reduced pulmonary function
is quite strong and suggestive of a causal relationship.

References

1.   Weiss, et al. No. 151 in Black Book (BB)

2.   Weiss, S.T., Tager, I.B., Speizer, F.E. Passive Smoking Its
     Relationship to Respiratory Symptoms, Pulmonary Function and
     Nonspecific Bronchial Responsiveness. Chest, 84(6) 651-652,
     1983.

3.   U.S. Public Health Service. The Health Consequences of Smoking.
     Chronic Obstructive Lung Disease. A Report of the Surgeon
     General. DHHS (PHS) 84-50205, 1984.

4.   U.S. Public Health Service. Report of Workshop on Respiratory
     Effects of Involuntary Smoke Exposure: Epidemiologic Studies.
     May 1-3, 1983. U.S. Department of Health and Human Services,
     1983.

5.   Rylander, R., Peterson, Y., Snella, M. Measurement and
     Estimation of Smoke Dosage to Non-smokers from Environmental
     Tobacco Smoke. Report from a Workshop on Effects and Exposure
     Levels, March 15-17, 1983. University of Geneva, Switzerland,
     1983.

6.   Aronow, W.S. Effect of Passive Smoking on Angina Pectoris.
     N. Engl. J. Med. 299:21-24, 1978.

7.   Dahms, T.E., Bolin, J.F., and Slavin, R.G. Passive Smoking.
     Effects on Bronchial Asthma Chest, 80, 5, 530-534, 1981.

8.   Shephard, R.J., Collins, R. and Silverman, F. Responses of
     Exercising Subjects to Acute "Passive" Cigarette Smoke Exposure.
     Environmental Research 19, 279-291, 1979.

9.   Shephard, R.J., Colling, R., and Silverman, F., "Passive
     Exposure" of Asthmatic Subjects to Cigarette Smoke.
     Environmental Research 20, 392-402, 1979.

10. Pimm, P.E., Silverman, F. Physiologic Effects of Acute Passive
     Exposure to Cigarette Smoke. Arch. Environ. Health 33:201-13,
     1978.

11. Sinzinger, H., Kefalides, A. Passive Smoking Severely Decreases
     Platelet Sensitivity to Antiaggregatory Prostaglandins. The
     Lancet 2, 392-393, 1982.

12. Burchfiel, C.M. Passive Smoking Respiratory Symptoms, Lung
     Function and Initiation of Smoking in Tecumseh, Michigan. Ph.D.
     Thesis. University of Michigan, 1984.

13. Burchfiel, C.M., Howatt, W., Keller, J., Butler, W., Higgins, I., and Higgins, M. Passive Smoking, Respiratory Symptoms and Pulmonary Function in the Pediatric Population of Tecumseh. Amer. Rev. Resp. Dis., 1983.

14. Kauffmann, F., Tessier, J.F., and Oriol, P. Adult Passive Smoking in the Home Environment: A Risk Factor for Chronic Airflow Limitation. Amer. J. Epidemiol. 117 No. 3 269-280, 1983.

15. White, J.R. and Froeb, H.F. Small-Airways Dysfunction in Nonsmokers Chronically Exposed to Tobacco Smoke. N. Eng. J. Med. 302; 720-723, 1980.

16. Hirayama, T. Passive Smoking and Lung Cancer. Presented at the Fifth World Conference on Smoking and Health, Winnipeg, Canada, July 1983.

17. Gillis, C.R., Hole, D.J., Hawthorne, V.M., and Boyle, P. "The Effect of Environmental Tobacco Smoke in Two Urban Communities in the West of Scotland. Eur. J. Resp. Dis. 65 (Supplement No. 133); 121-126, 1984.

18. Friedman, G.D., Pettiti, D.B., Bawol, R.D. Prevalence and Correlates of Passive Smoking. Am. J. Public Hlth. 73, 401-405, 1983.

19. Matsukura, S., Taminato, T., Kitano, N., Seino, Y., Hamada, H., Uchihashi, M., Nakajima, H., and Hirata, Y. Effects of Environmental Tobacco Smoke on Urinary Cotinine Excretion in Non-smokers: Evidence for Passive Smoking. N. Eng. J. Med. 311:828-832, 1984.

20, Jarvis, M.J., and Russell, M.A.H. Measurement and Estimation of Smoke Dosage to Non-smokers from Environmental Tobacco Smoke. Report from a Workshop on Effects and Exposure Levels, March 15-17, 1983. University of Geneva, Switzerland, 1983.

21. Tager, I.B., Weiss, S.T., Munoz, A., Rosuer, B., Speizer, F.E. Longitudinal Study of the Effects of Maternal Smoking on Pulmonary Function in Children. N. Eng. J. Med. 309 (12): 699-703, 1983.

22. Lebowitz, M.D., and Burrows, B. Respiratory Symptoms Related to Smoking Habits of Family Adults. Chest 69:48-50, 1976.

23. Dodge, R. The Effects of Indoor Pollution on Arizona Children. Archives of Environmental Health 37:151-155, 1982.

24.   Schilling, R.S.F., Letai, A.D., Hui, S.L., Beck, G.J.,
      Schoenberg, J.B., and Bouhuys, A.H. Lung Function, Respiratory
      Disease, and Smoking in Families. Amer. J. Epidemiol.
      106:274-283, 1977.

25.   Comstock, G.W., Meyer, M.B., Helsing, K.J., and Tockman, M.S.
      Respiratory Effects of Household Exposures to Tobacco Smoke and
      Gas Cooking. Amer. Review Resp. Dis. 124:143-148, 1981.

26.   Speer, F. Tobacco and the Nonsmoker. A study of Subjective
      Symptoms. Archs. Envir. Hlth. 16:443-446, 1968.

27.   Aronow, W.S., Ferlinz, J. and Glauser, F. Effect of Carbon
      Monoxide on Exercise Performance in Chronic Obstructive Pulmonary
      Disease. Am. J. Med. 63:904-908, 1977.

28.   Aronow, W.S., and Cassidy, J. Effect of Carbon Monoxide on
      Maximal Treadmill Exercise: A Study in normal persons. Ann.
      Intern. Med. 83:496-499, 1975.

29.   Aronow, W.S., and Isbell, M.W. Carbon Monoxide Effect on
      Exercise-induced Angina Pectoris. Ann. Intern. Med. 79:392-395,
      1973.

30.   Aronow, W.S., Stemmer, E.A. and Isbell, M.W. Effect of Carbon
      Monoxide Exposure on Intermittent Claudication. Circulation
      49:415-417, 1974.

31.   Anderson, E.W., Andelman, R.J., Strauch, J.M., Fortuin, N.J. and
      Knelson, J.H. Effect of Low-level Carbon Monoxide Exposure on
      Onset and Duration of Angina Pectoris. Ann. Intern. Med.
      79:46-50, 1973.

32.   Federal Register. 49:155, 31923-31926. August 9, 1984.

33.   National Center for Health Statistics.

34.   Higgins, M.W., Keller, J.B., Becker, M., Howatt, W., Landis,
      J.R., Rotman, H., Weg, J.G. and Higgins, I.: An Index of Risk
      for Obstructive Airways Disease. Am. Rev. Resp. Dis.
      125:144-151, 1982.

35.   Higgins, M.W.: The Tecumseh Community Health Study. In
      W.H.O./I.E.A. Monograph. Epidemiological Methods for
      Environmental Health Studies 1983.

Part Four

Combustion Products

19.

Part Four: Overview

David V. Bates

University of British Columbia
Vancouver, British Columbia

There is some evidence that children exposed to nitrogen oxides in the home may have an increased incidence of respiratory infections. This section on on Combustion Products was designed to explore this phenomenon, to examine the strengths of this association, and to examine new data concerning the nature of these exposures. Dr. Spengler's data gave us the context for our knowledge in this field. He reported that in kitchens of ordinary homes in the western world, the concentrations of NO_2 range to 80 $\mu g/m^3$; levels in the bedrooms are generally one-half that in kitchens. Where gas is not used for cooking, the levels are much lower, perhaps by as much as a factor of 12. As Dr. Spengler showed, the number of pilot lights in the home plays a part in determining annual average values; peak values in kitchens, however, are clearly more impressive.

The levels of NO_x that may be built up as a result of kerosene burners may also contribute to the high number of respiratory infections and morbidity. There is little doubt that biomass fuels used in restricted spaces without chimneys (in the developing world) may cause a considerable amount of respiratory disease, particularly in small children.

What experimental data exist to link nitrogen oxide exposure with morbidity? This question was explored in the report by Drs. Graham and Miller. Their data showed that spike exposures of NO_x of only about twice that of home levels have adverse effects on host defenses in animals. The concentration of NO_x seems more important than the time of exposure, but, of course, at any given concentration, the longer the exposure the greater the effect. NO_x also has complex effects on the immune system of the animal; however, it is not clear whether these play any part in the host response in normal circumstances. Their report, therefore, showed that there are animal experimental data of a kind similar to observed effects in humans.

Dr. Kleinerman reviewed the long-term reversible and irreversible changes in animals as a result of NO_2 exposure. At high-enough concentrations, changes in the airways of animals may be shown to persist for 9 months after removal from NO_x. The changes also appear to represent a permanent alteration in airway epithelium. Other changes on lung elasticity are less easy to document, but they certainly warn us that exposures to NO_x may not be without long-term effects. We thus have plenty of experimental evidence to warn us of the importance of this indoor air pollutant.

Dr. Vedal reviewed the epidemiological data relating childhood illness and pulmonary function to gas stove emissions. The great difficulty of these studies is the problem of eliminating the effect of other variables, particularly a misclassification of exposure and socioeconomic factors, which may also be important. Dr. Vedal's analysis of the power of these studies and of the accompanying confounding factors enables us to see the strengths and weaknesses of our present data.

Dr. Collier presented a different perspective; he and his colleagues have been following a group of children in a longitudinal study to determine the frequency of respiratory infections and whether there are long-term consequences from them. He reported that in the first 5 years of life, there are 6 to 8 episodes per year of respiratory infections, and as many as 20 episodes per year of lower respiratory infections occur in every 100 children. This will not come as news to parents of preschool children. Dr. Collier's findings also tell us that boys are more often affected than girls; furthermore, in 26% of the cases a virus can be isolated. It is possible that these episodes are related to the later development of pulmonary function.

The relationships between these phenomena appear to be generally present, but the total chain of evidence has certain weaknesses within it. The presentations in this section are provided to assist us in evaluating those parts of this "rope ladder" of causality that are strong and those that are weak. Although it is clear from the present state of the data that no definitive statement is yet possible, it is a good moment to review the present status of evidence from all of these fields.

20.

Emissions from Indoor Combustion Sources

John D. Spengler and Martin A. Cohen

Department of Environmental Science and Physiology, Harvard School of
Public Health, Boston, MA

INTRODUCTION

Biomass and fossil fuels are burned indoors for heating, cooking, and aesthetic purposes. Combustion often occurs at air-fuel
ratios different from stoichiometric conditions. The resulting
incomplete pyrolysis produces gaseous and particulate combustion
by-products as well as nitrogen oxides, carbon dioxide, and water
vapor.

Where combustion occurs indoors and the emissions are not
directly vented out, elevated concentrations of air contaminants can
occur. Numerous studies now offer clear evidence that nitrogen
dioxide, nitrogen oxide, carbon monoxide, carbon dioxide, sulfur
dioxide, formaldehyde, inorganic and organic particulate matter,
reduced sulfur compounds, various metals, and organic vapors can
reach concentrations substantially higher than outdoors. Because
combustion is often episodic, short-term concentrations are elevated
several fold over daily or weekly averaged concentrations. Concentration patterns indoors are quite variable between and within
rooms. Occupants also are mobile within a structure. Therefore, at
this time, measurements of indoor concentrations must be considered
as approximations of actual personal exposures. In fact, substantial between-person variation is expected even for those occupying
the same indoor environment.

This paper reviews literature reporting emissions of biomass
and fossil fuels likely to be used indoors. The emphasis is placed
on residential fuels. For more comprehensive reviews the reader is
referred to several published reports and books (1,2,3,4). The
focus of this article will be on emissions from gas appliances,
kerosene burners, sterno burners, and woodburning. New data are
presented on emission factors for camping equipment (stoves and lan-
terns) using liquid petroleum fuels. The resulting indoor concentrations are briefly summarized.

261

VARIABLES AFFECTING EMISSION COMPONENTS AND RATES

Laboratory and field studies characterizing the emission from combustion appliances show considerable variation in the emission species and emission rates. The principal factors affecting emissions are related to the fuel combustion conditions and the appliance or facility itself. These factors are, to some degree, interdependent. However, for discussion they will be treated separately. For each of these factors, there are elements of human interaction which will affect both the composition and rates of emissions.

Fuels. Indoor fuels are hydrocarbon based energy sources. The complex of hydrocarbon molecules and the physical state of the fuel (gas, liquid, or solid) will influence its combustion properties. Among the more common fuels are methane and propane gases, kerosene, animal or vegetable oils, and "clear" lead-free gasoline liquid fuels. Common solid fuels are coal, charcoal, wood, crop residue, and animal dung. The thermal energy content of these fuels varies. Many of these fuels, paticularly the solid fuels, contain organically bound or structurally bound molecules. Trace metals, sulfur, chlorine, potassium, nitrogen, calcium, silicon, and a variety of minerals are found in varying degrees in all liquid and solid fuels. Of course, gases and liquids have relatively lower concentrations of trace constituents unless deliberately added. Mercaptans are often added to methane and propane for leak detection. Sulfur compounds are formed during combustion of fuels containing mercaptans. Solid fuels contain incombustable materials generically referred to as ash content. The non-hydrocarbon content of fuels often end up in the effluent gas emissions.

Combustion Conditions. Emission rates and the constituents are directly influenced by combustion conditions. The typical air-fuel ratio for optimal burning is usually 12 to 1 by mass. However, the actual air-fuel ratio differs greatly by burning apparatus and operating conditions. In the oxygen-lean conditions poorer pyrolysis occurs resulting in partially burned hydrocarbon compounds. It is the oxygen-starved burning in coal and wood stoves that produces increased polycyclic and aromatic compounds. These conditions are characteristic of high CO emissions and yellow-colored flames for gas burning. Under oxygen-rich conditions when the fuel burns hot, both fuel and air are consumed quicker. The higher the flame temperature the greater the nitrogen oxide emissions.

The moisture content of fuel affects the burning conditions. Ayer measured the amount of unburned fuel emissions as a function of moisture in the wood. An optimal condition of 15 to 25% moisture in the wood resulted in about 10% unburned fuel emissions. In very dry wood the unburned fuel emissions rose to 30% (5).

Appliance. The design, use, and maintenance of the combustion facility will greatly influence the emission parameters. Primarily, these factors affect the oxygen/fuel relationships in the combustion zone. And, obviously, the equipment and operator are, to some degree, controlling the fuel charging rate. The published work on kerosene heaters illustrates this point. Convective and radiant kerosene heaters have substantially different NO/NO_2 emission ratios. Further, the CO emission rate is reported by different investigators as being a factor of 2 to 4 difference between the two types of kerosene heaters.

Lindstrom demonstrated that the emissions rate of BaP varied inversely with the size of the combustion facility. As the thermal rating of this facility increases from 10 to 10^6 kW/hr, BaP emissions decrease over 7 orders of magnitude (6). In part, this is due to better burning conditions and higher temperatures for the larger facilities.

The design of wood/coal burning stoves has undergone substantial changes. It is now possible to purchase stoves with a secondary combustion chamber with direct air feed, and stoves equipped with catalytic converters. These modifications reduce the emission rates of unburned hydrocarbons.

The maintenance of a combustion device should affect combustion efficiency and, hence, emission rates. Unfortunately, almost all of the experience with with emission testing has been in the laboratory. Because emission testing has not been performed on equipment and devices as they are actually used in homes, offices, etc., the variances in emission rates for indoor combustion has not been established.

EMISSION FACTORS AND CONCENTRATIONS

The background discussion of the previous section is a useful perspective when reviewing published emission factors. Almost all the emission results have been derived in laboratories. The variations in reported results reflect the variety of conditions that might affect the emission rates. So rather than report a single value, a range of reported emissions has been presented.

Table I is a compilation of emission factors for various combustion appliances, running under various conditions, using different fuels. In the original literature, the values were usually stated in terms of the mass of pollutant per unit mass of fuel, unit heat output or unit time (7-27). There were also two basic methods used to find the emissions factors. One entailed using a "chamber" and the other a ventilation system. Each method has its own advantages and disadvantages. In order to make the values comparable, some assumptions had to be made, in which case they were stated. Table II shows emission factors and carcinogenic activities for a variety of the components of woodsmoke (28), and Table III lists chemical and physical properties of different biomass fuels (29).

Table I   Emission Factors

	NO_2 (g/kg)	CO (g/kg)	PM^b (g/kg)	Remarks
Kerosene-Radiant Heater[a]				
Leaderer	0.192 – 0.256	1.85 – 3.17	–	high, low and med. flame setting
Caceres	2.26 – 4.52	0.136 – 0.208	–	–
Traynor	0.18 – 0.23	2.35 – 6.15	0.0003 – 0.0014	age & warm up time dp=0.005-0.4um
Yamanaka	0.479 ± 0.061	–	–	± SD
Woodring	0.2 – 0.24	3.80 – 8.35	0.057 – 0.127	RSP, 1-K & 2-K fuels
Weston	0.73 – 0.113	1.28 – 2.45	–	Old and New
Convective Heater[a]				
Leaderer	bkg – 0.0197	1.85 – 3.17	–	high, med, & low flame settings
Traynor	0.531 – 1.46	0.396 – 5.01	–	age & warm-up time dp=0.05-0.4 um
Yamanaka	2.61 ± 0.37	5.34 – 9.26	0.17 ± 0.07	± SD
Woodring	0.67 – 0.90	0.133 – 0.292	–	± SD 1-K & 2-K fuel
Weston	0.125 – 0.201		–	new
Caceres	0.18	0.61	–	wick type

[a] Assume rated output was fuel consumption rate, unless stated in the original work, kerosene 43.5 kJ/g.

[b] PM = Particulate Matter.

Table I cont.

	NO$_2$ (g/kg)	CO (g/kg)	PM (g/kg)	Remarks
Biomass				
Smith	–	1.3–7.5	0.11–1.78	Wood & cow dung back calculation from simulated hut
DeAngelo	–	11–40	1.8–2.9	Fireplace
Butcher	–	39–360	0.3–21.3	–
Butcher	–	63–158	1.6–6.4	Depends on wood type
Dasch	0.2–1.0(NOx)	58–200	1.5–20	Wood type
Knight	–	–	2–100	wood type & fuel consumption
Hayden	–	20–700	–	Fuel consumption & stove type
Ayer	–	20–125	–	Surface area to volume ratio of wood
Stove				
c Traynor	0.57	12.7 ± 0.96	0.0028	±SD, dp < 2.5 um
c Hollowell	0.40 – 0.80	8.0 – 17.3	–	–
c Cote	0.672– 0.975	7.36– 21.5	–	new and old
c Traynor	0.826	11.2 ± 1.9	0.0233	2 burners
Caceres	0.46	5.1	–	1 burner
Caceres	–	1.6	–	LPG
Cole	0.409–0.634	1.05– 10.8	–	lean, rich fuel
Yamanaka	0.957±0.142	–	–	"bath furnace", "gas burner", "hot & water supplier"

Table I cont.

Gas Oven	NO^2 (g/kg)	CO (g/kg)	PM (g/kg)	Remarks
Gas-Fired Space Heaters				
c Girman	–	0.014–14.7	–	well & poorly tuned, air intake open & closed
c Cote	0.586–0.619	4.26 – 8.42	–	high & low flames
c Traynor	0.361–1.12	0.597–34.9	–	high & low fuel consumption, excess air
Caceres	–	4.0 – 7.9	–	LPG

c Assume for gas 1000 BTU/SCF and rated output = fuel consumption rate unless stated.

Table II   Air Pollutants in Woodsmoke

Pollutant	Category	Carcinogenic Activity	Emission Factors (g/kg)		
			Stoves (1)	Fireplaces (2)	Ratio 1/2
Acenaphthylene	1		0.064	0.010	6.4
Flourene	1		0.020	0.0047	4.3
Anthracene/phenanthrene	1		0.096	0.0088	10.9
Phenol	1,3,4		0.1	0.02	5.0
Flouranthene	1,4		0.022	0.0016	13.7
Pyrene	1,4		0.019	0.0016	11.9
Benz(a)anthracene	1,2	+	0.0177	0.0019	9.3
Chrysene	1,2	+|			
Benzoflouranthenes	1,2		0.0135	0.0019	7.11
Benzo(b)flouranthene		++			
Benzo(j)flouranthene		++			
Benzopyrenes	2		0.009	0.0015	6.0
Benzo(a)pyrene	1,2	+++	0.0025	0.00073	3.4
Indeno(1,2,3-ed)pyrene	1,2	+			
Benzo(ghi)perylene	1		0.0059	0.0014	4.2
Dibenzanthracene	1,2		0.0010	0.00018	5.6
Dibenz[a,h]anthracene	1,2	+++			
Dibenz[a,c]anthracene	2	+			
Ancenapthene	1		0.0064	0.0012	5.3

Table II Air Pollutants in Woodsmoke (continued)

Pollutant	Category	Carcinogenic Activity	Emission Factors (g/kg)		
			Stoves (1)	Fireplaces (2)	Ratio 1/2
Ethyl benzene	1		0.041	0.0091	4.5
Phenathrene	1				
Dimethylbenzanthracene	2	+++			
Benzo[c]phenanthrene	2	+++	0.0025	0.008	0.31
Methylcholanthene	2				
3-methylcholanthene	2	++++			
Dibenzopyrenes	2		0.0007	0.0004	1.8
Dibenzo[a,l]pyrene	2	high			
Dibenzo[a,h]pyrene	2	+++			
Dibenzo[a,e]pyrene	2	+++			
Dibenzocarbazoles	2				
Dibenzo[a,g]carbazole	2	+/-			
Dibenzo[c,g]carbazole	2	+++			
Dibenzo[a,i]carbazole	2	+/-			
Formaldehyde	3		0.2	0.4	0.5
Propionaldehyde	3		0.2		
Acetaldehyde	3		0.1		
Isobutyraldehyde	3		0.3	0.5	0.6
Cresols	3		0.2	0.06	3.3
Catechol	4		0.01	0.014	0.7

Modified from Cooper (28)

1 = Priority pollutants measured in smoke from residential wood combustion sources.
2 = Carcinogenic compounds observed in smoke from residential wood combustion sources.
3 = Cilia toxic and mucus coagulating agents observed in smoke and flue gas from residential wood combustion sources.
4 = Initiating or cancer-promoting agents and co-carcinogenic compounds in smoke from residential wood combustion sources.

Table III   Chemical and Physical Properties of Biomass Fuels

(typical values dry basis, weight percent)[a]

Fuel	Carbon	Hydrogen	Nitrogen	Sulfur	Oxygen	Ash	Volatile matter	Gross heat MJ/kg
Hardwoods	51	6.3	0.1	0.0	42	0.8	–	19.7
Softwoods	52	6.0	0.1	0.0	41	1.1	85	20.8
Bark	54	5.9	0.1	0.0	38	1.5	75	20.2
Charcoal	93	2.5	1.5	–	3	1.0	10	33.7
Rice husks	39	5.7	0.5	0.0	40	16.0	–	15.4
Bagasse	45	6.0	0.0	0.0	35	11.0	80	18.3
Cattle dung	40	5.5	2.5	0.3	31	16.0	82	15.3
Bituminous coal	77	5.1	1.5	1.1	7.8	7.1	26	29.7
Anthracite coal	85	2.2	0.8	0.5	2.9	7.0	32	29.0

[a] Modified from Smith (29).

The use of the combustion devices indoors may lead to unhealthy pollutant concentrations. Some of the major factors that can affect the resulting levels are the emission rate, overall and local ventilation rates, spatial and temporal relationships, and certain use habits. The concentration measurements were carried out in various ways ranging from passive sampling to the use of direct reading instruments.

In homes using gas-fired stoves, ovens or heaters, NO_2 concentrations have been measured between 3 and 540 ppb (30,31),[2], CO between 0.9 and 40 ppm (32,33), and respirable suspended particles (RSP) between 5 and 1000 $\mu g/m^3$ (32). Some of the factors that have been investigated are gas versus electric stoves (32,34,35,36), spatial orientation with respect to the source (31,35,37), the number of pots on burners (33), the use of the oven for heating rooms (38), the use of multiple sources (30,33), seasonal variation (40), and the contributions from the pilot light (31). A study in a cafeteria where sterno was used to heat chafing dishes showed equilibrium concentrations of 50 ppm (56).

In homes using kerosene-fired heaters, concentrations of NO_2 have been measured between 7 and 380 ppb (40,41), CO up to 25 ppm (44), and total suspended particulate matter (TSP) between 117 and 350 $\mu g/m^3$ (42). These investigators looked at concentration relationships with respect to the age of the heater, (42) rated heat output (41,42), and spatial orientation with respect to the heater (40).

In homes using biomass (wood and/or dung) as a fuel, concentrations of NO_2 have been measured between 3 and 76 ppb (43,44), CO between 0.4 and 152 ppm (43,45), and TSP between 0.54 and 83 mg/m^3 (45). Some of the environmental factors that affected the concentrations of the indoor pollutants from burning biomass are; the air exchange rate (46), the type of house (46,47), the use of heat exchangers (44), and the frequency of stoking (48). These results come from a mixture of studies on biomass stoves, furnaces, and fireplaces. In Smith's book (29), Table 4-7 ha concentrations of CO and TSP from about 10 studies in "developing countries." Under primative wood burning indoors TSP has been recorded between 1 and 20 mg/m^3. BaP measurements during cooking often exceed a microgram per cubic meter.

It should be noted that these lists are not meant to be comprehensive, but just provide an overview of the factors that affect the indoor concentrations of pollutants. Not only have field measurements been taken, but chamber and modeling studies have been done to predict concentrations under certain conditions (3). Some other studies using personal monitoring have also been done (49,50,51).

EXPOSURES TO CAMPERS

As stated previously, the use of cooking, lighting, and heating equipment may result in NO_2, CO, CO_2, and particulate exposures to campers if these devices are used indoors (in tents). Emission factors for two commonly used camping combustion appliances were determined in a laboratory study and are reported here.

The appliances were an Optimus 123 stove (Svea) and a model 288-700 Coleman Lantern, both burning white gas (Coleman fuel). The stove was self-pressurizing after initial priming with fuel, and the lantern required the pumping of a plunger for 25 strokes to achieve enough pressure to light. Both the stove and lantern were about three months old and were used primarily in the lab.

A ventilation system was built that vented off the combustion gases to be analyzed. The system consisted of a hood supported by cinder block, a blast gate and dilution port to control the dilution air, a copper probe with teflon tubing inserted in it, a thermometer, a venturi and inclined manometer to indirectly measure the volumetric air flow rate, and a fan exhausting outside of the building.

The copper/teflon probe was attached to the sampling train by a teflon-T, where one side led to an interference filter for CO_2 and the CO analyzer and the other side to a particulate filter and the NO-NOx analyzer. The instruments used were an Electrochemical CO Ecolyzer Model 2000 and a Thermo-Electron Corporation 10A Chemiluminescent NO-NOx Analyzer (converter efficiency measured at >99%). They were calibrated with 56 ppm CO and 1.45 ppm NO, respectively. Carbon dioxide and water vapor concentrations were measured, revealing that neither was at a high enough level to interfere with either analyzer.

The emission rate (g pollutant/hr) was determined by multiplying the average mass concentration per run (after steady state was reached) by the volumetric flow rate of air in the ventilation system during that run. The fuel consumption rate was found by measuring the net change in mass of the appliance and dividing by the time the fuel valve was open. The change in mass was found using an Ohaus triple-beam balance with a tolerance of $\pm$ 300 mg and the time was measured with a wrist watch.

Table IV summarizes the results of 9 trials with the stove and 8 with the lantern. Both appliances were run at their highest heat or light output during the entire burn. The stove was tested with an aluminum pot on the burner holding about 2 liters of water initially at 14°C (SC = 3.2).

Table IV   Emission Factors for Stove and Lantern
(mean values $\pm$ standard deviation)

	Number of Trials	Fuel (kg/hr)[a] Consumption	CO (g/kg)	NO_2 (g/kg)	NO (g/kg)	NO/NO_2
Stove	9	0.124 ± 0.005	2.71 ± 4.28	1.03 ± 0.040	0.349 ± 0.092	0.339 ± 0.018
Lantern	8	0.056 ± 0.002	b	0.711 ± 0.092	5.33 ± 0.462	7.58 ± 1.00

a   white gas has 49 kJ/g

b   Preliminary measurements of CO for the lantern were made indicating levels that
were not measurably different from the background level, therefore CO measurements
for the lantern were not taken.

It should be noted that when the stove was run without the pot of water, the CO concentrations decreased to just above background, NO increased about two-fold, and NO_2 remained unchanged. When the stove was tested with the pot of cold water and was turned down to "low", CO and NO_2 output were cut in half and NO went down to barely above the background level. No changes in the concentrations were detected after the water in the pot boiled. When the lantern was turned down to low, both NO_2 and NO went down to just above their background levels.

Past experiments in the 1940's by Henderson et al. (52) and Pugh (53) have shown approximate emission factors for CO between 13-86 g/kg fuel and 13-18 g/kg fuel respectively (the values presented here weren't calculated in the experiments, only rough concentrations, times, and fuel consumption rates were in print). Even though the procedures and stoves were different (theirs burnt kerosene) the values agree with those reported here. Future studies may be done using snow in the pot on the stove, older appliances and different types of appliances.

Concentrations of CO inside tents with stoves running have been measured by Pugh (53) at 30 ppm and by Gallagher (54) between 10 and 100 ppm. Air exchange rates were determined for a small two-person tent (1.2 m^3) by plotting the decay of CO after the source was removed. Air exchange rates of 85 to 130 hr^{-1} were noted. Applying a one-box indoor air pollution model based on conservation of mass, equilibrium concentrations of 25 ppm CO, 580 ppb NO_2 and 360 ppb NO are calculated if a stove of the type tested is used. These concentrations are very sensitive to the air exchange rate and, therefore, the environment. Future field work will be conducted to verify these values and under different conditions to see if in fact, unhealthy pollutant levels can be reached.

ACKNOWLEDGEMENTS

This paper includes the original contributions of Martin Cohen's investigation of emissions from camping equipment. We appreciate the assistance of Yukio Yanagisawa, William Turner, and George Allen with experimental design, set up and analysis. Partial support for these activities was derived from a contract with the Gas Research Institute (5082-251-0739).

REFERENCES

1. National Research Council, <u>Indoor Pollutants</u> National Academy Press, Washington, D.C.1981.

2. Walsh, P.J., Dudney, C.S., Copenhaver, E.D., <u>Indoor Air Quality</u>, C.R.C. Press 1984.

3. <u>Indoor Air Quality Environmental Information Handbook: Combustion Sources</u>, prepared with assistance from Mueller Assoc., Baltimore, MD, Supcon Corp., Washington, D.C., Brookhaven National Laboratory, Upton, NY, and U.S. Department of Energy, Draft June 1984.

4. Moschandreas, D., Relwani, S., O'Neil, H.J., Cole, J.T., Elkins, R.H., Macriss, R.A., "Characterization of Emission Sources into the Indoor Environment",Gas Research Institute, IITRI project C08675, Draft March 1985.

5. Ayer, F.A., Ed. <u>Proceedings of Conference on Wood Combustion Environmental Assessment</u>, New Orleans 1981, NTIS.

6. Lindstrom, O., "Gasification/Combustion of Wood," in <u>Residential Solid Fuels: Environmental Impacts and Solutions</u>, Eds., Cooper J.A. and Malek D., proceedings from a conference held in Portland, OR, June 1-4, 1981, pp. 789-807.

7. Leaderer, B.P., "Air Pollutant Emissions from Kerosene Space Heaters," <u>Science</u> 218:1113(1982).

8. Caceres, T., Soto, H., Lissi, E., Cisternas, R., "Indoor House Pollution: Appliance Emissions and Indoor Ambient Concentrations, "<u>Atm. Env.</u> 17:1009(1983).

9. Turner, G.W., Allen, J.R., Apte, M.G., Girman, J.R., Hollowell, C.D., "Pollutant Emissions from Portable Kerosene-Fired Space Heaters," <u>Env. Sci. & Technol.</u>, 17:369(1983).

10. Yamanaka, S., Hirose, H., Takado, S., "Nitrogen Oxides Emissions from Domestic Kerosene-Fired and Gas-Fired Appliances," <u>Atm. Env.</u> 13:407(1979).

11. Woodring, J.L., Duffy, T.L., Davis, J.T., Bechtel, R.R., "Measurements of Emission Factors of Kerosene Heaters," Presented at American Industrial Hygiene Conference, Philadelphia, PA 1983.

12. Weston, as cited in <u>Indoor Air Quality Environmental Information Handbook: Combustion Sources</u> prepared with assistance from Mueller Assoc., Baltimore, MD Supcon Corp., Washington, D.C., Brookhaven National Laboratory, Upton, NY and U.S. Department of Energy, Draft June 1984, p. 2-11.

13. Traynor, G.W., Anthkon, D.W., Hollowell, C.D., "Techniques for Determining Pollutant Emissions from a Gas-Fired Range," LBL-9522, Dec. 1981, Preprint to Atm. Env.

14. Hollowell, C.D., Traynor, G.W., "Combustion-Generated Indoor Air Pollution," LBL-7832 Presented at the 13th International Colloquium on Polluted Atmospheres, Paris, France, April 25-28, 1978.

15. Cote, W.A., as cited in Caceres, T., Soto, H., Lissi, E., Cisternas, R., "Indoor House Pollution: Appliance Emissions and Indoor Ambient Concentrations," Atm.Env. 17:1009(1983), p. 2-19.

16. Cole, J.T., as cited in Indoor Air Quality Environmental Information Handbook: Combustion Sources, prepared with assistance from Mueller Assoc., Baltimore, MD, Supcon Corp., Washington, D.C., Brookhaven National Laboratory Upton, NY, and U.S. Department of Energy, Draft June 1984, p. 2-29.

17. Girman, J.R., Apte, M.G., Apte, G.W., Traynor, G.W., Allen, J.R., Hollowell, C.D., "Pollutant Emission Rates from Indoor Combustion Appliances and Sidestream Cigarette Smoke," LBL-12562, Env. Int. Vol. 8, pp. 213-221, May 1982.

18. Cote, W.A., as cited in Indoor Air Quality Environmental Information Handbook: Combustion Sources, prepared with assistance from Mueller Assoc., Baltimore, MD, Supcon Corp., Washington, D.C., Brookhaven National Laboratory, Upton, NY, and U.S. Department of Energy, Draft June 1984, p. 8.

19. Traynor G.W., as cited in Indoor Air Quality Environmental Information Handbook: Combustion Sources, prepared with assistance from Mueller Assoc., Baltimore, MD, Supcon Corp., Washington, D.C., Brookhaven National Laboratory, Upton, NY, and U.S. Department of Energy, Draft June 1984, p. 2-29.

20. Smith, K.R., Apte, M., Menon, P., Shrestha, M., "Carbon Monoxide and Particulates from Cooking Stoves: Results from a Simulated Village Kitchen," Presented at the Third International Conference on Indoor Air Quality and Climate, Stockholm, Sweden, Aug. 1984.

21. DeAngelo, D.G., as cited in Smith, K.R., Apte, M., Menon, P., Shrestha, M., "Carbon Monoxide and Particulates from Cooking Stoves: Results from a Simulated Village Kitchen," Presented at the Third International Conference on Indoor Air Quality and Climate, Stockholm, Sweden, Aug. 1984.

22. Butcher, S.S., Rao, U., Smith, K.R., Osborn, J., Azrima, P., Fields, H., "Emission Factors and Efficiencies for Small-Scale Open Biomass Combustion: Toward Standard Measuring Techniques," Presented at the Annual Meeting of the American Chemical Society Division of Fuel Chemistry, Philadelphia, PA, Aug. 1984.

23. Butcher, S.S., Ellenbecker, M. J., "Particulate Emission Factors for Small Wood and Coal Stoves," J. Air Pollut. Control Assoc., 32:380(1982).

24. Dasch, J.M., "Particulate and Gaseous Emissions from Wood-Burning Fireplaces," Env. Sci.&Tech., 16:380(1982).

25. Knight, C.V., from K.R. Smith, Traditional Biomass Fuels and Air Pollution, Draft fig. 2.4.

26. Hayden, A.C.S., from K.R. Smith, Traditional Biomass Fuels and Air Pollution, Draft, fig. 2.6-8.

27. Ayer, F.A., from K.R. Smith, Traditional Biomass Fuels and Air Pollution, Draft, fig 2.11.

28. Cooper, J.A., "Environmental Impact of Residential Wood Combustion Emissions and its Implications," J. Air Pollut. Control Assoc., 30:855(1980).

29. Smith, K.R., Traditional Biomass Fuels and Air Pollution, Draft.

30. Palmes E.D., as cited in Indoor Air Quality Environmental Information Handbook: Combustion Sources, prepared with assistance from Mueller Assoc., Baltimore, MD, Supcon Corp., Washington, D.C., Brookhaven National Laboratory, Upton, NY, and U.S. Department of Energy, Draft June 1984, p. 4-32.

31. Hollowell C.D., as cited in Indoor Air Quality Environmental Information Handbook: Combustion Sources, prepared with assistance from Mueller Assoc., Baltimore, MD, Supcon Corp., Washington, D.C., Brookhaven National Laboratory, Upton, NY, and U.S. Department of Energy, Draft June 1984, p. 4-50.

32. Lebowitz, M.D., Holberg, C.J., O'Rourk, M.K., Corman, G., Dodge, R., "Gas Stove Usage, CO and TSP, and Respiratory Effects," APCA #83-9.1.

33. Sterling, T.D., Sterling, E., "CO Levels in Kitchens and Homes with Gas Cookers," J.A.P.C.A. 29:238(1979).

34. Palmes, E.D., Tonczyk, C., DiMattio, "Average NO_2 Concentrations in Dwellings with Gas or Electric Stoves," Atm. Env. 11:869(1977).

35. Spengler, J.D., Duffy, C.P., Letz, R., Tibitts, T.W., Ferris, B.G., "Nitrogen Dioxide Inside and Outside 137 Homes and Implications for Ambient Air Quality Standards and Health Effects Research," Env. Sci. & Tech. 17:164(1983).

36. Florey, C., as cited in Indoor Air Quality Environmental Information Handbook: Combustion Sources, prepared with assistance from Mueller Assoc., Baltimore, MD, Supcon Corp., Washington, D.C., Brookhaven National Laboratory, Upton, NY, and U.S. Department of Energy, Draft June 1984, p. 4-42.

37. Boley, S.M., Brunekreef, B., Lebret, E., Biersteker, K., "Indoor NOx Pollution," International NOx Symposium 1982, Mastricht, The Netherlands.

38. Sterling, T.D., Kobayashi, D., "Use of Gas Ranges for Cooking and Heating in Urban Dwellings," J.A.P.C.A. 31:162(1981).

39. Wade, W.A., as cited in Indoor Air Quality Environmental Information Handbook: Combustion Sources, prepared with assistance from Mueller Assoc., Baltimore, MD, Supcon Corp., Washington, D.C., Brookhaven National Laboratory, Upton, NY, and U.S. Department of Energy, Draft June 1984, p. 4-44-47.

40. Ryan, P.B., Spengler, J.D., Letz, R., "The effects of Kerosene Heaters on Indoor Pollutant Concentrations: A Monitoring and Modelling Study," Atm. Env. 17:1339(1983).

41. CPSC, as cited in Indoor Air Quality Environmental Information Handbook: Combustion Sources, prepared with assistance from Mueller Assoc., Baltimore,MD, Supcon Corp., Washington, D.C., Brookhaven National Laboratory, Upton, NY, and U.S. Department of Energy, Draft June 1984, p. 4-17.

42. Ritchie, I.M., as cited in Indoor Air Quality Environmental Information Handbook: Combustion Sources, prepared with assistance from Mueller Assoc., Baltimore, MD, Supcon Corp., Washington, D.C., Brookhaven National Laboratory, Upton, NY, and U.S. Department of Energy, Draft June 1984, p. 4-22.

43. Traynor, G.W., Allen, J.R., Apte, M.G., Dillworth, J.F., Girman, J.R., Hollowell, C.D., Koonce, J.F., "Indoor Air Pollution from Portable Kerosene- Fired Space Heaters, Wood-Burning Stoves, and Wood-Burning Furnaces," LBL-14027 presented at A.P.C.A. specialty meeting on residential wood and coal combustion, Louisville, KY, March 1-2, 1982.

44. Grimsrud, D.T., Lipshitz, R.D., Girman, J.R., "Indoor Air Quality in Energy Efficient Residences," Indoor Air Quality, Eds., P.J. Walsh, C.S. Dudney, E.D. Copenhover, C.R.C. Press 1984.

45. Dollar, A.M., as cited in Smith, K.R., <u>Traditional Biomass Fuels and Air Pollution</u>, Draft, Tables 4.1 and 4.4.

46. Smith, K.R., Aggarwal A.L., Dave R.M., "Air Pollution and Rural Biomass Fuels in Developing Countries: A Pilot Village Study in India and Implications for Research and Policy," <u>Atm. Env.</u> 17:2343 (1983).

47. Aggarwal, A.L., Raigani, C.V., Patel, P.D., Shah, P.G., Chaterjee, S.K., "Assessment of Exposure to Benzo(a)Pyrene in Air for Various Population Groups in Ahmedabad," <u>Atm. Env.</u> 16:867 (1982).

48. Moschandreas, D., as cited in <u>Indoor Air Quality Environmental Information Handbook: Combustion Sources</u>, prepared with assistance from Mueller Assoc., Baltimore, MD, Supcon Corp., Washington, D.C., Brookhaven National Laboratory, Upton, NY, and U.S. Department of Energy, Draft June 1984, p. 4-37.

49. Nitta, H., Maeda, K., "Personal Exposure Monitoring to Nitrogen Dioxide," not published.

50. Yanagisawa, Y., Matsuki, H., Osaka, F., Kasuga, H., Nishimura, H., "Annual Variation of Personal Exposure to Nitrogen Dioxide," not published.

51. Kim, Y.S., Spengler, J.D., Yanagisawa, Y., Chung, Y., Kwon, S.P., "Indoor Nitrogen Dioxide Pollution in Korea," not published.

52. Henderson, Y., McCullough, J., "Carbon Monoxide as a Hazard of Polar Exploration," <u>Nature</u>, 145:92(1940).

53. Pugh, L. G. C. E., "Carbon Monoxide Hazard in Antarctica," <u>British Medical Journal</u>, 1:192(1959).

54. Gallagher, K., Mountain Safety Research, Seattle, WA, private communication (1984).

55. Murray, T.J., "Carbon Monoxide Poisoning from Sterno," <u>Canadian Medical Assoc. Journal</u>, 118:800(1978).

21.

Interference with Lung Defenses by Nitrogen Dioxide Exposure

Judith A. Graham and Frederick J. Miller

Toxicology Branch, Health Effects Research Laboratory, Environmental Protection Agency, MD-82, Research Triangle Park, NC 27711

Disclaimer: This report has been reviewed by the Health Effects Research Laboratory, U.S. Environmental Protection Agency, and approved for publication. Mention of trade names or commercial products does not constitute endorsement or recommendation for use.

INTRODUCTION

It has been established for several years that nitrogen dioxide (NO_2) affects host defenses against pulmonary infection in animals, but the mechanisms of toxicity involved have not yet been fully identified. Controversies exist regarding the lowest effective concentrations of NO_2 and whether or not man might experience some of the effects observed in animals. This review will initially discuss the effects of NO_2 on host defenses in the experimental animals using models in which exposed animals are challenged with viable microbes and mortality is measured. A presentation of the effects of NO_2 on more specific host defenses (e.g., mucociliary escalator, alveolar macrophages) follows. Studies of the effects of NO_2 on systemic humoral and cell-mediated immunity will be summarized in an attempt to speculate on immunological effects that might occur in the lung. Reviews of pulmonary defense mechanisms are available [1,2], as is information on the effects of NO_2 upon these defense mechanisms [3-7].

INCREASED SUSCEPTIBILITY TO INFECTION

Animal infectivity models have proven to be useful probes in toxicological studies on the effects of airborne environmental contaminants [8-17]. In these model systems, animals are usually exposed to a pollutant and then are challenged with an aerosol of viable microorganisms, such as Klebsiella pneumoniae, Diplococcus pneumoniae, Streptococcus (Group C), and influenza PR-8 virus. Increased mortality, compared to corresponding controls, is determined over a 14- to 15-day post-exposure observation period. Host resistance to respiratory infection assessed in this manner may reflect an integration or summation of effects (edema, inflammation, alteration of macrophage function, alterations in mucociliary function, etc.) of the pollutant on the lung [4,17].

The toxicological data base that has evolved from infectivity model studies using mice demonstrates that variations in concentration, length of exposure, and pattern of exposure are important factors in the determination of effects of NO_2 [3]. However, in the absence of NO_2 pulmonary dosimetric comparisons between mice and human subjects, the implications of these studies relative to human exposures are unclear, especially since increased mortality in animal-bacterial models would more likely need to be compared to increased morbidity in man. Moreover, as one compares the exposure regimens of the various studies to diurnal profiles of NO_2 in urban air [3], it is apparent that more studies are needed employing realistic exposure profiles. In urban air, there are often low background levels of NO_2 on which are superimposed higher diurnal peaks. These peaks are usually of short duration and irregular occurrence. Also, epidemiological studies comparing NO_2 exposures in homes using gas stoves to those using electric stoves report NO_2 exposures with peak patterns [18,19].

What comparisons can be made concerning differences in host susceptibility to infection as a result of the interaction between NO_2 concentration (C) and length (time) of exposure (T)? If there is no interaction between these two factors, no statistical difference in mortality response should be seen when either factor is varied, provided that the product of these factors remains constant. Table 1 indicates that in the mouse infectivity model using Streptococcus (Group C), concentration has a greater influence on the magnitude of the response than does the length of exposure.

However, length of exposure cannot be totally dismissed as evidenced from studies [20] in which the length of continuous exposure varied from a few minutes to 21 days (Table 2). For any given concentration of NO_2 between 2.8 mg/m^3 (1.5ppm) and 526.7 mg/m^3 (28 ppm), there is a statistically significant (p < 0.05) positive increasing linear relationship between length of exposure and the resulting mortality. Moreover, with increasing concentrations of NO_2, the rate of increase in mortality also increases. Using Klebsiella pneumoniae, Ehrlich and Henry [21] found significant mortality increases with continuous exposure to 0.94 mg/m^3 (0.5 ppm). However, McGrath and Oyervides [10] found no increase in mortality consequent to K. pneumoniae challenge in mice exposed for 3 or 8 months to 0.94 mg/m^3 (0.5 ppm).

Comparing the responses of animals to intermittent (7 hr/day for various numbers of days) versus continuous exposure to 6.6 mg/m^3 (3.5 ppm) and 2.8 mg/m^3 (1.5 ppm) of NO_2, show that the results [20,22] are concentration dependent. Continuous exposure to 2.8 mg/m^3 (1.5 ppm) initially resulted in a greater mortality response

Table 1.   Interaction of NO_2 Concentration and Length of Exposure on Enhancement of Mortality [a]

Concentration (C), ppm	Time (T) hr	CxT	Mortality, % (NO_2 - Control)
1.5	4.7	7	6.4
3.5	2.0		18.7
7.0	1.0		30.2
14.0	0.5		21.7
28.0	0.25		55.5
1.5	9.3	14	10.2
3.5	4.0		27.0
7.0	2.0		41.8
14.0	1.0		44.9
28.0	0.5		67.2
1.5	14.00	21	12.5
3.5	6.00		31.9
7.0	3.00		48.6
14.0	1.50		58.5
28.0	0.75		74.0

[a]These are predicted values obtained from Figure 1 of Gardner et al. [20]. The table is adapted from Gardner et al. [20].

than with intermittent exposure; but after 2 or 3 weeks of the intermittent exposure regimen, the response was the same as for continuous exposure. When comparisons are made on the basis of CxT (concentration x time), the intermittent exposure appears to yield a higher mortality than does continuous exposure only when the CxT reaches 220.5 (Table 3).   This corresponds to 21 days of inter- mittent exposure to 2.8 mg/m^3 (1.5 ppm). From these data, one can only speculate as to whether this trend would continue with longer exposure times to 2.8 mg/m^3 (1.5 ppm). When the concentration was increased to 6.6 mg/m^3 (3.5 ppm), the effect observed for the two exposure modes was not different over the time periods studied.

While one might consider the above intermittent (7 hr/day) exposures to be "broad peaks", the NO_2 peaks in ambient air and in the gas stove epidemiological studies are shorter in duration but more frequent in occurrence. Studies described by Graham et al. [23] have examined the importance of the variation in the length and persistence of a spike concentration of NO_2 on increased suscepti- bility of the host to infection. The effect these peaks have when superimposed on a lower continuous background concentration of NO_2 was also examined. In the latter studies, the ratio of the peak to the basal NO_2 concentration was 3:1, a ratio frequently seen in urban air.

Table 2.   Influence of Length of Continuous Exposure on
Susceptibility of Mice to Infection for Various
Concentrations of NO_2[a]

A.   Experimental Data

NO_2 Level, ppm	Exposure length, hr	Mortality, % (NO_2-control)
1.5	2	6.7
	5	0.0
	8	24.4
	18	25.0
	24	15.6
	96 (6)[b]	10.8
	126	35.0
	168 (4)	25.0
	222	25.0
	336 (4)	31.3
	504 (3)	45.0
3.5	0.5 (5)	10.0
	1.0 (10)	8.0
	2.0 (2)	12.5
	3.0 (10)	19.0
	5.0 (2)	37.5
	7.0 (14)	34.3
	14.0 (4)	45.0
	24.0 (7)	49.3
	48.0 (8)	56.3
	96.0	75.0
	168.0	85.0
	384.0	65.0
7.0	0.5 (6)	20.8
	1.0 (6)	29.2
	1.5 (6)	28.3
	2.0 (6)	49.2
14.0	0.5 (3)	23.3
	1.0 (3)	38.3
	1.5 (3)	66.7
	2.0 (3)	65.0
28	0.10 (5)	41.6
	0.25 (5)	53.6
	0.42 (5)	60.6
	0.58 (5)	73.6

[a]Table and regression results based on Gardner et al. [20]
[b]The number of replicate experiments is indicated in parentheses.

Table 2. (Continued)   COMBUSTION PRODUCTS   283

B.   Regression model: Percent Mortality (NO_2 - Control) =
$$\alpha + \beta \log_{10} Time$$

==

NO_2 Level, (ppm)	Intercept (α)	Slope (β)
1.5	-2.14	12.76
3.5	10.42	27.58
7.0	30.17	38.66
14.0	44.94	77.12
28.0	78.88	38.85

All regressions are significant at the 0.05 probability level

---

Table 3. A Comparison of Intermittent (7 hr/day) and Continuous Exposure Effects on Susceptibility of Mice to Respiratory Infection[a]

==

	1.5 ppm NO_2			3.5 ppm NO_2	
	Mortality, % (NO_2 - Control)			Mortality, % (NO_2 - Control)	
CxT	Intermittent	Continuous	CxT	Intermittent	Continuous
10.5	6	9	24.5	40	32
21.0	5	12	49.0	37	42
31.5	6	15	73.5	43	47
42.0	6	16	98.0	55	50
73.5	13	19	171.5	55	57
147.0	26	23	269.5	75	62
220.5	33	26	367.5	60	66

[a]Data are based upon Gardner et al. |20,22|.

In the single peak exposure experiments [23], mice were exposed to 8.1 mg NO_2/m^3 (4.5 ppm) for either 1, 3.5, or 7 hrs prior to being challenged with Streptococcus either immediately or 18 hrs after exposure. Animals challenged immediately after the NO_2 exposure showed a mortality pattern that was linearly related to the length of the peak exposure, whereas no mortality increase over controls occurred in the 18 hr recovery groups. When a "lead in" basal exposure for 64 hrs to 2.8 mg/m^3 (1.5 ppm) preceded the peak exposure, no significant enhancement in mortality over the peak exposure alone was seen in animals challenged immediately after cessation of the NO_2 spike. However, if the baseline exposure was continued during the 18 hr period after the peak exposure, there was a significant increase in mortality for those mice receiving peak exposures of 3.5 or 7 hrs.

Studies using multiple peak exposures on a continuous background concentration of NO_2 were also performed [23]. With the exception of the weekends, there were morning and afternoon peaks of 1 hr duration. The mice were challenged with Streptococcus either before or after the morning peak. When the challenge was given before the morning peak, there was no consistent mortality trend over the 15 day experimental regimen. However, on 3/4 of the days examined, the mortality increase was significantly greater than zero, and a number of days were statistically significant even when a Bonferroni correction [24] was applied for multiple t-tests. For animals challenged after the morning peak, mortality was consistently enhanced during the second week of exposure and approached the mortality level of continuous exposure to 2.8 mg/m³ (1.5 ppm) found by Gardner et al. [20]. These results demonstrate the complexity in determining toxicological responses associated with various, but realistic exposure patterns. While NO_2 exposure results in significant increases in mortality, the pattern and magnitude of the response are influenced by the specific exposure regimen.

In order to assess the contribution of peaks compared to continuous exposure to the background concentration alone at levels of NO_2 "nearer to ambient", we have exposed mice for 16, 32, or 52 weeks to the following regimens: clean air, 0.38 mg/m³ (0.2 ppm) for 23 hr/day for 7 days/wk, or the baseline of 0.38 mg/m³ plus two 1 hr peaks of 1.5 mg/m³ (0.8 ppm) Monday through Friday. The study was essentially negative in that no significant increase in mortality over control was seen at any time point for either of the two NO_2 exposure regimens.

Very few studies have been conducted using viral infectivity models. Henry et al. [15] observed that 18.8 mg/m³ (10 ppm) NO_2 increased mortality (6 of 6 monkeys) after infection with influenza A/PR/8 virus; at 9.4 mg/m³ (5 ppm), 1 of 3 monkeys died. Mice challenged with influenza A/PR/8 virus also experience enhanced mortality following a 39-day continuous exposure to 0.94-1.88 mg/m³ (0.5-1 ppm) or a 2 hr/day (for 1, 3, and 5 day) exposure to 18.8 mg/m³ (10 ppm) [25]. Thus, it appears that antiviral defenses can be affected by NO_2, but due to the limited data base, no firm conclusions can be drawn.

MUCOCILIARY CLEARANCE

Microbes deposited within the tracheobronchial region can be cleared via the mucociliary escalator. Generally, when bacteria are inhaled, about 10-20% are cleared rapidly, most probably by this mechanism [26]. Alveolar macrophages (AM) can also be cleared from the lung via the mucociliary escalator. These AM may have phagocytized bacteria or viruses and may contain some viable microbes if their microbiocidal function has been altered. Thus, slowing of tracheobronchial clearance would be expected to increase the residence time of microbes in the lung. The only major study below 10 ppm NO_2 showed that a 6-week exposure to 6 ppm NO_2 decreased mucous clearance [27].

ALVEOLAR MACROPHAGES

Alveolar macrophages are the cells primarily responsible for maintaining sterility of the lower respiratory tract [1]. They have the capacity to phagocytize and kill microbes, kill tumor cells, produce mediators (e.g., interferon and chemotactic factors) which influence other host defenses, and release proteolytic enzymes that may be involved in the development of chronic lung disease. They are the predominant cell recoverable by lung lavage. Alveolar macrophage function, structure, and biochemistry have been studied following exposure to NO_2 (Table 4).

Using different approaches and different bacterial species, Goldstein et al. [28] and Ehrlich et al. [29] have shown that acute and subchronic exposure to NO_2 decrease the ability of mouse lungs to clear viable bacteria. These effects are thought to be due to decrements in the function of AM.

Alveolar macrophages inactivate bacteria via a multistage process [1], each step of which may be susceptible to effects of NO_2 exposure. Bacterial attachment to the AM membrane is facilitated by opsonizing antibody, but this process has not been studied after NO_2 exposure. Membrane alterations have been observed, but it is not known to what degree the changes influence bactericidal activity. Subchronic exposure causes morphological changes in the surface of AM as shown by scanning electron microscopy [30]. In vitro exposure increases the agglutination of rat AM, illustrating membrane damage [31]. Concanavalin A-induced membrane receptors were found to be altered by NO_2 exposure, as was the ability of the cells to recognize heterologous red blood cells [32].

The next major phase in the bactericidal process is phagocytosis, internalization of the bacteria within the cell. Gardner et al. [33] found that acute exposure to NO_2 decreases in vivo phagocytosis of bacteria; others found no such effect studying in vitro phagocytosis of inert particles [34]. Acute, but not repeated, exposures to very high levels of NO_2 increased in vitro phagocytosis by AM, suggesting an activation [35]. Under the same conditions, the AM had an increased cytotoxic response towards syngenic tumor cells and were more sensitive to activating agents [35]. Subchronic exposure to NO_2 increased AM congregation on epithelial cells in vitro, a response that might reflect activation, although the precise mechanism is unknown [36].

We are unaware of any studies of the effects of NO_2 on lysosomal enzymes. Ozone, which has many effects in common with NO_2, decreases activities of these enzymes, so the next stage of the bactericidal process (within the secondary lysosome) might be sus-

Table 4. Effects of NO_2 on Alveolar Macrophages

NO_2, ppm	Time of Exposure	Species	Effects	Reference
0.5	3 hr/day, 5 days/wk, 1,2,3 mo	Mouse	Decreased phagocytosis at 2 mo, no change in viability or % macrophages, decreased clearance of bacteria (Streptococcus sp.) at 1 and 2 months.	29
0.5	continuous, 3 wk	Mice	Increased in vitro congregation of macrophages on epithelial and spindle target cells.	36
2; 0.5 with spikes to 2	continuous; continuous to 0.5 ppm with spikes 1 hr/day 5 days/wk up to 33 wk	Mouse	Morphological changes in cell membrane.	30
1-20 ppm (3.9-51 ppm/hr)	2-3 hr	Rat	Dose-response (linear) decrease in the production of superoxide anion radicals.	37
1.9-14.8	4 hr	Mouse	Decreased pulmonary bactericidal (S. aureus) activity $\geq$ 7 ppm.	28
3.6, 12.1	1-2 hr	Rat	Increased agglutination induced by Conconavalin A	31

COMBUSTION PRODUCTS   287

Table 4. Effects of NO_2 on Alveolar Macrophages (Cont.)

NO_2, ppm	Time of Exposure	Species	Effects	Reference
7	24 hr	Rabbit	Increased rosette formation in macrophages treated with wheat germ lipase.	32
10	3 hr	Rabbit	Decreased phagocytic activity.	33
10	24 hr	Rat	No change in phagocytosis.	34
40	4 hr, 1, 7, 14 days	Rat	Effects only at 1 day. Increased phagocytosis, cytotoxicity to adenocarcinoma cells, and sensitivity to activating agents.	35

ceptible to NO_2 exposure. The production of superoxide anion radicals also appears to play a significant role in bactericidal activity. In vitro exposure of AM to NO_2 caused a significant linear dose-response decrease in the production of superoxide anion radicals [37].

Little is known about antiviral activities of AM following exposure to NO_2. Earlier research showed that a 3 hr exposure to 25 ppm NO_2 increased the attachment of parainfluenza-3 virus to macrophages and enhanced their penetration [38].

IMMUNE SYSTEM

The immune system is active in the lung [2]. It may be hypothesized that local lung immunity is at more risk to NO_2 exposure than is the systemic immune system since the lung would receive a greater dose. However, to date, there have been no reports of such pulmonary studies. (See Table 5 for summary of immunological effects).

Acute exposure to high levels of NO_2 decreases the primary humoral antibody response [39]. Longer exposure to lower levels decreases the primary response and slightly increases the secondary antibody response [40]. Longer exposure to NO_2 causes a biphasic response in which immunostimulation proceeds to no change and then proceeds to immunosuppression of primary humoral immunity. A somewhat similar pattern is observed for graft-vs-host reactions, but for the response of splenic cells to PHA, depression is observed at all time points [41]. Concentration of NO_2 also plays a role in immunologic responses. At 1 ppm, monkeys showed increased antibody responses to a vaccine of influenza [42]; at 5 ppm, serum neutralizing titers were decreased [43]. NO_2 also decreases serum neutralizing titers in mice and generally increases serum immunoglobulin titers [44]. Morphological and weight changes also occur in the spleens of mice after subchronic NO_2 exposure [45]. These observations of the immune system after NO_2 exposure indicate the complexities of the responses. The only clear interpretation is that the results are highly dependent on concentration and time of exposure, with time being a modulator (e.g., increase or decrease) of the response.

SUMMARY AND CONCLUSIONS

The studies discussed here clearly show the propensity of NO_2 to cause decrements in pulmonary host defenses against bacterial infections in animals. Concentration is more important than time of exposure in causing effects, although time plays a role. Most of the supporting information is derived from the mouse streptococcal infectivity model which uses mortality as an endpoint. The mouse model system is not intended to portray potential effects on human mortality since the combination of circumstances of the mouse model would not be expected to occur in man. However, the mouse model reflects the net effect on several host defense systems which also operate in man. The animal data also illustrate the complexities

Table 5. Effects of NO_2 on the Systemic Immune System

NO_2, ppm	Time of Exposure	Species	Effects	Reference
0.35	Continuous, 6 wk	Mice	In spleen: increase in weight as a percent of body weight; smaller increase in cell number per spleen weight increment; increase in size of lymphoid nodules; apparent greater predominance of red cells in the red pulp.	45
0.4, 1.6	Continuous, 4 wk	Mice	Primary antibody response (splenic plaque forming cells to sheep erythrocytes) decreased. Secondary antibody response increased at 1.6 ppm. No difference between splenic B and T cells.	40
2 ppm; 0.5 ppm with 2 ppm spikes	Continuous; Continuous with 1 hr 5 days/wk spikes 40 wk	Mice	Mice exposed for 12 wk, vaccinated with influenza vaccine, and re-exposed for another 28 wk. Decreased serum neutralization titer in 0.5/2 ppm group shortly after vaccination. No changes 12 wks after vaccination. No effect on hemagglutination inhibition titers. Variety of changes in serum Ig titers.	44
1	493 days	Monkey	Monkeys challenged with influenza virus. No change in hemagglutination inhibition titers. Increased serum neutralizing titers.	42

Table 5. Effects of NO_2 on the System Immune System (Cont.)

NO_2, ppm	Time of Exposure	Species	Effects	Reference
5	169 days	Monkey	Monkeys challenged with influenza virus. Initial depression in serum neutralizing titers; return to normal by 133 days. Hemagglutination inhibition titers unchanged.	43
5, 20, 40	12 hr	Mice	At 5 ppm: decreased number of thymus cells. At 20 and 40 ppm: decreased primary antibody response (splenic plaque-forming cells and hemagglutination titer to sheep erythrocytes); decreased number of spleen and thymus cells.	39
10	2 hr/day 5 days/wk 30 wk	Mice	Hemagglutination titers to red blood cells increased at 10 wk, did not change at 20 wk, and decreased at 30 wk. No changes in titers to T-dependent antigen (PVP). Response of spleen cells to PHA depressed after 10, 15, 23, and 26 wk. Graft-vs-host response of spleen cells increased at 10–23 wk, decreased at 26 wk. Decreased survival when injected with tumor cells.	41

when attempting to hypothesize whether or not man might experience effects on host defenses. If human alveolar macrophages encounter a sufficient dose-rate of NO_2, there is no apparent reason why they would not be affected. Quantitative data on dosimetric and sensitivity relationships across species (including man) are needed to extrapolate toxicological responses between animals and man. Nevertheless, it may still be suggested that the principles derived from the animal studies are likely to hold for man. That is, the dose-rate of NO_2 has a predominant influence on responses. This makes detailed exposure assessment especially important in assessing risk.

Unfortunately, we know very little about the effects of NO_2 on antiviral defenses. The high-concentration studies available must be interpreted cautiously due to the importance of concentration in causing effects. The single study [25] at a lower level is difficult to interpret accurately. Considering the clinical and societal importance of lung viral infections, this deficiency in knowledge is of concern.

There is no information on the effects of NO_2 on the pulmonary immune system. This system has a degree of compartmentalization from the rest of the body and is important to lung defense. For example, local immunity can be more important than systemic immunity in defending the lung. The systemic humoral and cell-mediated immune system is susceptible to NO_2. However, the current difficulties in interpreting the magnitude of the immunological changes in animals in terms of human health inhibit hazard assessment. Additional studies are needed to help elucidate the effects of NO_2 on lung host defenses.

REFERENCES

1.   Green, G. M., G. J. Jakab, R. B. Low, and G. S. Davis. "Defense Mechanisms of the Respiratory Membrane," Am. Rev. Resp. Dis. 115:479-514 (1977).

2.   Bice, D. E. "Methods and Approaches for Assessing Immunotoxicity of the Lower Respiratory Tract," in Toxicology of the Immune System, J. A. Dean, A. Munson, and M. Luster, Eds. (New York: Raven Press, in press, 1984).

3.   EPA, "Studies of the Effects of Nitrogen Compounds on Animals," in Air Quality Criteria for Oxides of Nitrogen, Ch. 14, EPA-600/8-82-026, (Research Triangle Park, NC: ECAO, 1982), 126 pp.

4.   Gardner, D. E. and J. A. Graham. "Increased Pulmonary Disease Mediated Through Altered Bacterial Defenses," in Pulmonary Macrophage and Epithelial Cells. R. P. Schneider, G. E. Doyle and H. A. Ragan, Eds. (Richland: Proc. 16th Annual Hanford Biology Symposium, 1976), pp. 1-21.

5.   Graham, J. A. and D. E. Gardner. "Immunotoxicity of Air Pollutants," in Toxicology of the Immune System, J. A. Dean, A. Munson, and M. Luster, Eds. (New York: Raven Press, in press, 1984).

6.   Green, G. M. "Similarities of Host Defense Mechanisms Against Pulmonary Infectious Diseases in Animals and Man," in Fundamentals of Extrapolation Modeling of Inhaled Toxicants: Ozone and Nitrogen Dioxide. F. J. Miller and D. B. Menzel, Eds. (New York: Hemisphere Publishing Corp., 1984), pp. 291-298.

7.   Jakab, G. J. "Nitrogen Dioxide-Induced Susceptibility to Acute Respiratory Illness: a Perspective," Bull. N.Y. Acad. Med. 56: 847-855 (1980).

8.   Coffin, D. L., and E. J. Blommer. "Acute Toxicity of Irradiated Auto Exhaust," Arch. Environ. Health 15:36-38 (1967).

9.   Coffin, D. L., Blommer, E. J., Gardner, D. E., and R. S. Holzman. "Effect of Air Pollution on Alteration of Susceptibility to Pulmonary Infection," in Proceedings of 3rd Annual Conference on Atmospheric Contamination in Confined Space, (Dayton, OH: Aerospace Medical Research Lab, 1968), pp. 71-80.

10.   McGrath, J. J. and J. Oyervides. "Response of NO_2-exposed mice to Klebsiella challenge," Adv. Modern Environ. Toxicol., S. D. Lee, M. G. Mustafa, and M. A. Mehlman, Eds. (Princeton Sci. Pub., Princeton), 5:475-485 (1983).

11. Gardner, D. E., F. J. Miller, J. W. Illing, and J. M. Kirtz. "Alterations in Bacterial Defense Mechanisms of the Lung Induced by Inhalation of Cadmium," Bull. Eur. Physiopathol. Resp. 13:157-174 (1977).

12. Maigetter, R. Z., R. Ehrlich, J. D. Fenters, and D. E. Gardner. "Potentiating Effects of Manganese Dioxide on Experimental Respiratory Infection," Environ. Res. 11:386-   (1976).

13. Miller, F. J., J. W. Illing, and D. E. Gardner. "Effect of Urban Ozone Levels on Laboratory-Induced Respiratory Infections," Toxicol. Lett. 2:163-169 (1978).

14. Purvis, M. R. and R. Ehrlich. "Effect of Atmospheric Pollutants on Susceptibility to Respiratory Infection. II.   Effect of NO_2," J. Infect. Dis. 113:72-76 (1963).

15. Henry, M. C., J. Findlay, J. Spangler, and R. Ehrlich. "Chronic Toxicity of NO_2 in Squirrel Monkeys," Arch. Environ. Health 20:566-570 (1970).

16. Henry, M. C., R. Ehrlich, and W. L. Blair. "Effect of Nitrogen Dioxide on Resistance of Squirrel Monkeys to Klebsiella pneumoniae infection," Arch. Environ. Health 18:580-587 (1969).

17. Gardner, D. E. "The Use of Experimental Airborne Infections for Monitoring Altered Host Defense," Environ. Health Perspect. 43:99-107 (1982).

18. Speizer, F. E., B. G. Ferris, Jr., Y. M. M. Bishop, and J. Spengler. "Respiratory Disease Rates and Pulmonary Function in Children Associated with NO_2 Exposure," Am. Rev. Res. Dis. 121:3-10 (1980).

19. Spengler, J. D., B. G. Ferris, Jr., and D. W. Dockery. "Sulfur Dioxide and Nitrogen Dioxide Levels Inside and Outside Homes and the Implications on Health Effects Research," Environ. Sci. Technol. 13:1276-1271 (1979).

20. Gardner, D. E., F. J. Miller, E. J. Blommer, and D. L. Coffin. "Influence of Exposure Mode on the Toxicity of NO_2," Environ. Health Perspect. 30:23-29 (1979).

21. Ehrlich, R., and M. C. Henry. "Chronic Toxicity of Nitrogen Dioxide. I.   Effect on Resistance to Bacterial Pneumonia," Arch. Environ. Health 17:860-865 (1968).

22. Gardner, D. E., F. J. Miller, E. J. Blommer, and D. L. Coffin. "Relationships Between NO_2 Concentration, Time, and Level of Effort Using an Animal Infectivity Model," in Proceedings International Conference on Photochemical Oxidant Pollution and

its Control, Vol. 1.   (Washington, DC: EPA-600/3-77-001a, 1977), pp. 513-525.

23. Graham, J. A., D. E. Gardner, E. J. Blommer, and F. J. Miller. "Influence of Exposure Patterns of Nitrogen Dioxide and Modifications by Ozone in Affecting Susceptibility to Bacterial Infectious Disease in Mice," (in preparation, 1984).

24. Dayton, C. M., and W. D. Schafer, "Extended Tables of t and chi Square for Bonferroni Tests with Unequal Error Allocation," J. A. S. A. 68:78-83 (1973).

25. Ito, K. "Effect of Nitrogen Dioxide Inhalation on Influenza Virus Infection in Mice," Jap. J. Hygiene 26:304-314 (1971).

26. Green, G. M. and E. H. Kass. "The Role of the Alveolar Macrophage in the Clearance of Bacteria from the Lung," J. Exp. Med. 119:167-172 (1967).

27. Giordano, A. M., and P. E. Morrow. "Chronic Low-Level Nitrogen Dioxide Exposure and Mucociliary Clearance," Arch. Environ. Health 25:443-449 (1972).

28. Goldstein, E., M. C. Eagle, and P. D. Hoeprich. "Effect of Nitrogen Dioxide on Pulmonary Bacterial Defense Mechanisms," Arch. Environ. Health 26:202-204 (1973).

29. Ehrlich, R., J. C. Findlay, and D. E. Gardner. "Effects of Repeated Exposures to Peak Concentrations of Nitrogen Dioxide and Ozone on Resistance to Streptococcal Pneumonia," J. Toxicol. Environ. Health 5:631-642 (1979).

30. Aranyi, C., J. Fenters, R. Ehrlich, and D. Gardner. "Scanning Electron Microscopy of Alveolar Macrophages after Exposure to O_2, NO_2, and O_3," (Abst.) Environ. Health Perspect. 16:180 (1976).

31. Goldstein, G. D., S. J. Hamburger, G. W. Falk, and M. A. Amoruso. "Effect of Ozone and Nitrogen Dioxide on the Agglutination of Rat Alveolar Macrophages by Concanavalin A," Life Sci. 21: 1637-1644 (1977).

32. Hadley, J. G., D. E. Gardner, D. L. Coffin, and D. B. Menzel. "Effects of Ozone and Nitrogen Dioxide Exposure of Rabbits on the Binding of Autologous Red Cells to Alveolar Macrophages," in Proceedings International Conf. Photochemical Oxidant Pollution and its Control, Vol. 1. EPA-600/3-77-001a. (Research Triangle Park, NC: U.S. Environmental Protection Agency, 1977), pp. 505-511.

33. Gardner, D. E., R. S. Holzman, and D. L. Coffin. "Effects of Nitrogen Dioxide on Pulmonary Cell Population," J. Bacteriol. 98:1041-1043 (1969).

34. Katz, G. V., and S. Laskin. "Pulmonary Macrophage Response to Irritant Gases," in Air Pollution and the Lung, E. F. Aharaonson, A. Ben-David, and M. A. Klingberg, Eds. (New York: John Wiley and Sons, 1975), pp. 82-99.

35. Sone, S., L. M. Brennan, and D. A. Creasia. "In vivo and In vitro NO_2 Exposures Enhance Phagocytic and Tumoricidal Activities of Rat Alveolar Macrophages," J. Toxicol. Environ. Health 11:151-163 (1983).

36. Richters, V., G. Elliott, and R. S. Sherwin. "Influence of 0.5 ppm Nitrogen Dioxide Exposures of Mice on Macrophage Congregation in the Lungs," In Vitro 14:458-464 (1978).

37. Amorusco, M. A., G. Witz, and B. D. Goldstein. "Decreased Superoxide Anion Radical Production by Rat Alveolar Macrophages Following Inhalation of Ozone or Nitrogen Dioxide." Life Sci. 28:2215-2221 (1981).

38. Williams, R. D., J. D. Acton, and Q. N. Myrvick. "Influence of Nitrogen Dioxide on the Uptake of Parainfluenza-3 Virus by Alveolar Macrophages," J. Reticuloendothel. Soc. 11:627-636 (1972).

39. Fujimaki, H., and F. Shimizi. "Effects of Acute Exposure to Nitrogen Dioxide on Primary Antibody Response," Arch. Environ. Health 36:114-119 (1981).

40. Fujimaki, H., F. Shimizu, and K. Kubota, "Effect of Subacute Exposure to NO_2 on Lymphocytes Required for Antibody Responses," Environ. Res. 29:280-286 (1982).

41. Holt, P. G., L. M. Findlay-Jones, D. Keast, and J. M. Papidimitrou, "Immunological Function in Mice Chronically Exposed to Nitrogen Oxides (NO_x)," Environ. Res. 19:154-162 (1979).

42. Fenters, J. D., J. P. Findlay, C. D. Port, R. Ehrlich, and D. L. Coffin. "Chronic Exposure to Nitrogen Dioxide: Immunologic, Physiologic and Pathologic Effect in Virus-Challenged Squirrel Monkeys," Arch. Environ. Health 27:85-89 (1973).

43. Fenters, J. D., R. Ehrlich, J. Spangler, and V. Tolkacz. "Serologic Response in Squirrel Monkeys Exposed to Nitrogen Dioxide and Influenza Virus," Am. Rev. Resp. Dis. 104:448-451 (1971).

44. Ehrlich, R., E. Silverstein, R. Maigetter, J. D. Fenters, and D. E. Gardner, "Immunologic Response in Vaccinated Mice During

Long-term Exposure to Nitrogen Dioxide," Environ. Res. 10:217-223 (1975).

45. Kuraitis, K. V., A. Richters, and R. P. Sherwin. "Spleen Changes in Animals Inhaling Ambient Levels of Nitrogen Dioxide," J. Toxicol. Environ. Health 7:851-859 (1981).

22.

Structure and Function of Airways in Experimental Chronic Nitrogen Dioxide Exposure

Jerome Kleinerman, Ronald E. Gordon, Michael P. C. Ip, and Alexander Collins

Mt. Sinai Medical Center, New York, NY

We are all aware of the acute injurious effect of NO_2 inhalation on the lung. Most of us, however, have been impressed with the acute injury affecting the pulmonary vascular bed and occurring with either a short or prolonged delay to produce pulmonary edema. In this chapter, I will focus on a different aspect of NO_2 injury to the lung and discuss the effect of NO_2 from the vantage point of the experimental pathologist. Our goal is to produce models of human disease, and, with this goal in mind, we have, for some time, studied the effect of relatively high concentrations of gaseous NO_2 on the lung. The concentrations I refer to are about 20 to 30 ppm, a level 100 times that present in the urban environment. Most animals so exposed not only survive but thrive in such an environment, and so our exploration of the effects of such exposures has continued for days, months, and even years. The effects of these exposures have taken numerous forms. We have examined certain aspects of the function of the isolated lungs and airways at the termination of the exposure. The static compliance, the total pulmonary resistance (TPR) of the airways, and the upstream airways resistance (UAR) were measured. In addition, lung volumes were measured and a histopathologic study of the airways by transmission and scanning electron microscopy was performed. In this discussion we have purposefully omitted description and analysis of changes in the alveolar parenchyma and airspaces.

Let us first discuss a study in which hamsters were exposed to NO_2 in concentrations of 2 to 4 ppm for 20 to 22 h daily 7 d/week for 14 months. They received simultaneous exposure to fly ash in total concentrations of 4.72 ± 184 mg/m^3 of air. The respirable fraction of this dust load was 0.85 mg/m^3. One of the most interesting observations resulting from this study was the survival of the exposed and control groups. This study had only three groups: (1) dust only, n = 36 at start; (2) dust-NO_2, n = 36 at start; and (3) ambient air, n = 26 at start (see Table 1).

Table 1.   Survival in NO_2 - Dust Studies (14 months)

Group 1, Dust only	19/36 - 52.8%
Group 2, Dust - NO_2	19/36 - 52.8%
Group 3, Ambient air	9/26 - 34.6%

This result suggests that there may be a biological advantage to exposure to NO_2 and dust, or to dust alone in the environment of the laboratory. The causes of death in all groups did not commonly include pulmonary complications. The major observed causes of death were facial abscesses, sinusitis and inflammatory lesions of the CNS, and gastrointestinal disturbances characterized by diarrhea and cachexia (wet tail). These lesions were observed in both the dust only and dust-NO_2-exposed hamsters in similar proportions, but were less frequent in the ambient air group.

Pulmonary function studies were performed at the time of the termination of the study, but within 24 h of the last NO_2 or dust exposure. These studies were performed on isolated lungs and airways within a plethysmograph by inflating the lungs with known volumes of air using a calibrated syringe and imposing pressure gradients across the inflated lungs to determine TPR and UAR. The static compliance values in the dust-NO_2 groups are listed in Table 2.

Table 2.   Pulmonary Function Studies Dust - NO_2 - (14 Months)[a]

	Dust Only	NO_2 - Dust	Ambient Air
N	15	15	9
Static compliance[b]	0.515 ± 0.045	0.505 ± 0.044	0.526 ± 0.097
Values for TPR and UAR			
TPR, 25% TV[b]	0.231 ± 0.039	0.218 ± 0.045	0.227 ± 0.076
UAR, 25% TV[b]	0.147 ± 0.042	0.127 ± 0.029	0.147 ± 0.051

[a]No significant differences are observed in both parameters between groups.

[b]No significant differences were observed among these groups.

Exposure to NO_2 in higher concentrations for periods of 12 months demonstrated similar results; namely, no significant changes in compliance, TPR, or UAR in hamsters. These exposures were to concentrations of 30 ± 5 ppm NO_2. These data (Table 3) also indicate a lack of any significant compromise in static compliance, TPR,

Table 3. Effects of NO_2 on Lung Function[a]

	Control			NO_2		
	1 month n = 2	6 month n = 6	12 month n = 6	1 month n = 4	6 month n = 6	12 month n = 6
Static compliance	0.40 ± 0.02	0.414 ± 0.007	0.465 ±0.001	0.342 ± 0.014	0.411 ± 0.018	0.502 ± 0.021
TPR, 25% TV	0.152 ± 0.001	0.147 ± 0.005	0.126 ± 0.003	0.184 ± 0.013	0.149 ± 0.008	0.140 ± 0.009
UAR, 25% TV	0.122 ± 0.002	0.115 ± 0.003	0.098 ± 0.003	0.129 ± 0.006	0.123 ± 0.007	0.110 ± 0.010
Lung volume, n = 6	3.93 ± 0.134	4.67 ± 0.34	5.72 ± 0.36	5.28 ± 0.12	5.74 ± 0.22	6.80 ± 0.38

[a]No significant differences between NO_2 and control groups in any parameters.

or UAR in hamsters exposed for varying times up to 12 months to NO_2 concentrations of 30 ± 5 ppm.

These data suggest that there is no measurable impairment of function resulting from prolonged exposure to NO_2 in moderate or relatively high concentrations for prolonged periods. There are, however, well defined and consistent morphological changes that develop within the airways in association with prolonged NO_2 exposure. These changes are most intense in the peripheral airways but can be seen even in the proximal and larger airways. They are more frequent and larger in hamsters exposed for periods of 5 months or longer, and are observed only during histologic evaluation, particularly when transmission and scanning electron microscopy are used. When the airways of hamsters chronically exposed to NO_2 are studied, a number of unusual degenerative and proliferative lesions are observed in the epithelial lining of the airways. These lesions are not the logical sequel to those observed as a result of acute or short-term exposures to NO_2. One of the changes is the progressive erosion and destruction of the cilia. This lesion is seen after several months of NO_2 exposure. It appears to increase in number and extent with increasing length of exposure. After cessation of exposure, the repair of the cilium occurs and normal appearing cilia are regenerated.

The injury, erosion, and repair of the surface cilia is in contrast to a bizarre observation within the deeper layer of the epithelial cells. In this locus, after several months of NO_2 exposure, intraepithelial cilia appear. These cilia protrude from cell surfaces within the epithelial cells as deep as the basal lamina and at each level of the epithelium. This disorientation of ciliary growth occurs at all levels of the airways but is more frequent in the smaller airways. While the surface ciliary injury tends to disappear after NO_2 exposure ceases, the intraepithelial cilia persist even 9 months after cessation of NO_2, and may remain in the intraepithelial locus permanently.

Within 5 months after the initiation of NO_2 exposure, nodules and papillomata develop in the bronchial and bronchiolar epithelium. These nodules may be sessile or papillary and are seen in every aspect of the bronchial epithelium, but appear more frequent in the bronchioles. The nodules are composed of nonciliated or indeterminate cells without definite evidence of secretory granules or centrioles. These cells were present in clusters, nodules, or ridges (without cilia) and would certainly compromise clearance of epithelial debris in these regions.

In many areas within the epithelial layers, cystic lesions are present. Many of these cysts contain epithelial debris, but others hold elongate, normal appearing cilia that arise from adjacent cells. The cysts are lined by cytoplasmic layers and bilamellar cell membranes that appear to be residual fragments of the affected cells. These epithelial ridges, nodules or papillomata, and the intraepithelial cyst do not disappear even 9 months after cessation of NO_2 exposure and, at least for the period of time we have studied them, appear to be permanent epithelial alterations of the airways.

These focal and persistent alterations of the bronchial and bronchiolar epithelium prompted the question concerning changes in the remainder of the epithelium exclusive of nodules and papillomata. To investigate this question, 1000 consecutive epithelial cells in the nonnodular epithelial regions were classified by cell types. Four basic cell types were defined: ciliated, secretory, indeterminate, and basal. The characteristics defining each of these cell types is self-explanatory. We observed that after 5 months of NO_2 exposure the proportion of secretory cells was significantly decreased in bronchiolar epithelium as compared with controls. This alteration in cell distribution persisted for the remainder of the NO_2 exposure and even thereafter for as long as our study lasted. This observation suggests that this alteration in cell distribution and maturation is not reversible and may be permanent. Whatever the changes in the potential for epithelial maturation in the bronchioles, we have never observed the development of any malignant or progressive epithelial proliferation in our NO_2 studies.

In summary, exposure of hamsters to chronic NO_2 inhalation of 30 ± 5 ppm for periods of 9 to 12 months produces no significant changes in resistance to the flow of air in small or large airways. These prolonged exposures do, however, produce significant injury to surface cilia and stimulate the growth of cilia between epithelial cells below the epithelial surface. Focal nodules, papillomata, and ridge-like projections develop on the epithelial surface after prolonged and relatively high NO_2 exposures. The nodules are composed predominantly of epithelial cells without distinctive differentiation, and they persist long after cessation of NO_2 exposure and may be permanent. There is a simultaneous alteration in the distribution of epithelial cells of the airways during prolonged NO_2 exposure. This change is characterized by a decrease in the proportion of secretory cells as compared to controls. The deviation from the normal persists for 9 months after NO_2 exposures have ceased and may be permanent. Despite this extensive epithelial proliferation and alterations in maturation, not a single evidence of malignant epithelial proliferation has been noted. We conclude that (1) our techniques to study functional changes in airflow are insensitive to the development of extensive proliferative bronchiolar epithelial changes; (2) prolonged NO_2 exposure does not appear to increase susceptibility to infection despite potential compromise of clearance mechanisms and severe airway injury; and (3) despite evidence of active epithelial proliferation, the prolonged exposure to NO_2 does not initiate malignant changes in the bronchiolar epithelium.

23.

Epidemiological Studies of Childhood Illness and Pulmonary Function Associated with Gas Stove Use

Sverre Vedal

University of British Columbia, Vancouver, Canada

INTRODUCTION

At issue in this paper is whether home gas stove use is harmful to respiratory health. Much data has been collected which bears directly or indirectly on the issue, either from epidemiological studies or experimental studies involving nitrogen dioxide (NO_2) exposure. The discussion will be limited to the epidemiological data, since only such data have been able to address the question from the standpoint of actual gas stove use in the home.

As an initial look at the epidemiological evidence, it is useful to tally the actual number of studies which have either found or not found adverse respiratory health effects associated with gas stove use. Details of all epidemiological studies which contribute pertinent data are listed in Table 1. Results from studies of both children and adults are outlined, but the analysis will concentrate only on the children's studies. Respiratory health effects are divided into respiratory symptoms and levels of pulmonary function. The study size was defined as the actual number of subjects who provided information for the gas stove analysis, which was often smaller than the size of the entire study. The participation rate was calculated as the percentage of those eligible for participation who actually participated. A tally of the results from Table 1 shows that whereas five studies of children have seen an association between either respiratory symptoms or illness and gas stove use, four have not; and, whereas two have seen an association between lower levels of pulmonary function and gas stove use, five have not. From this superficial look at the pertinent data it is apparent why there has been no consensus on the respiratory health effects of gas stoves.

Table 1. Epidemiological Studies of Gas Stove Use.

Study	Age range, size and participa- tion rate	Findings in gas stove homes		
		Respiratory symptom prevalence	Pulmonary function	Controlled factors
Children				
Melia et al. [1] cross-sectional	6-11 5758 (65%)	Increased respiratory illness in urban girls only.		SES*, urbaniza- tion,out- door SO_2 and smoke
Hosein and Bouhuys [2,3] cross-sectional	7-14 576 (90%)	No increase	Similar FVC,FEV_1 PEFR Vmax 50, Vmax 75	parental smoking, parental symptoms. SES not a confounder
Melia et al.[4] cross-sectional and longitudinal	5-10 4827 (62%)	Rate ratio↑ for respir. illness (1.25 in boys, 1.19 in girls). This ratio declined with decreasing age.		as for [1] and parent- al smoking
Florey et al.[5] cross-sectional and longitudinal	6-7 ∿480 (60%)	Increased respiratory illness assoc. with bedroom NO_2 but not kitchen NO_2.	Similar $FEV_{0.75}$ PEFR, MMEF	as for Melia [4]
Keller et al.[6] cross-sectional and longitudinal	6-15 770	No increased incidence of acute respir. illness	Similar FVC, $FEV_{0.75}$	crowding. SES uniform
Speizer et al.[7]	See Ware et al.[13]			
Hasselblad et al. [8] cross-section.	6-13 11,211 (43%)		Lower $FEV_{0.75}$ only in 9-13 year old girls	SES, parental smoking, ambient air pollution

*socioeconomic status

Table 1. Epidemiological Studies of Gas Stove Use. (Continued)

Dodge[9] longitudinal	8-12 419 (47%)	Increased cough	Similar change in FEV_1	Neither parental smoking nor SES were confonders
Ekwo et al.[10] cross-sectional	6-12 1139 (64%)	2.4 rate ratio for respiratory hospitalization before age 2		Parental smoking.
Schenker et al.[11] cross-sectional	5-14 4071 (93%)	No increase		SES, parental smoking, parental symptoms
Vedal et al.[12] cross-sectional	5-14 3175 (80%)		Similar FVC, FEV MMEF, Vmax 50 Vmax 75	SES, parental smoking
Ware et al.[13] cross-sectional	6-9 10,106	1.1 rate ratio for respir. illness before age 2 (p=0.14)	Lower FVC and FEV_1	SES, parental smoking, parental symptoms

Adults

Hosein and Bouhuys[2]	~ 6500	No increase	Similar FVC, FEV_1 PEFR, Vmax 50, Vmax 75	cigarette smoking. SES not a confounder
Keller et al.[6]	1182	No increase	Similar FVC and FEV_1	SES, cigarette smoking
Comstock et al.[14]	1724	Increased symptoms in nonsmoking men	Lower FEV_1 and FEV_1/FVC in nonsmoking men	SES, cigarette smoking

A more thorough assessment of these studies is needed for several reasons. It is possible that most of the studies finding an association between gas stoves and ill health are seriously flawed, or alternatively, that most of the studies showing no association are flawed. If either were the case, then a consensus might be possible. More likely, all of the studies will be found to have flaws. The burden will be to identify those flaws which might invalidate the study results, and then to determine whether identifying the studies with critical flaws will help in reaching a consensus. If no such consensus can be reached, it may be possible through this process to determine features of the study designs which made the results inconclusive, and to recommend designs for future studies.

EVALUATION CRITERIA

Of primary concern in evaluating an epidemiological study is to determine whether the study results were due to bias or could potentially have been the result of bias. Bias can result in the creation of an association when in fact one is not present. It can also cause the obliteration of a real association. A useful classification of biases which are prevalent in epidemiological studies distinguishes three types: selection bias, information bias, and confounding bias [15]. Selection bias occurs when the groups to be compared are selected into the study according to different criteria. Information bias is present when the data collected from the groups to be compared differ in quality. Confounding bias, probably the most common type of bias present in epidemiological studies, occurs when an extraneous factor varies in the comparison groups and is itself associated with the outcome to be compared in the groups.

Biases are not the only concerns however. It is important to know whether studies are powerful enough, in a statistical sense, to detect any differences between the comparison groups, if differences in fact exist. Power is a measure of the ability of a study to detect true differences and is partly a function of the number of subjects included in the study and the size of the difference which is present. It is also necessary that the assignment of individuals into comparison groups results in the groups being distinct with respect to the exposure. If the comparison groups cannot be characterized by truly distinct exposures, then misclassification is said to exist. Random misclassification results in underestimation of the association and a resultant inability to identify differences.

APPLICATION OF CRITERIA

Misclassification of Exposure

A major limitation of all studies which have attempted to determine whether there are respiratory health effects associated with gas stoves is that simple classification of exposure by either gas stove use or no gas stove use may result in a great deal of misclassification of the exposure of primary interest, presumably NO_2. There are large overlaps in NO_2 concentrations measured in homes using gas stoves and homes where no gas stove is used, although mean concentrations of NO_2 in gas stove homes as a group are dramatically higher than in homes with no gas stove (Table 2).

Table 2. Kitchen NO_2 levels in homes with gas stoves and electric stoves

	NO_2 Concentration (ppb)	
	Electric	Gas
mean	18.0	112.2
range	6-188	5-317

(Goldstein et al. [16])

This overlap is due to there being other sources of NO_2 in the home besides gas stoves, including gas heaters, pilot lights, and outdoor NO_2, and to differences in ventilation and amount of stove use [17]. The result of the overlap is that some subjects in homes without a gas stove will have exposure to higher concentrations of NO_2 than some subjects in gas stove homes. These individuals will therefore be misclassified into the wrong exposure group when exposure is defined simply by gas or no gas stove use.

Another source of misclassification results from the use of children as the study subjects. All of the studies which address the question of health effects of gas stove use were not designed primarily to address that question. Instead, the study of gas stove effects has been incorporated into studies on childhood respiratory illness in general or studies designed to evaluate the effects of ambient air pollution. Children are sometimes ideal study subjects because the investigators need not contend with the extraneous effects of cigarette smoking or occupational exposures and, if exposure is defined by residence, as it often is in air pollution studies, because children tend not to have changed area of residence often. Children may also represent a more susceptible group. However, children's exposure to gas stoves can vary by the amount of time they spend indoors and the amount of time spent in the kitchen. This is then another source of variability in exposure which is not well reflected by the dichotomy, gas stove or no gas stove use.

As was noted above, the effect of misclassification will be to underestimate the true association, resulting in a tendency to find no significant association. The situation is not so clear when multiple regression techniques are used in an attempt to control for the effects of extraneous factors which might distort the association of interest. In these instances, because of the complex correlations between the several factors included in the regressions, and because of less than perfect reliability in measurement of the covariates, misclassification may result in either underestimation or overestimation of the true degree of association [18,19].

Florey et al.[5] attempted to obtain a more direct measure of exposure in order to minimize misclassification. NO_2 was measured in kitchens and in a 25% sample of bedrooms. Respiratory illness prevalence increased with increases in bedroom NO_2 concentration (p=0.10) as seen in Table 3, but not with increases in kitchen NO_2 concentration. Keller et al.[6] also measured indoor NO_2 on

Table 3.   Prevalence (%) of respiratory illness
in children from homes with gas stoves.

	Bedroom NO_2 levels (ppb)		
	0-	20-	40-
Prevalence (%)	44	59	71
Number of children	48	34	21

a small random sample but the monitoring location was not specified. No differences in respiratory illness rates were found in homes with high and low concentrations of NO_2.

Power

Since the size of the effects of gas stove use on respiratory symptoms and levels of pulmonary function have been small in those studies in which effects have been found, the sample sizes required to identify these effects need to be large. Assuming that a 20% increase from a baseline prevalence of 20% for respiratory symptoms is to be identified, and assuming a significance level of 0.05 and a power of 0.80 (an 80% probability of detecting the assumed 20% difference if such a difference exists), each comparison group would need to be composed of approximately 1,300 children. If the baseline symptom prevalence were 10%, the number required would be 2,800. For level of FEV_1, for the same levels of significance and power assumed above, and assuming a standard deviation of 500 ml and a desired detectable difference of 25 ml (1% of 2.5 L), then approximately 4,900 would be required for each group. This latter is an overestimate of required sample size because the variability of pulmonary function measurements is decreased by controlling for differences due to height, age, and sex. Therefore, assuming a standard deviation for FEV_1 of 10% of predicted (with "predicted" calculated from a prediction equation controlling for height, age, and sex), and a difference to be detected of 1%, then approximately only 1230 would be needed for each group. With the decrease in precision of estimates which often results from the control of extraneous factors, even these calculated sample sizes may be underestimates.

Only the studies of Ware et al.[13] from the Six Cities Study, Hasselblad et al.[8],and Vedal et al. [12] clearly included enough subjects to avoid overlooking an association, and all observed an association between gas stove use and lower levels of pulmonary function. However, after control for SES, the association in the studies of Ware et al. and Vedal et al. was no longer statistically significant. Studies which clearly had an inadequate number of children include the studies of Florey et al.[5], Keller et al.[6], Dodge [9], and Ekwo et al.[10]. Nevertheless, the study by Ekwo et al.[10] found an increased risk of respiratory illness hospitalization before age 2. This risk was detected because the risk of hospitalization in gas stove homes was 2.4 times that in no gas stove homes.

Florey et al.[5] suggested a dose response relationship between bedroom NO_2 concentration and respiratory symptoms (Table 3) in a study with inadequate numbers of children according to the previous calculations. However, the calculations above were based on expected increased risk based on exposure defined by gas stove use. With less exposure misclassification, a result of having actual NO_2 concentration measurements, the expected increase in risk due to exposure may be much higher. Much smaller sample sizes would then be required to detect increased risk.

Selection Bias

Most of the epidemiological studies on gas stoves have used variations of a cross-sectional design, where subjects were selected from the general population, and had measurements of respiratory outcomes performed at one point in time. It seems difficult to imagine, with this design, that subjects in the group where a gas stove was used in the home for cooking were selected by different criteria than those in the group from homes where no gas stove was used. However, a way in which a selection bias might occur in this situation is if the number of subjects actually participating in the study is significantly less than that selected for participation and if nonparticipation in one group is associated with either respiratory symptoms or with level of pulmonary function.

For example, if children from homes with gas stoves who had more symptoms or lower levels of pulmonary function for some reason were less likely to take part in the study, then one might incorrectly conclude that gas stove exposure was not associated with either respiratory symptoms or lower levels of pulmonary function. Since studies were often performed at schools, school absences due to illness would result in nonparticipation. Inability to perform adequate pulmonary function tests might also be related to illness [20].

Participation rates varied among the studies. Studies with "good" participation rates are not susceptible to significant bias due to selection, whereas studies with "poor" participation rates, although not necessarily biased, are at least susceptible to it. If good participation is arbitrarily defined as 80% or better, then the Six Cities Study [7,13] and the Chestnut Ridge Study [11,12] are the only studies which are not susceptible to selection bias (Table 1). In neither of these studies was an association between gas stove use and respiratory symptoms found. Both found lower levels of pulmonary function in gas stove homes which were no longer detectable after adjustment for SES.

It is not possible, for the most part, to determine whether non-participants were less healthy than participants, since they would have needed to have participated in the study for this to be determined. For at least two studies, however [2,10], an attempt was made to sample nonparticipants. Hosein et al. [2,21] found that adults in nonparticipating families tended to smoke more than those participating but did not consistently have more respiratory symptoms. Ekwo et al. [10] found no differences in either adult smoking habits or children's respiratory symptoms in participating homes and nonparticipating homes. Although no information was available from either study on relative gas stove use in nonparticipants, it is unlikely that significant selection bias exists in studies even with relatively poor participation rates.

Information bias

Bias could exist in these studies if subjects in one group were more likely to report symptoms on the questionnaire or if the performance or measurement of pulmonary function varied in the exposure groups. The latter is unlikely unless gas stove use was unusually common or rare in some study regions and different equipment or techniques were used in these regions. However, differential reporting of symptoms is more complicated. It has been shown that symptomatic adults tend to report more symptoms in their children [11]. If adults in gas stove homes are more symptomatic, then increased childhood illness rates might be reported. In this case, gas stoves would be correctly implicated, but in the wrong population. However, the presence of a gas stove in the home may be associated with cigarette smoking in the home [4,9], in which case, even if adults were not more symptomatic as a result of a gas stove, adults symptomatic from cigarette smoking might report more symptoms in their children. Such differential reporting will result in a confounding bias due to cigarette smoking, with incorrect implication of gas stoves.

Of the many respiratory symptoms and illnesses which have been compared, the only partly consistent finding has been that children from gas stove homes were more likely to have been either hospitalized for or had a severe respiratory illness before the age of two years (Table 1). Although this may be an important observation, especially since hospitalizations at least should be more objectively reported than symptoms, recall of events occurring three or more years in the past is especially sensitive to bias.

Confounding bias

An association between gas stove use and either respiratory symptoms or levels of pulmonary function will be created if some extraneous factor or factors associated with gas stove use is also associated with the respiratory outcomes. Parental cigarette smoking may be one such extraneous factor as noted above because it can be associated with both gas stove use and with increased respiratory symptoms and lower levels of pulmonary function in children. Socioeconomic status (SES) has also been implicated as another factor which might confound the association, with children from a low socioeconomic background being more likely to live in homes with gas stoves and to have more respiratory symptoms and worse lung function. Since SES obviously cannot be directly associated with any health outcome, it must be a marker for other factors which might cause respiratory illness, such as crowding, for example. Other potential confounding factors include indoor temperature and humidity and outdoor air pollution. Florey et al.[5] have argued that humidity may be higher in gas stove homes due to water produced from the burning of gas and because gas stoves were often used to help dry clothing. Also, with increased humidity and the resultant condensation, windows might be opened, causing lower room temperatures. Cigarette smoking by the children themselves has been considered as a possible intervening variable. If children tend to smoke at an earlier age if their parents smoke, and if parental smoking is associated with gas stoves, then the children's cigarette smoking may itself cause the respiratory outcomes. However, it is unlikely that the number of children smokers is large enough or the effects of such smoking great enough to be of importance.

In the studies cited in Table 1, various methods were used to attempt to control for the effect of some of the above mentioned factors. Most commonly, control was attempted through the use of multiple regression models in which the potentially confounding factors were considered as covariates. Less commonly, the analyses were stratified by levels of one or two of the factors, or it was argued that the population sample was homogeneous with respect to the important factors and therefore did not require control. For example, Keller et al. [6] had a population of only upper middle class families and therefore felt no need to control for SES. Dodge [9] found that SES in his population in the southwestern U.S. was not associated with either respiratory symptoms or level of pulmonary function and therefore did not need to be controlled. Florey et al.[5] also found no association between SES and respiratory symptoms. In other populations, SES has been associated with both respiratory symptoms and lung function and has also been associated with gas stove use [13]. Most studies controlled for SES (Table 1). If those studies which did not control for it are ignored, there is still no consensus among the remaining studies as to the presence of health effects with gas stove use.

It has been suggested that control for SES may not be appropriate in these studies because SES functions as a surrogate for gas stove use, and control of a surrogate for gas stoves would have the effect of obscuring any effect of gas stoves [13]. In two studies where gas stove use was associated with slightly lower (less than a 1% decrease) levels of pulmonary function [12,13], adjustment for SES decreased the association, causing it to be no longer statistically significant. However, it is clear that SES, no matter what it represents, is a predictor of the respiratory outcomes in those populations where adjustment for SES is considered. Because it is also associated with gas stove exposure, but is more than a surrogate, it needs to be considered a likely confounder.

Parental cigarette smoking was controlled in most of the studies. Dodge [9] did not control for parental smoking, but parental smoking was not predictive of the respiratory outcomes in that population. Parental smoking was also not controlled in the study of Keller et al.[6] and the first study of Melia et al.[1]. When these studies are ignored, there is still no good consensus.

Outdoor air pollution was controlled in the British studies [1,4,5] and by Hasselblad et al. [8], but did not alter the results. No studies have attempted to control for any effects of home humidity or temperature.

CONCLUSIONS AND RECOMMENDATIONS

Initial review of the studies which have analyzed the association between gas stove use and respiratory health effects in children suggested a lack of consensus. A more detailed review showed that most of these studies suffered from a misclassification of the exposure, with NO_2 concentrations overlapping in gas stove and no gas stove homes. Such misclassification results in underestimation of associations and a consequent increase in statistical power requirements. If studies which were too small to identify statistically significant associations are ignored, there is still no consensus as to the association between gas stove use and respiratory symptoms, but there is some consensus that gas stove use is associated with slightly lower levels of pulmonary function. Most biases which might result in spurious associations are unlikely to have contributed significantly to the finding. However, control for differences in socioeconomic status (SES), as one such potentially biasing factor, critically reduced the degreee of association between gas stove use and level of pulmonary function in 2 studies [12,13].

What recommendations can be made to improve the likelihood of obtaining conclusive results from future epidemiologic studies? Obviously improvement in the classification of study subjects in terms of exposure is needed. Stationary indoor measurements of NO_2 concentrations improve exposure classification, as would personal monitoring. Both increase the expense and complexity of studies, although these would be in part offset by the smaller sample sizes required. As an alternative to directly measuring NO_2 exposure, it might be possible to predict individual exposure based on activity patterns, cooking patterns, and home characteristics. [22]. Although this would improve on the dichotomous exposure classification of gas stove versus no gas stove, collection of the data required to apply an adequate prediction model may prove to be as complicated as actually obtaining the NO_2 measurements directly.

Added to the difficulty of classifying exposure is determining the nature of the exposure of interest. For example, high peak concentrations rather than mean concentrations may be the only important exposure. Also, concentrations of combustion products other than NO_2 might be the critical exposure. In either case, direct measurements of NO_2 might still result in significant misclassification.

The role of SES as a confounding factor in the association between gas stove use and level of pulmonary function is controversial. As currently measured, there is undoubtedly significant unreliability in the measurement of SES, and this unreliability is compounded when this measure of SES is in turn used as a measure of that unknown feature of SES which is a predictor of level of pulmonary function. This measurement error can result in serious errors in estimation of the association of interest [19]. A partial solution would be to restrict the study population to a relatively homogeneous group with respect to SES, such as was done by Keller et al. [6], who studied a population of upper middle class families.

The only partly consistent finding with respect to the questionnaire outcomes in the studies reviewed was the association between gas stove use and an increased prevalence of severe, or hospitalization requiring, chest illness before age 2. The susceptibility of this association to an information bias due to varying recall of past events was noted. It would be of interest, therefore, to see a prospective study of severe chest illness in this age group related to indoor NO_2 concentrations to minimize any effect of this potential information bias. This age group might also be expected to have more distinct exposures to NO_2 in gas stove and no gas stove homes than older children because of their more regular confinement indoors.

Finally, if one is primarily interested in studying the effects of home gas stove use, then a population which has a large range of exposure should ideally be used. Housewives are an obvious group. Difficulties with use of housewives as opposed to children are that they have been exposed to more extraneous influences on respiratory health, including cigarette smoking, and may not be as susceptible to gas stove effects. Comstock et al. [14], in a study of adults, found an association between gas stove use and respiratory health only in nonsmoking males, and argued that females might have a natural resistance. Also, the effect of gas stoves in children diminished as the children aged in the study of Melia et al. [4]. Limitation of a study to nonsmoking housewives of uniform SES would, it might be argued, also limit the generalizability of any findings. Given the lack of consensus as to the respiratory health effects of gas stoves, a conclusive study on a population in which an effect should be detectable, if it exists, should take precedence over any concern over generalizability.

REFERENCES

[1]   Melia, R.J.W., Florey, C.du V., Altman, D.G., Swan, A.V.
      Association between gas cooking and respiratory disease in
      children. Br. Med. J. 2: 149-152 (1977).

[2]   Hosein, H.R. and Bouhuys, A. Possible environmental hazards of
      cooking (letter). Br. Med. J. 1: 125 (1979).

[3]   Schilling, R.S.F., Letai, A.D., Hui, S.L. et al. Lung function,
      respiratory disease, and smoking in families. Am. J. Epidemiol.
      106 (2): 274-283 (1977).

[4]   Melia, R.J.W., Florey, C. du V., Chinn, S. The relation between
      respiratory illness in primary school children and the use of
      gas for cooking.I.Results from a national survey. Int. J.
      Epidemiol. 8 (4): 333-338 (1979).

[5]   Florey, C. du V., Melia, R.J.W., Chinn, S. et al. The relation
      between respiratory illness in primary school children and the
      use of gas for cooking. III. Nitrogen dioxide, respiratory
      illness and lung infection. Int. J. Epidemiol. 8 (4): 347-353
      (1979).

[6]   Keller, M.D., Lanese, R.R., Mitchell, R.I., Cote, R.W.
      Respiratory illness in households using gas and electricity for
      cooking. 1. Survey of incidence. Environ. Research 19: 495-503
      (1979).

[7]   Speizer, F.E., Ferris, B.G., Jr., Bishop, Y.M.M., Spengler, J.
      Respiratory disease rates and pulmonary function in children
      assoicated with NO_2 exposure. Am. Rev. Respir. Dis. 121:3-10
      (1980).

[8]   Hasselblad, V., Humble, C.G., Graham, M.G., Anderson, H.S.
      Indoor environmental determinants of lung function in children.
      Am. Rev. Respir. Dis. 123:479-485 (1981).

[9]   Dodge, R. The effects of indoor pollution on Arizona children.
      Arch. Environ. Health 37(3): 151-155 (1982).

[10] Ekwo, E.E., Weingerger, M.M., Lachenbruch, P.A., Huntley, W.H.
      Relationship of parental smoking and gas cooking to respiratory
      disease in children. Chest 84(6): 662-668 (1983).

[11] Schenker, M.B., Samet, J.M., Speizer, F.E. Risk factors for
      childhood respiratory disease: the effect of host factors and
      home environmental exposures. Am. Rev. Respir. Dis. 128:
      1038-1043 (1983).

[12] Vedal, S., Schenker, M.B., Samet, J.M., Speizer, F.E. Risk
      factors for childhood respiratory disease: analysis of
      pulmonary function. Am. Rev. Respir. Dis. 130: 187-192 (1984).

[13] Ware, J.H., Dockery, D.W., Spiro, A. et al. Passive smoking,
      gas cooking, and respiratory health of children living in six
      cities. Am. Rev. Respir. Dis. 129:366-374 (1984).

[14] Comstock, G.W., Meyer, M.B., Helsing, K.J., Tockman, M.S.
      Respiratory effects of household exposure to tobacco smoke and
      gas cooking. Am. Rev. Respir. Dis. 124: 143-148 (1981).

[15] Kleinbaum, D.G., Kupper, L.L., Morgenstern, H. Epidemiologic
      research: principles and quantitative methods. (Belmont,
      California: Lifetime Learning Publications, 1982). pg.190-191.

[16] Goldstein, B.D., Melia, R.J.W., Chinn, S. et al. The relation
      between respiratory illness in primary school children and the
      use of gas for cooking. II. Factors affecting nitrogen dioxide
      levels in the home. Int. J. Epidemiol. 8 (4): 339-345 (1979).

[17] Spengler, J.D., Duffy, C.P., Letz, R. et al. Nitrogen dioxide
      inside and outside 137 homes and implications for ambient air
      quality standards and health effects research. Environ. Sci.
      Technol. 17 (3): 164-168 (1983).

[18] Cochran, W.G. Errors of measurement in statistics.
      Technometrics 10 (4): 637-666 (1968).

[19] Kupper, L.L., Effects of the use of unreliable surrogate
      variables on the valididy of epidemiologic research studies.
      Am. J. Epidemiol. 120 (4): 643-648 (1984).

[20] Eisen, E.A., Wegman D.H., Louis, T.A. Effects of selection in
      a prospective study of forced expiratory volume in Vermont
      granite workers. Am. Rev. Respir. Dis. 128: 587-591 (1983).

[21] Mitchell, C.A., Schilling, K.S.F., Bouhuys, A. Community
      studies of lung disease in Connecticut: organization and
      methods. Am. J. Epidemiol. 103 (2): 212-225 (1976).

[22] Sexton, K., Letz, R.,Spengler, J.D. Estimating human exposure
      to nitrogen dioxide: an indoor/outdoor modeling approach.
      Environ. Research 32: 151-166 (1983).

24.

Significance of Childhood Lower Respiratory Infections

Frederick W. Henderson and Alan M. Collier

Department of Pediatrics, University of North Carolina School of Medicine, Chapel Hill, North Carolina

INTRODUCTION

Identification of the determinants of respiratory health in childhood and assessment of their relative importance as risk factors are complex problems. Knowledge must be available in several domains to be confident that risk estimates for individual parameters are not confounded. More important factors include:

1) degree of allergic immune responsiveness,
2) degree of airway reactivity,
3) lower respiratory infection experience,
4) level of passive exposure to tobacco smoke and products of fossil fuel combustion and,
5) tendency to mucus hypersecretion.

Each of these factors is related to long-term respiratory health, but they are also interrelated. The purpose of this review is to characterize the parameter lower respiratory infection experience. Clinical evidence of lower respiratory tract disease (LRD) occurs in a subset of individuals during the course of acute respiratory infections; the risk of lower respiratory involvement is highest in young children. In large part, clinical disease is a direct reflection of infection of cells lining the trachea, larger bronchi, intrapulmonary airways, and alveoli. However, the clinical pathophysiology of respiratory infection is modulated by patterns of host immunologic response to infectious and non-infectious antigens, host physiologic response to airway insults, and by antecedent or concurrent exposure to inhaled irritant air pollutants. In fact, the health effects of exposure to indoor air pollutants are most readily reflected in children as an altered risk for acute LRD and as an altered prevalence of lower respiratory symptoms (1). Herein, we review

the clinical patterns of lower respiratory infection (LRI)
observed in childhood and describe the etiology and epidemiology
of childhood LRI. The importance of age, sex, and etiology as
determinants of LRI epidemiology will be described. Patterns of
recurrent LRI among children who experience LRI in the first 2
years of life will also be presented. Finally, a brief summary of
studies linking childhood LRI to later patterns of pulmonary
function will be provided together with a short review of studies
relating the occurrence of acute LRI in infancy to exposure to air
contaminants in the home.

MATERIALS AND METHODS

The data to be described were collected in two populations of
children studied in Chapel Hill, NC. From July 1, 1964, to
June 30, 1975, all children seeking care for LRI at the only
pediatric practice in Chapel Hill were recruited for study (2-5).
When parents consented, a description of the clinical illness was
recorded, epidemiologic data were collected, and upper respiratory
secretions were cultured for viruses and mycoplasmas. Data from a
16-year study of children followed longitudinally from early
infancy in the day care program of the Frank Porter Graham Child
Development Center provide a second source of information on
childhood LRI (6, 7). Children were admitted to this program at
6-12 weeks of age and followed continuously through elementary
school. During the pre-school years, they attended a research day
care center 5 days a week. Our group provided continuous medical
care for the children. Clinical features of all respiratory
infections were recorded, and viral and mycoplasmal cultures of
upper respiratory secretions performed at the onset of illness.

RESULTS

Clinical Syndromes of LRI

There are four clinical syndromes of LRI in children, three of
which are quite distinctive. Table 1 shows the syndrome
definitions in common use in pediatrics and the ones used in our
work. Croup is an illness in which children with signs of acute
respiratory infection develop evidence of upper tracheal and
perilaryngeal disease with partial airflow obstruction. Clinical
features include hoarseness, barking-seal cough, and inspiratory
stridor. Wheezing-associated respiratory infections (WARI) are
readily identified clinically. As the name implies, this complex

TABLE 1

CLINICAL SYNDROMES OF CHILDHOOD
LOWER RESPIRATORY INFECTIONS

Syndrome	Signs
Croup:	Hoarseness, Croupy Cough, Stridor
Tracheobronchitis:	Cough, Coarse Rhonchi
Wheezing Associated Respiratory Infections:	Wheezing (Including Bronchiolitis)
Pneumonia:	Rales, Evidence of Pulmonary Consolidation

includes infant WARI - frequently termed bronchiolitis and
post-infancy WARI. Some infants with WARI have hyperreactive
airways and could be considered as manifesting first episodes of
asthma. This subset of infants cannot be reliably identified at
the time of their first wheezing episode. With increasing age,
the likelihood that WARI is occurring in a host with altered
airway reactivity increases, but at no age can one presume that
all infection-associated wheezing is attributable to hyperreactive
airways. There are children who are not prone to recurrent
wheezing who express wheezing as a manifestation of illness with a
single respiratory infection after infancy. In our studies, the
diagnosis of pneumonia was based on the findings of rales or
clinical evidence of pulmonary consolidation. X-rays were usually
not obtained. The fourth syndrome of LRI, tracheobronchitis, is
less distinctive clinically. Its diagnosis relies on the
occurrence of cough and coarse rhonchi indicative of mucus
hypersecretion in larger airways. Unfortunately, cough occurs in
40% of children with predominantly upper respiratory illness
symptoms. Deciding when URI with cough crosses the boundary to
tracheobronchitis is a clinical one. In older children and
adults, this is the predominant syndrome of lower respiratory
involvement in patients with acute respiratory infections. We

think tracheobronchitis is a clinical entity in young children as well, but we recognize that various observers would differ in their assignment of this diagnosis.

Age and LRI Incidence

    The incidence of acute respiratory illness is relatively stable over the first 5 years of life at approximately 6-8 respiratory illnesses per child year. The occurrence of lower respiratory disease, however, is age dependent; its incidence is related inversely to age. In the pediatric practice study, the incidence of LRI was highest in the first year of life at an attack rate of 18-26 illnesses per 100 children per year. By age 5, the incidence had fallen to 10-13 LRI/100 children/year, and in school children the incidence ranged from 4-7 LRI/100 children/year. The attack rate of LRI was higher in boys than in girls, with greater sex related differences in LRI incidence in infancy and early childhood than in later childhood. If one accepts 7 respiratory illnesses per child per year as a reasonable estimate of total respiratory illness incidence, then our information from the pediatric practice suggests that 2-3% of respiratory illnesses include a sufficient degree of lower respiratory involvement to elicit physician visits. Our data from the day care population, where illnesses were monitored daily, show that children experienced an average of 1 lower respiratory illness per child per year in the first year of life and 0.4 to 0.5 LRI's per child per year over the next four years of life. In this study, lower respiratory involvement was documented in 8% of respiratory illnesses in pre-school children. The importance of sex as a determinant of LRI incidence varies for the different LRI syndromes, and this point will be clarified later.

Age and Syndrome of LRI

    Age is not only an important factor determining LRI incidence, it is an important determinant of clinical expression of illness. Data from the pediatric practice study showed that in infants, wheezing was observed in one-half of LRI studied. After age 3, the proportion of LRI with wheezing declined to 20% and remained stable in that range through the early years of high school. Tracheobronchitis was diagnosed in one-fourth of infants with LRI. This syndrome emerged as the predominant clinical pattern of LRI in school age children, as it is in adults. Croup accounted for 20% of LRI in 1-3 year old children; the maximum contribution of the pneumonia syndrome (28-32%) was observed in the 2-5 year age group.

Sex and Syndrome of LRI

As described earlier, boys experience LRI more frequently
than girls and this difference is most apparent in pre-school
children. Sex related differences in LRI incidence are better
understood when the LRI syndromes are examined separately. Croup
was the syndrome whose incidence was most strongly influenced by
sex of the affected child. Averaged across the pre-school years,
the male:female incidence ratio for croup was 1.5, while the
male:female incidence ratio in pre-school children was 1.3 for
each of the other LRI syndromes. The maximum differences in
illness incidence by sex were observed in 6-12 month old children
with croup and WARI; the attack rates of these illnesses were
1.7-1.8 times higher in boys in this age bracket. In school
children, croup and WARI continued to occur more frequently in
males, while the attack rates of tracheobronchitis and pneumonia
were equal for the sexes.

Etiology of LRI

Viruses and bacteria are the principal classes of organisms
responsible for lower respiratory tract infections in previously
healthy individuals. Our studies concern the etiologic roles of
respiratory viruses and a single bacterium, Mycoplasma pneumoniae
in childhood LRI. Investigators have had difficulty defining the
importance of S. pneumoniae and H. influenzae as etiologic agents
of outpatient LRD since isolation of these bacteria from blood,
pleural fluid or pulmonary tissue is usually required to establish
their etiologic role in LRD patients with certainty. We shall not
discuss these agents further.

A respiratory virus or M. pneumoniae was isolated from 26% of
LRI patients in the pediatric practice study. Respiratory
syncytial virus, parainfluenza virus types 1 and 3 and M.
pneumoniae were recovered most frequently; these agents accounted
for 70% of all isolates. Our data probably underestimate the
importance of the influenza A viruses as causes of childhood LRI.
These viruses vary in the ease with which they can be recovered in
tissue culture; had embryonated eggs been employed in our studies,
the influenza A viruses would have been isolated more frequently.
In studies from Houston, it appears that influenza A viruses and
respiratory syncytial virus may be nearly equivalent in their
capacity to elicit LRD severe enough to warrant hospitalization in
infants (8), in years with major influenza epidemics.

Each of the principal etiologic agents of childhood LRI has a
unique biology that is reflected epidemiologically in:

1) their association with certain syndromes,
2) the age of host when infection is linked closely to LRI and,
3) the season of their occurrence.

The tendency for particular agents to be associated with specific syndromes is most dramatic for croup and infant WARI. Among 360 croup patients from whom an isolate was recovered, 50% of the cases were associated with parainfluenza virus type 1 infection, 18% were attributed to infection with parainfluenza virus type 3, and 10% related to respiratory syncytical virus infection. In 396 patients with WARI of identified etiology, 32% of isolates were respiratory syncytial virus, 10% parainfluenza virus type 1, and 10% parainfluenza virus type 3. M. pneumoniae accounted for 30% of isolates from pneumonia patients while respiratory syncytial virus and the parainfluenza viruses accounted for 25% and 20% of isolates in this patient population. One can also study this relationship by examining the patterns of LRI observed in children known to have infections with different agents. Forty percent of 343 children with respiratory syncytial virus infections had WARI; pneumonia and tracheobronchitis were the next most common syndromes of LRI in patients with respiratory syncytical virus infection. Sixty percent of 304 children with parainfluenza virus type 1 LRI had croup. No LRI syndrome was predominant in patients with parainfluenza virus type 3 LRI. M. pneumoniae infections most often presented as pneumonia or tracheobronchitis, while influenza virus infections were associated closely with tracheobronchitis. The various agents of childhood LRI manifest clinical tropism for different regions of the lower respiratory tract. We presume that the clinical tropism is a close correlate of cellular and tissue tropism of the infecting agent, and we consider this a pathogenetically important biologic characteristic of each agent.

Specific agents tended to cause LRI in children of different ages. Respiratory syncytial virus, the parainfluenza viruses, and the adenoviruses caused lower respiratory disease primarily in infants and preschool children. Influenza virus LRI's were identified throughout childhood; while M. pneumoniae LRI was observed predominantly in school aged children. These varying patterns of LRI incidence with age, all have immunologic bases. All children are infected with respiratory syncytial virus and the parainfluenza viruses in early childhood and usually have additional infections with these viruses during the preschool and early school years. Since antigenic variation in these agents is minimal, children become partially immune to infection with these viruses and resistant to illness in association with reinfection. Adenovirus infections also occur in infancy and early childhood and probably elicit long-lasting type specific immunity. Antigenic variation is a major factor influencing a population's susceptibility to influenza virus infections. Persons of any age

may become infected with these viruses if the degree of antigenic
change in circulating viruses is sufficient. The magnitude of the
immunologic response to infection may be a determinant of illness
severity in children with M. pneumoniae infections of the
respiratory tract. Older children may have an enhanced
capacity to make the vigorous immune responses which appear to be
required for manifesting clinical disease.

     Each of the agents of LRI tends to cause infections in
different seasons of the year. Furthermore, since different
agents are associated more closely with different LRI syndromes,
the different clinical patterns of LRI have a distinctive seasonal
occurrence. As examples, the incidence of bronchiolitis is
highest from December through March, the months when respiratory
syncytial virus epidemics occur most commonly. The incidence of
croup peaks in September through November corresponding to the
fall peak in incidence of parainfluenza virus type 1 infections.
Other syndrome-outbreak associations are less dramatic, but there
remain important seasonal aspects to the occurrence of the other
agents of LRI. Parainfluenza virus type 3 infections may occur as
outbreaks in fall or spring or they may behave in a more endemic
pattern. M. pneumoniae infections cluster in the fall of the year
and continue into early winter. These infections may also become
endemic in a community. Influenza virus infections occur as
outbreaks usually between December and April in non-pandemic
years.

Recurrent LRI

     Among children who experience lower respiratory infections in
early life, a subset proceeds to manifest repeated bouts of LRD in
later childhood. We compared information from the pediatric
practice study and the day care study on the problem of recurrent
LRI in the first 6 years of life, among children who experienced
at least one LRI before age 2 years. As indicated previously,
children in the day care study were monitored closely during all
respiratory illnesses and the occurrence of LRI signs recorded.
The day care data are influenced by the inclusion of milder forms
of LRD. 1,314 children with LRI in the first 2 years of life were
identified in the pediatric practice. The size of the population
at risk is not known with precision; however, we estimate that
approximately 30–35% of children made physician visits for LRD in
the first 2 years of life. In the day care study, 65% of 173
children followed longitudinally experienced LRI in the first 2
years of life. Among those with LRI in the first 2 years of life:
61% of those in pediatric practice and 26% of those in day care
experienced a single LRI in the first 6 years of life.
Forty-eight percent of LRI patients in pediatric practice

experienced at least one WARI while wheezing was observed during LRI in 68% of day care children. Eighteen percent of practice LRI patients made 3 or more visits for LRD in the first 6 years of life compared to 51% of day care children with any LRD. At least one recurrence of wheezing illness occurred in 21% of practice patients who ever wheezed and in 47% of day care children with a wheezing associated respiratory illness. Three or more total LRI's occurred more frequently in children who had at least one wheezing illness in their histories than in children who were free of wheezing illness.

In one sense, the day care data may give a more complete picture of the extent of LRI involvement during the course of common respiratory infections since illnesses were monitored so thoroughly. However, the day care information probably reflects, in addition, enhanced exposure of children to the agents of LRD in the day care setting. We think knowledge of the origins of recurrent LRI will prove particularly relevant, for we consider it likely that chronic respiratory symptoms and decrements in pulmonary function will be most prevalent in this subset of children who experience early childhood LRI.

DISCUSSION

        The influence of lower respiratory tract infections in early
life on the subsequent growth, development, and function of the
lung is not completely understood. Certain viral lower
respiratory tract infections, usually occurring in infancy, can
cause severe lung injury with long-lasting pathophysiologic
changes demonstrable after the event. Obliterative adenovirus
bronchiolitis is an established example of this type (9). These
infections are uncommon, and the acute illness is always severe,
frequently life-threatening. Respiratory syncytial virus
infection is the single most important cause of lower respiratory
disease in young children, including those requiring
hospitalization for LRD and those with illness cared for as
outpatients. Several investigators have identified decrements in
lung function in school-aged children who were hospitalized with
respiratory syncytial virus bronchiolitis in infancy (10, 11, 12).
Nothing is known of the long term implications of the great
majority of early childhood LRI which is of insufficient severity
to warrant hospitalization. The data we have presented today
provide a foundation of knowledge regarding the incidence,
clinical characteristics, and causes of these relatively common
illnesses of childhood. We have demonstrated the complexity of
the LRI history variable for investigators trying to define the
determinants of lower respiratory health in children. One must
have knowledge of the frequency of LRI, the age of occurrence of
first and recurrent bouts of LRI, the region of the lower
respiratory tract involved by infection, and the etiology of
illness. The LRI problem has almost as many components as the
problem, air pollution. Having characterized LRI history it
becomes imperative to establish the extent to which the variable
is a measure of infection experience or a reflection of the
patterns of host response to infection determined by the
non-infections variables, atopy, airway reactivity, and exposure
to environmental pollutants. Atopy, airway hyper-reactivity, and
recurrent wheezing illness are clearly associated in childhood.
This is not to imply that all children with recurrent wheezing are
allergic or that all children who report wheezing have
hyper-reactive airways. We need to define the origins of
recurrent wheezing illness with greater precision. The problem of
recurrent non-wheezing LRI in childhood has not been investigated.
The importance of tobacco smoke exposure as a respiratory health
hazard has been addressed thoroughly by other authors in this
symposium. The correlation between passive exposure to cigarette
smoke and an enhanced risk of acute LRI in infancy is becoming
rather convincingly established. Exposure to NO_2, SO_2 and carbon
monoxide in the home are important areas for continued
investigation.

Acute lower respiratory infections and recurrent lower respiratory disease are important causes of acute and chronic illness morbidity in childhood. Their importance as determinants of long term respiratory health needs to be investigated carefully. Prevention of infections with respiratory syncytial virus, the parainfluenza and influenza viruses, and M. pneumoniae has proven a difficult task; continued work in this area is certainly warranted.

REFERENCES

1. Weiss, S. T., I. B. Tager, M. Schenker, and F. E. Speizer. "The Health Effects of Involuntary Smoking," Am. Rev. Respir. Dis. 128: 933-942 (1983).

2. Henderson, F. W., W. A. Clyde, A. M. Collier, and F. W. Denny. "The Etiology and Epidemiologic Spectrus of Bronchiolitis in Pediatric Practice," J. Pediatr. 95: 183-190 (1979).

3. Murphy, T. F., F. W. Henderson, W. A. Clyde, A. M. Collier, and F. W. Denny. "Pneumonia: An Eleven Year Study in a Pediatric Practice," Am. J. Epidemiol. 113: 12-21 (1981).

4. Chapman, R. S., F. W. Henderson, W. A. Clyde, A. M. Collier, and F. W. Denny. "The Epidemiology of Tracheobronchitis in Pediatric Practice," Am. J. Epidemiol. 114: 786-797 (1981)

5. Denny, F. W., T. F. Murphy, W. A. Clyde, A. M. Collier, and F. W. Henderson. "Croup: An 11-Year Study in a Pediatric Practice" Pediatr. 71: 871-876 (1983).

6. Henderson, F. W., A. M. Collier, W. A. Clyde, and F. W. Denny. "Respiratory Syncytial Virus Infections, Reinfections and Immunity," N. Engl. J. Med. 300: 530-534 (1979).

7. Loda, F. A., W. P. Glezen, and W. A. Clyde. "Respiratory Disease in Group Day Care," Pediatr. 49: 428-437 (1972).

8. Glezen, W. P., A. Paredes, and L. H. Taber. "Influenza in Children," J. Am. Med. Ass. 243:1345-1349 (1980).

9. Becroft, D. M. "Bronchiolitis Obliterans, Bronchiectasis, and Other Sequelae of Adenovirus Type 21 Infection in Young Children," J. Clin. Pathol. 24: 72-82 (1971).

10. Pullan, C. R., and E. N. Hey. "Wheezing, Asthma and Pulmonary Dysfunction 10 Years after Infection with Respiratory Syncytial Virus in Infancy," Br. Med. J. 284: 1665-1669 (1982).

11. Kattan, M., T. G. Keens, J. G. Lapierre, H. Levison, A. C. Bryan, and B.J. Reilly. "Pulmonary Function Abnormalities in Symptom-free Children after Bronchiolitis," Pediatr. 59: 683-688 (1977).

12. Sims, D. G., M. A. P. S. Downham, and P. S. Gardner. "Study of 8-year-old Children with a History of Respiratory Syncytial Virus Bronchiolitis in Infancy," Br. Med. J. 1: 11-14 (1978).

Part Five

Organics

25.

Part Five: Overview

Lance A. Wallace

U.S. Environmental Protection Agency
Research Triangle Park, North Carolina 27711

Organic compounds indoors present a rich and bewilderingly complex picture. Over 800 different compounds from a single class --
the volatile vapors -- have been observed in four buildings studied
by the U.S. Environmental Protection Agency (EPA). Semivolatile
vapors (pesticides and PCBs) and organics adsorbed to particulates
(PAHs) have not been so extensively categorized indoors, but are
likely to be equally complex.

To categorize this variety requires new methods of collection
and analysis suitable to indoor conditions and concentrations. For
example, samplers must be relatively small, portable, and low-volume
to be acceptable in homes or offices. Dr. Sheldon describes the
advances in these areas of sampling and analysis made by the
Research Triangle Institute and other organizations under EPA sponsorship. Her paper concentrates on volatile organics collected on
Tenax. The indications are that levels of contamination on blank
samples of Tenax, percent recoveries following 10 weeks of storage,
and precision are acceptable for all purposes except those requiring
extremely high accuracy. One area of promise is the use of passive
badges with diffusionally controlled rates of sampling (with either
very clean activated carbon or Tenax sorbents) to measure personal
exposures in an unobtrusive and dependable way.

The paper by Riggin and Petersen focuses on the semivolatile
and nonvolatile classes of compounds which include pesticides, PAHs,
and PCBs. The importance is emphasized of considering <u>both</u> the particulate and vapor phase components for certain compounds that are
on or near the borderline (vapor pressure approximately 10^{-7} torr)
between these two classes. Research needs to include additional
study of polyurethane foam (PUF) as an adsorbent for semivolatiles.
Another research need is the study of interactions of organics with
react inorganic gases such as ozone and nitrogen dioxide.

Once the instruments are available, studies to characterize the presence, concentrations, and sources of organics become possible. Drs. Sterling and Wallace describe many studies, most conducted during the last 5 years, on VOCs (including formaldehyde). (Notable by their absence are studies of the two other large groups -- semivolatiles and particulate-bound organics.) We conclude from these reviews that:

1. Indoor median concentrations of volatile organics are consistently greater, by factors of 2 to 5, than outdoor medians.

2. At higher concentrations the indoor-outdoor ratio increases, often beyond factors of 10.

3. Concentrations are extremely variable, covering 3 to 4 orders of magnitude, indicating the presence of intense indoor sources.

4. These sources are many, including paints, adhesives, cleansers, cosmetics, and other consumer products and building materials; but also common activities, such as visiting the dry cleaner shop or even taking a hot shower!

Perhaps our most fruitful area of research, based on these two reports, lies in identifying the important sources in consumer products and building materials. Once identified, it will be possible for manufacturers to employ substitutes, architects, and consumers to select materials and products with low-emission characteristics. Even lacking this research, homeowners can reduce their exposure by storing cleansers, paints, cosmetics, etc., in areas separated from living space.

Two reports on health effects of organics conclude this section. Reed and Frigas present a rigorous test of the hypothesis that formaldehyde vapor causes asthma, with a clearly negative result, at least for the 13 asthmatic and formaldehyde-exposed patients tested. Mølhave presents an equally rigorous test of the hypothesis that mixtures of organic vapors in the amounts commonly seen in new Danish housing can cause the symptoms associated with the "sick-building syndrome," at least in persons who have previously displayed such symptoms. The positive findings from this double-blind test appear to be among the first objective proofs that a major component of the sick-building syndrome may be caused by common indoor organic vapors at levels far below the occupational health standards.

In summary, these reports present a complete outline of a chapter of indoor air research that has been initiated and has made great strides in the last 5 years - the characterization of the presence, concentrations, and effects of volatile organic compounds in indoor air. This is a remarkable accomplishment, but it should not blind us to the necessity, first, of taking practical steps to reduce the potential effects of volatile organics on public health, and second, of applying similar determination to elucidating indoor concentrations and sources of the remaining categories of organics – the semivolatile and the particle-bound organics.

26.

Review of Analytical Methods for Volatile Organic Compounds in the Indoor Environment

Linda S. Sheldon, Charles M. Sparacino,
  and Edo D. Pellizzari

Analytical and Chemical Sciences
   Research Triangle Institute
      Post Office Box 12194
Research Triangle Park, NC 27709

Early research efforts related to indoor air quality concentrated on several criteria pollutants (CO, NO_2, particulates) and several uniquely indoor pollutants (radon, formaldehyde, asbestos). More recently, researchers have begun to investigate indoor air quality impacts of noncriteria pollutants including volatile organics. Several studies have identified and quantitated large numbers of volatile organics in the indoor environment at concentrations ranging from 100 ppt in a new unoccupied hospital to 10-20 ppb in occupied offices [1,2], homes [3,4,5], and schools [6]. In several of these studies, specific organics were related to building materials or products present in the indoor environment [2,3,4,6].

Successful methods for the analysis of volatile organics in the indoor environment must fulfill several criteria. They must be capable of detecting pollutants at ambient levels (i.e., ppt-ppb). They must utilize collection/measuring devices which are lightweight, compact, and quiet. They should be easy to calibrate and use in the field. They should provide accurate and reproducible analysis with a minimum of artifactual and contamination problems. Finally, the method must allow for sampling periods which are compatible with monitoring needs. For example, if a correlation between indoor air concentrations and air exchange rates is being tested, then the sampling period for both parameters should be identical and should be sufficiently long to minimize large fluctuations in the exchange rate. On the other hand, if pollutant concentrations are to be related to a specific episodic source emission, then a short monitoring period coincident with the emission would be desirable. A number of devices which have been laboratory and/or field tested has been developed to meet these criteria.

Methods for monitoring volatile organics in air may be classified as one of two types [7]: analytical methods which detect and quantitate pollutants on site, and, collection techniques which concentrate organics on some type of sorbent for later analysis. Each of these two categories can be further subdivided into active and passive methods. Active methods employ a power source to pull air across a sensor or collector, while passive methods rely upon permeation or diffusion to bring the analyte in contact with the collector or detector.

Analytical methods are capable of providing concentration profiles which describe both short-term peak concentrations and time-weighted-average concentrations. Unfortunately, these are generally the most complex methods, utilize the most expensive equipment, and require the greatest amount of calibration and maintenance in the field. On the other hand, collection methods, although simpler for field use, can only provide concentration data as time weighted averages.

The analytical performance of the currently available methods in each of these categories will be reviewed. Information on instrumentation, detection limits, interferences, precision and accuracy will be given in order to allow for a comparison between methods.

COLLECTION METHODS

Solid sorbents are the materials most commonly employed for collecting vapor phase organics. Any method based on these sorbents involves three basic steps: Adsorption of the vapor phase organics onto the sorbent, desorption of organics from the sorbent material, and finally analysis of the desorbed organics. Table 1 lists the type of sorbents available and the techniques generally used to accomplish the adsorption, desorption and analysis steps.

TABLE 1.   CHARACTERISTICS OF COLLECTION METHODS

Sorbent	Application	Desorption	Analysis
Polymeric resin	Sample bp 60-300°C except highly polar	Thermal	GC/MS
Inorganic	Strongly sorbs water; not generally useful	-	-
Activated carbon	Sample bp 0-300°C, polar and nonpolar	Solvent	GC/FID GC/ECD GC/NPD
Carbon molecular sieve	Sample bp 0-70°C, i.e., vinyl chloride	Thermal	GC/MS

Organic polymer adsorbents include materials such as Tenax GC and the XAD resins. Tenax GC (2,6-diphenyl-p-phenylene oxide) has been tested extensively and has been shown to adsorb and desorb a wide variety of volatile (bp>60°C) organic compounds. Many polymeric sorbents have a low affinity for water; therefore collection efficiency is generally not strongly dependent on humidity. Further, sorbed water does not significantly interfere with desorption and subsequent analysis. For the porous polymers, adsorption is a reversible process and collection efficiency is dependent on temperature. Low collection efficiencies have been reported for very volatile compounds, both apolar (vinyl chloride, C_1-C_4 alkanes) and polar (methanol, ethylene oxide, acrolein). At high temperatures, collected organics can be thermally desorbed for subsequent analysis. Tenax GC has good thermal stability compared to most porous polymers allowing thermal desorption at high temperatures (300°C) which extends the upper volatility range of analysis compared to other resins.

Inorganic adsorbents include silica gel, alumina, Florisil, and molecular sieves. Since these materials efficiently sorb water which rapidly deactivates their surface sites, they are not commonly used for collecting volatile organics in air. As a special application of these sorbents, a molecular sieve has been used for collecting formaldehyde in ambient air samples [8].

Carbon adsorbents are relatively nonpolar and, hence, water adsorption is not a significant problem. Organics are usually adsorbed through chemisorption allowing efficient collection of all volatile organics including vinyl chloride and the low molecular weigh alkanes; however, adsorption is primarily irreversible and requires solvent extraction techniques for efficient desorption. Typically, methanol, acetone, or CS_2 (or combinations of these solvents) are used for desorption.

Carbonaceous polymeric adsorbents have a spherical macroporous structure with properties intermediate between activated carbon and organic polymeric adsorbents. These materials have a greater adsorption affinity than Tenax GC for the very volatile and polar organics. In addition, thermal desorption may be possible for very volatile organics, although desorption efficiency decreases rapidly for compounds with boiling points greater than 100°C [9].

Desorption of collected organics from any sorbent is accomplished using either thermal or solvent desorption techniques. Generally, the polymeric resins which adsorb through reversible interactions may be thermally desorbed. In these cases, GC/MS is the preferred method of analysis since it provides positive identification of target organics as well as broad spectrum analysis. On the other hand, activated carbon which adsorbs organics through chemical interactions requires solvent desorption for efficient recovery of collect organics. For low molecular weight organics, solvent desorption will usually preclude GC/MS analysis since the solvent front will interfere with analysis. In these cases, solvents such as CS_2 or acetone are used and gas chromatographic detectors (i.e., FID, ECD, NPD) which are relatively insensitive to the eluting solvent are employed. When solvent desorption is used only a fraction of the sample (1/100 or 1/500) is introduced for analysis which limits the sensitivity of the technique compared to thermal desorption when similar detection systems are used.

Details of specific active and passive sampling methods will now be given.

Active Sampling

Tenax GC

As described earlier, active sampling methods rely upon a pump to pull air through a sorbent cartridge. The use of Tenax GC as a sorbent has been extensively tested in the laboratory as well as the field by Research Triangle Institute, and has been used for the collection of field samples [1,5,10,11,12].

Using this method, volatile organics are collected from air samples by passing air through a glass sampling cartridge (10 × 15 cm) containing ∿1.5 g of Tenax GC. Flow rates through the sorbent range from ∿0.01 to 1 L/min depending upon the compounds to be analyzed and the sampling period. Usually, ∿0.025 m^3 of air are collected. Recovery of volatile organics is accomplished by thermal desorption with helium into a liquid nitrogen cooled trap. Vapors are then introduced into a high resolution fused silica chromatographic column and analyzed by electron impact mass spectrometry. The instrument may be calibrated from a known mass of standards loaded onto Tenax cartridges which are thermally desorbed, or through direct liquid injections into the GC/MS system.

The limits of detection (LOD) and quantitation limits (QL) for analysis depend upon several factors. First, the breakthrough volume for each specific compound defines the maximum sample size which can be used. Table 2 lists calculated breakthrough volumes for selected organic compounds [13]. Second, method detection limits are directly related to instrumental detection limits. Response factors calculated relative to a fixed amount of external standard, e.g., bromopentafluorobenzene, generally give a linear response over the range 5 to 1,000 ng per compound, although most organics can be detected at ∿1 ng injected on column. With an instrumental detection limit of 1 ng and a 25 L sample, the method LOD would be 40 ng/M^3 or ∿20 ppt. Quantitation limits are approximately five times higher. For compounds which have high breakthrough volumes, significantly larger samples may be collected to give correspondingly lower LODs and QL. For example, for chloro-benzene which has a breakthrough volume of 473 L at 70°F, a 450 L sample could be collected which would then give a method LOD of 0.5 ppt. Finally, background contamination on Tenax cartridges will increase the amount of sample component which must be collected for accurate determination. Table 3 gives levels of volatile organics found on both field and laboratory blanks during a recent indoor air study.

TABLE 2.   TENAX GC BREAKTHROUGH VOLUMES FOR SELECTED VOLATILE COMPOUNDS[a]

Compound	Breakthrough Volume (L)	
	70°F	80°F
Benzene	54	38
Ethylbenzene	693	487
Chloroform	9.1	6.6
Carbon tetrachloride	20	14
1,1,1-Trichloroethane	15	12
Chlorobenzene	473	303
Tetrachloroethylene	196	144
Trichloroethylene	50	38
o-Dichlorobenzene	1,463	1,139
m-Dichlorobenzene	1,291	948
Bromodichloromethane	45	34

[a]Based on 1.4 g Tenax GC.
For a bed of Tenax 1.5 × 6.0 cm.

TABLE 3.   BACKGROUND CONTAMINATION ON BLANK SAMPLES
(ng per cartridge)

Compound	Field Blanks	Laboratory Blanks
1,1,1-Trichloroethane	6 ± 1	4 ± 3
Benzene	12 ± 3	5 ± 1
Trichloroethylene	2 ± 1	ND
n-Butylacetate	1 ± 1	ND
Tetrachloroethylene	1 ± 0	ND
Ethylbenzene	2 ± 1	ND
m-Xylene	2 ± 1	ND
Styrene	9 ± 4	1 ± 1

Accuracy of this method using Tenax GC combined with GC/MS analysis has been assessed in a variety of ways. First, recovery of volatile organics desorbed from Tenax cartridges versus liquid injections has been evaluated. Results in Table 4 shows percent

TABLE 4.   PERCENT RECOVERY OF FIELD CONTROLS

Compound	% Recovery $\pm$ S.D.
1,2-Dichloroethane	107 $\pm$ 17
Trichloroethane	92 $\pm$ 29
Benzene	101 $\pm$ 24
Carbon tetrachloride	94 $\pm$ 17
Bromodichloromethane	91 $\pm$ 24
Trichloroethylene	97 $\pm$ 9
m-Butylacetate	82 $\pm$ 15
Tetrachloroethylene	91 $\pm$ 10
Chlorobenzene	87 $\pm$ 14
Styrene	94 $\pm$ 26
2-Ethoxyethyl acetate	51 $\pm$ 18
o-Xylene	84 $\pm$ 11
Tetrachloroethane	88 $\pm$ 14
α-Pinene	65 $\pm$ 12
1,2,4-Trimethylbenzene	84 $\pm$ 13
m-Dichlorobenzene	91 $\pm$ 13
n-Decane	90 $\pm$ 18
n-Undecane	85 $\pm$ 7
n-Dodecane	85 $\pm$ 7
Epichlorohydrin	68 $\pm$ 15
m-Cresol	72 $\pm$ 22

recovery of volatile organics loaded onto field controls after a 10-week storage period. Second, sets of performance evaluation cartridges have been prepared by EPA and analyzed in our laboratory. Results of these evaluations are shown in Table 5. Finally, in a recent study, three colocated samples are being collected at each site at three different flow rates, 10, 20 and 30 L/min. It is assumed that only reported levels found on a sample cartridge which increase with increasing sample volume are real and above the method detection limits. Results from the study are not yet available.

Method precision has been evaluated by inter- and intralaboratory analysis of 85 duplicate field samples collected at a 10% rate of the total number of samples. Results of these analyses are given in Table 6. For duplicate pairs for which both samples contained measurable quantities, the inter- and intra-laboratory precision (RSD) for indoor air was 20 to 30%. Benzene exhibited the poorest agreement, chlorobenzene and trichloroethylene showed the best precision.

TABLE 5.   PERFORMANCE EVALUATION RESULTS

Compounds	Number of Samples	Bias	Precision
1,2-Dichloroethane	9	15	+47
1,1,1-Trichloroethane	9	10	+20
Benzene	9	-6	+31
Carbon tetrachloride	9	-3	+26
Trichloroethylene	9	-10	+17
Tetrachloroethylene	9	-3	+15
Chlorobenzene	9	-4	+19
Ethylbenzene	9	-4	+17
o-Xylene	9	-2	+16

TABLE 6.   MEAN PERCENT RELATIVE STANDARD DEVIATION (%RSD) BETWEEN
ANALYTE CONCENTRATIONS REPORTED FOR INDOOR AIR FIELD
AND DUPLICATE SAMPLES

Analyte	% RSD by Sample/Duplicate Type			
	F1/D1	(N)	F1/Q1	(N)
Chloroform	30.0	(24)	39.0	(12)
1,1,1-Trichloroethane	32.0	(43)	43.0	(25)
Benzene	45.0	(57)	43.0	(32)
Carbon tetrachloride	28.0	(18)	58.0	(9)
Trichloroethylene	16.0	(23)	21.0	(18)
Tetrachloroethylene	28.0	(55)	33.0	(30)
Chlorobenzene	17.0	(2)	33.0	(2)
Styrene	25.0	(48)	33.0	(28)
p-Dichlorobenzene	29.0	(45)	33.0	(30)
Ethylbenzene	31.0	(49)	30.0	(31)
o-Xylene	29.0	(45)	--	(0)
p-Xylene	36.0	(58)	40.0	(34)

$$\% \text{ RSD} = \frac{\text{Standard Deviation Between Field and Duplicate Content}}{\text{Average of Field and Duplicate Content}} \times 100$$

Mean %RSD = Σ %RSD for analyte over all samples/N. Duplicate sets with
one or both samples containing trace or not detected amounts of analyte
are not included.

The major problem with the use of Tenax GC is low breakthrough volumes for volatile organics. A number of solid sorbents has been substituted or added to sampling cartridges to improve retention capacity. A cartridge system containing silica gel, Chromosorb 101 and activated carbon in addition to Tenax GC has been reported [14]. Work in our laboratory showed Chromosorb 104 to have better retention capacity [11]. Work by Hillenbrand and Riggin demonstrated that, although absorption capacity of Chromosorb 101 was good, kinetics of adsorption were slow and therefore collection efficiency would not be high [15]. The major drawbacks of these other sorbents used in conjunction with thermal desorption are water uptake during sampling and poor desorption efficiency.

Activated Carbon

Activated carbon sorbents were developed and used for sampling airborne volatiles in industrial environments. NIOSH approved methods are available for ethylene oxide, vinyl chloride, 2-butanone, n-propanol, and n-butanol [16]. These methods use sample sizes which are small (2 L) compared to sample sizes for ambient monitoring (~50 L). Detection limits depend upon both the detection system as well as sample size. Representative detection limits are given in Table 7 assuming an 18 L sample size, solvent desorption using 1 mL of solvent, and analysis of 1 µL of the extract using GC/FID or GC/ECD. Precisions of the NIOSH methods are also listed in Table 7.

TABLE 7.   METHOD PERFORMANCE FOR NIOSH METHODS

Compound	Range (mg/m^3)	Analysis Method	L.O.D. (mg/m^3)[a]	% RSD
Ethylene oxide	20-270	GC/FID	<5.5	10.3
Vinyl chloride	0.008-5.2	GC/ECD	<0.089	8.0
2-Butanone	70-1500	GC/FID	<40	7.2
n-Propanol	50-1000	GC/FID	<40	7.5
n-Butanol	30-900	GC/FID	<40	6.5

[a]For an 18-liter sample.

Passive Monitoring

Passive monitors are either permeation or diffusion controlled. With diffusion-limited devices the collector is isolated from the environment by a porous barrier containing a well-defined series of channels or pores. The purpose of these channels is to provide a geometrically well-defined zone of undistrubed air through which mass transport is achieved by diffusion. The mass flow of pollutant (m) may then be given by Ficks first law;

$$m = (DA/L) \ (C_\infty - C_o),$$

where D is the diffusion coefficient for the compound of interest, A is the total channel area, L is channel length, C_∞ is the ambient concentration of the species and C_o is the concentration at the collector surface [17]. For activated carbon where efficiency of sorption is high, C_o is generally considered as zero. The quantity DA/L which is defined as the sampling rate is usually 30 to 100 cm^3 for most volatile organics and should be determined experimentally for each passive sampling device.

Several passive diffusion devices using activated carbon as the sorbent have been developed for monitoring the industrial environment. Coutant et al. [18-20] have investigated the application of these devices to ambient air monitoring. In an initial investigation [20], background contamination of commercial devices was evaluated. Results in Table 8 show high background for several chlorinated organics, with the DuPont badges showing lowest contamination. Further studies were then performed with the DuPont badges to determine effects of humidity and wind velocity on collection efficiency [18,19]. From the results of these studies (Tables 9 and 10), it was concluded that the DuPont badges had potential use for ambient air monitoring provided that: (1) they are manufactured with extreme care and under special conditions to prevent low-level contamination, (2) they are desorbed with the minimum solvent volume (1 mL) and (3) their use is limited to relative humidities below 80% and wind velocities above 15 cm/sec. A few measurements of the drift velocities of room air inside single-family homes have been reported recently [21]; it is significant that in most instances the air movement was less than 15 cm/sec.

TABLE 8.   CONTAMINATION LEVELS ON COMMERCIALLY
AVAILABLE PASSIVE MONITORS [21]

Compound	Median Concentration and (Range)*		
	DuPont	3M	Abcor
Chloroform	<5.0 (<5-108)	15 (<5.0-16)	<5.0
1,2-Dichloroethane	<170	<170	<170
1,1,1-Trichloroethane	22 (<1.6-68)	240 (178-292)	443 (400-760)
Carbon tetrachloride	6.0 (<0.36-14)	5.3 (4.8-7.4)	230 (220-1720)
Bromodichloromethane	<0.78	<0.78	<0.78
Trichloroethylene	4.0 (<3.6-45)	230 (194-305)	120 (110-430)
Tetrachloroethylene	24 (<0.7-41)	15 (9.6-141)	44 (20-50)
Benzene	<10	<10	<10
Chlorobenzene	<18	<18	<18

*
 ng per monitor.

TABLE 9.   EFFECT OF AIR VELOCITY ON DuPONT BADGES [19]

Compound	Apparent TWA Concentrations (ppbv)	
	At 60 cm/s	At $\cong$ 0 cm/s
Chloroform	0.8	-
1,1,1-Trichloroethane	815	554
Carbon tetrachloride	0.9	0.5
Trichloroethylene	27	18
Tetrachloroethylene	8.6	4.5
Benzene	7.0	2.0
Chlorobenzene	<0.3	<0.3

TABLE 10.   EFFECT OF HIGH HUMIDITY ON
RESPONSE OF DuPONT BADGES [19]

Chemical	Mean Relative Response (corrected)[a]	
	At 7-72% RH $\bar{R}$	At 80-92% RH $\bar{R}$
Chloroform	0.85	0.44
1,1,1-Trichloroethane	1.09	0.69
Carbon tetrachloride	0.67	0.27
Trichloroethylene	1.00	0.58
Tetrachloroethylene	0.89	0.39
Benzene	0.79	0.35
Chlorobenzene	0.84	0.69

[a]Corrected by exclusion of two badges which gave abnormally high responses.

Monsanto Research Corporation has developed a passive monitoring device using Tenax GC as the sorbent [19]. The major advantage of this monitor over charcoal based commerical devices is the enhanced sensitivity achieved by thermal desorption. In this device, adsorption is reversible, hence, C_o cannot be assumed to be zero and the gas phase concentration at the surface of the sorbent becomes more important. This phenomenon will cause decreased adsorption at either high loadings (>1 μg) or increased sampling times. Theoretical calculations have shown that after one hour the apparent sampling rate for 1,1,1-trichloroethane decreased by ~20%. Experiments have demonstrated that temperature, relative humidity, and high wind speed have no effect on collection efficiency. Samples at low linear face velocities (<5 cm/s) resulted in reduced absorption.

For permeation devices, a polymeric membrane serves as a barrier to ambient atmospheres. Gaseous contaminants contact the membrane, dissolve, and are transported through the membrane to a collection medium. The permeation constant, k, for the membrane used and for a given organic permits the determination of a time weighted average

$$c = wk/t$$

where: c = concentration,
         w = weight of organic,
         t = exposure time.

These devices have been applied to the analysis of vinyl chloride using activated carbon as a sorbent [23,24]. It was claimed by the authors that the method is not affected by temperature, relative humidity, or wind velocity, however, reported detection limits are ∿20 ppb which may be too high for monitoring indoor air. The effects of improving sensitivity by increasing the sampling time have not been investigated.

ANALYTICAL METHODS

Analytical methods rely upon the use of automated, portable instrumentation which can provide real time or quasi-real time pollutant measurements in the field. Most of the existing methods and instrumentation used for automated determinations have been developed for inorganic species (e.g., NO_x, CO, particulate), although some specific organics measurement devices (e.g., formaldehyde, acrylonitrile) are available. Many of the commercially available devices were designed for workplace monitoring, and lack the requisite sensitivities for ambient, indoor air levels. Infrared spectroscopy can be useful for obtaining qualitative information, but spectral overlays of major inorganic and organic components prevent readily obtained quantitative data on trace level species.

Portable, automated gas chromatographs are the most common method for field monitoring organic pollutants. In addition to the prime advantage of providing compound separation, GC has the added power of selective detection.

Two portable GCs are available for field use (HNu and Photovac) both of which utilize a photoionization detector for increased selectivity and sensitivity [25,26]. In EPA field tests one of the units was capable of sensitivity levels (for chlorinated hydrocarbons) in the low ppb range for 1 mL air samples [27]. The use of cryogenic preconcentration using Nafion tubing is currently being investigated as a means increasing sample size to 1-10 L and thereby lowering detection limits to levels comparable to sorbent column techniques [28]. Driscoll et al. [26] have described the HNu model 501 which, through micro-processor control, can operate unattended for 24 hours. The use of a far UV detector (FUV) and an FID provide for the detection of "virtually any possible combination of contaminants."

Gas chromatographs of this type have been utilized in a number of studies. For example, a 40-home study [29] utilized a packed column GC/PID for measuring volatile organics. A variation of the normal field GC approach has been reported by DiNandi et al. [30] where a single GC was used in a ten-point air sampling system installed in a single residence.

A different air monitoring device has been developed at the Argonne National Laboratory [31] that utilizes multiple sensors, operating in different modes, and pattern recognition to analyze results. The prototype instrument uses an array of four electro-chemical sensors that respond to toxic gases, and heated noble-metal filaments that serve to pyrolyze or oxidize air components. The combination of four sensors and four operating modes provides sixteen measured parameters, i.e., sixteen independent data channels. Histograms from these data channels provide a "finger-print" which, compared to data from standards, can serve to identify and quantitate compounds at or below threshold level values.

One final method is the mobile MS system developed by SCIEX [32]. Instruments are either single or triple quadropole mass spectrometers equipped with an atmospheric pressure (APCI) and a low pressure chemical ionization (LPCI) source. Air samples are admitted directly into the inlet system for real time analysis. Air flow is adjusted to levels sufficient to minimize memory effects and interactions in the sampling time.

A series of controlled experiments has been performed by EPA to evaluate the system's ability to identify unknown organics. During the tests, 13 compounds were correctly and unambiguously identified, 7 were identified as either/or, one was identified incorrectly, and two were missed. Concentrations at 1 ppm and 20 ppb were tested. Average standard deviations were 17 and 36%, respectively. In the atmospheric pressure chemical ionization (APCI) source, depletion effects were noted at concentrations greater than 500 ppb. Recommendations for an indoor air monitoring approach included the use of computer search systems and some preseparation if unambiguous identifications are needed.

REFERENCES

1. Pellizzari, E.D., Sheldon, L.S., Sparacino, C.M., Bursey, J.T., Wallace, L. and S. Bromberg. "Volatile Organic Levels in Indoor Air," in Indoor Air, Volume 4, Chemicals Characterization and Personal Exposure, B. Berglund, T. Lindvall, and J. Sundell, Eds., Proceedings of the 3rd International Conference on Indoor Air Quality and Climate, August 20, 1984, p. 303.

2. Miksch, R.R., Hollowell, C.D., and H.E. Schmidt. "Trace Organic Chemical Contaminants in Office Spaces," Environment International 8:129-137 (1982).

3. DeBortoli, M., Knoppel, Pecchio, E., Peil, A., Rogora, L., Schauenburg, _., Schlitt, H., and H. Vessers. "Integrating 'Real Life' Measurements of Organic Pollution in Indoor and Outdoor Air of Homes in Northern Italy. Op. cit., B. Berglund, T. Lindvall and J. Sundell, Eds., p. 21 (1984).

4. Lebret, E., van de Wikl, H.J., Bos, H.P., Noy, D., and J.S.M. Baley. "Volatile Hydrocarbons in Dutch Homes." Ibid., p. 57.

5. Pellizzari, E.D., Sparacino, C.M., Sheldon, L.S., Leininger, C.C., Zelon, H., Hartwell, T.D., and L. Wallace. "Sampling and Analysis for Volatile Organics in Indoor and Outdoor Air in New Jersey." Ibid., p. 221.

6. Berglund, B., Johansson, I., and T. Lindvall. "A Longitudinal Study of Air Contaminants in a Newly Built Preschool." Environment International 8:111-115 (1982).

7. Wallace, L.A., and W.R. Ott. "Personal Monitors: A State-of-the-Art Survey." J.AP.C.A. 32(6): 601-610 (1982).

8. Matthews, T.G., and T.C. Howell. "Solid Sorbent for Formaldehyde Monitoring." Anal. Chem. 54(9):1495-1498 (1982).

9. Riggin, R. "Compendium of Methods for the Determination of Toxic Organic Compounds in Ambient Air," Publication Number EPA-600/4-84-041 (1984).

10. Pellizzari, E.D. "Development of Analytical Techniques for Measuring Ambient Atmospheric Carcinogenic Vapors." Publication No. EPA-600/2-75-075 (1975).

11. Pellizzari, E.D. "The Measurement of Carcinogenic Vapors in Ambient Atmospherics." Publication No. EPA-600-7-77-055 (1977).

12. Pellizzari, E.D., Bunch, J.E., Berkley, R.C., and J. McRae. "Determination of Trace Hazardous Organic Vapor Pollutants in Ambient Atmospheres by Gas Chromatography/Mass Spectrometry/Computer." Anal. Chem. 48:803-807 (1976).

13. Krost, K.J., Pellizzari, E.D., Walburn, S.G., and S.A. Hubbard. "Collection and Analysis of Hazardous Organic Emissions." Anal. Chem. 54:810-814 (1982).

14. Hawthorne, A.R. "A Review of Indoor Air Quality Research at Oak Ridge National Laboratory." Proceedings: National Symposium on Recent Advances in Pollutant Monitoring of Ambient Air and Stationary Sources, May 1984, in press.

15. Hillenbrand, L.J., and R.M. Riggin. "Evaluation of Solid Sorbents for Collection of Volatile Organics in Ambient Air." Proceedings: National Symposium on Recent Advances in Pollutant Monitoring of Ambient Air and Stationary Sources, May 1983. Publication No. EPA-600/9-84-001, 344-357 (1984).

16. National Institute of Occupational Safety and Health (NIOSH), Manual of Analytical Methods, Volume 3, Method 5156.

17. Lautenberger, W.J., King, E.V., and J.A. Morello. "A New Personal Badge Monitor for Organic Vapors." Am. Ind. Hyg. Assoc. J. 41(10):737-747 (1980).

18. Coutant, R.W., and D.R. Scott. "Applicability of Passive Dosimeters for Ambient Air Monitoring of Toxic Organic Compounds." Environ. Sci. Technol. 16:410-416 (1982).

19. Lewis, R.G., Coutant, R.W., Wooten, G.W., McMillin, E.R., and J.D. Mulik. "Applicability of Passive Monitoring Devices to Measurement of Volatile Organic Chemicals," in Ambient Air. Paper presented at 1983 Spring National Meeting, American Institute of Chemical Engineers, Houston, TX, March 27, 1983.

20. Coutant, R.W. "Laboratory Evaluation of Commercially Available Passive Organic Personal Monitors." Publication Number EPA-600/54-82-031.

21. Gammage, R.B., and Hawthorn. "Current Status of Measurement Techniques and Concentrations of Formaldehyde in Residences" in Formaldehyde, V. Turoski, Ed. (A.C.S. Advances in Chemistry Series No. 210), in press.

22. Coutant, R.W., Lewis, R.W., and J.D. Mulik. "Passive Sampling Devices with Reversible Adsorption: Mechanics of Sampling." Proceedings, National Symposium on Recent Advances in Pollutant Monitoring of Ambient Air and Stationary Sources, May, 1984, in press.

23. West, P.   "Passive Monitoring of Personal Exposures to Gaseous Toxins." American Laboratory 3:35-38 (1980).

24. Nelms, L.H., Reiszner, K.D., and P.W. West.   "Personal Vinyl Chloride Monitoring Device with Permeation Technique for Sampling." Anal. Chem. 49:994-998 (1977).

25. Barker, J., and R.C. Levenson.   "A Portable Photoionization GC for Direct Air Analysis." Amer. Lab. 76-79 (1980).

26. Disscoll, J.N., Welshire, A.G., and J.W. Bodenrader.   "New Continuous Monitoring Systems for Measurement of Hazardous Pollutants." Proceedings, National Symposium on Recent Advances in Pollutant Monitoring of Ambient Air and Stationary Sources, May 1984, in press.

27. Treitman, R.D., and J.D. Spengler.   "Equipment for Personal and Portable Air Monitoring: A State-of-the-Art Survey and Review. Op. Cit. B. Berglund, T. Lindvall, and J. Sundell, Eds., p. 63 (1984).

28. McClenny, W.A., and J.D. Peel.   "Reduced Temperature Preconcentration of Volatile Organics for Gas Chromatographic Analysis: System Automation." Proceedings, National Symposium on Recent Advances in Pollutant Monitoring of Ambient Air and Stationary Sources, May 1984, in press.

29. Gammage, R.B., White, D.A., and K.C. Gupta.   "Residential Monitoring of High Volatility Organics and Their Sources." Op. Cit. B. Berglund, T. Lindvall, and J. Sundell, Eds., p. 157 (1984).

30. DeNardi, S.R., Ludwig, J.F., Tastaglia, M.S., and M.W. Abromovitz.   "A Systems Approach to the Monitoring of Indoor Air Pollutants. Ibid., p. 197.

31. Zaromb, S.   "Portable Instrument for the Detection and Identification of Air Pollutants." Proceedings, National Sympsoium on Recent Advances in Pollutant Monitoring of Amient Air and Stationary Sources, May 1984, in press.

32. Thompson, B.A.   "Mobil Air Monitoring by MS/MS - A Study of the Mobil TAGA 6000 System." Ibid., 1984.

27.

Sampling and Analysis Methodology for Semivolatile and Nonvolatile Organic Compounds in Air

Ralph M. Riggin and Bruce A. Petersen

Analytical Chemistry Section, Battelle Columbus Laboratories
Columbus, OH

INTRODUCTION

Volatility of organic compounds is a fundamental property which determines the type of sampling and analysis technique used for each compound. Operationally we have chosen to divide organic compounds into three classes —volatile, semivolatile, and nonvolatile.

Volatile compounds have vapor pressures greater than ∿1 mm Hg at ambient temperature and exist entirely in the vapor phase. Semivolatile compounds have vapor pressures of 10^{-7} to 1 mm Hg and are present in both the vapor and particle-bound state. Nonvolatile compounds have vapor pressures less than 10^{-7} mm Hg and are found exclusively in the particle-bound state. While these classifications are approximate, since other factors affect the vapor/particle distribution of a compound, they provide useful guidance for selecting appropriate sampling and analysis methodology.

Other factors to be considered in selecting methodology are chemical reactivity, photochemical reactivity, and required detectability for the target compounds. The last factor impacts (1) the volume of sample which must be collected and (2) the analytical method to be employed. Indoor air studies, in particular, must be carefully designed with respect to analytical sensitivity because only a limited volume of sample may be available.

Subsequent sections of this review will address the classes of semivolatile and nonvolatile compounds of interest, the state of the art is sampling methodology, and available analytical methodology.

IMPORTANT COMPOUND CLASSES

Some of the classes of semivolatile and nonvolatile organic compounds of particular interest include:

351

- Polycyclic aromatic hydrocarbons (PAHs)
- Organochlorine pesticides
- Organophosphorous pesticides
- Polychlorinated biphenyls (PCBs)
- Chlorinated dibenzodioxins and furans

These compounds are of primary concern due to their widespread presence in the environment and toxicological effects.

PAHs and polar (oxygenated and nitro-substituted) derivatives have received a great deal of attention. Two recent reports have reviewed the various physical and chemical considerations for sampling this group of compounds (1,2). Vapor pressure data for a sample of PAHs are shown in Table 1. According to the class definitions provided early, one would consider compounds less volatile than chrysene to exist primarily in the particle-bound state (i.e., nonvolatile components). Table 2 provides experimental data illustrating the vapor/particle distribution of various PAHs, under ambient sampling conditions. As shown in the table, chrysene is predominately collected in the particle phase, but still has significant distribution into the vapor phase. Consequently PAHs as volatile as chrysene require that provisions be made to capture <u>both</u> the particle and vapor phase components in order to obtain an accurate estimate of total concentration.

TABLE 1.   VAPOR PRESSURES OF SELECTED PAHs

Compound	P at 25°C, torr
Phenanthrene	1.7×10^{-4}
Anthracene	2.5×10^{-5}
Pyrene	6.8×10^{-6}
Fluoranthrene	4.9×10^{-6}
Benzo(a)anthracene	1.1×10^{-7}
Chrysene	9.0×10^{-9}
Benzo(a)pyrene	5.5×10^{-9}
Benzo(e)pyrene	5.5×10^{-9}
Perylene	4.3×10^{-9}
Benzo(g,h,i)perylene	1.0×10^{-10}
Coronene	1.5×10^{-12}

TABLE 2. PAH VAPOR/PARTICLE DISTRIBUTION RATIOS[a]

Compound	Vapor/Particle Ratio
Phenanthrene and Anthracene	37
Methylphenanthrene and Methylanthracene	11
Fluoranthene	3.8
Benzofluorenes	0.80
Benzo(a)anthracene and Chrysene	0.32
Benzopyrenes and Perylene	0.13
Benzofluoranthenes	0.09

(a) Taken from Reference 3.

Lewis and coworkers (4,5) have conducted considerable research in the sampling and analysis of PCBs and organochlorine/organophosphorus pesticides in ambient air. Such compounds exist predominately in the vapor state. Using highly sensitive electron capture and flame photometric detectors they have been able to determine ng/m^3 levels of these compounds. Most of this research has employed high or medium volume filtration samplers modified to include polyurethane foam or porous polymeric adsorbents to capture the vapor phase compounds.

Chlorinated dioxins and furans have only recently been studied with regard to the sampling and analysis methodology for ambient air (6). Although more work needs to be conducted, present information indicates that these compounds can be collected using the same approach as PAHs or PCBs. However, highly selective analytical techniques are required to detect the pg/m^3 levels of concern.

SAMPLING METHODOLOGY

By far the most widely used sampling strategy for semivolatile and nonvolatile compounds is based on the sampler shown in Figure 1 (4). An expanded view of the sampling head is shown in Figure 2. This sampler, which has recently become commercially available, employs a standard "Hi-Vol" housing. However, the conventional blower is replaced with a bypass motor which can better tolerate the reduced flow conditions. The sampling head (Figure 2) contains a

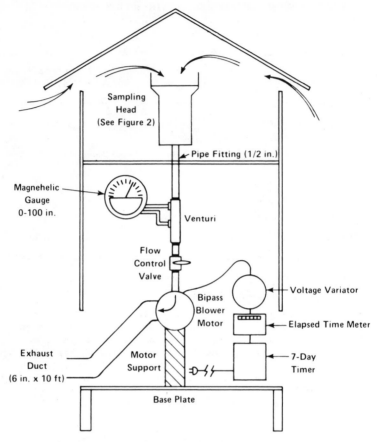

FIGURE 1.   MODIFIED HI-VOL SAMPLER

four-inch diameter glass fiber (or quartz fiber) filter backed by a
glass cartridge containing a polyurethane foam (PUF) plug.

PUF has a number of advantages relative to other adsorbents in
this application, including: low restriction to flow, ease of
purification and handling, and low cost.   However, PUF has been
shown (4) to have poorer retention for the more volatile pesticides
and PCBs compared to XAD-2   or Tenax   resins.   These materials were
used in a sandwich form, surrounded by PUF plugs to hold the
adsorbents in place.   Another disadvantage of PUF is the apparent
formation of mutagenic artifacts during sampling, thus reducing its
usefulness as a collection medium for bioassay studies.

Unfortunately, the assembly shown in Figure 1 is much too noisy
for use indoors.   A low volume sampler, suitable for indoor use, has

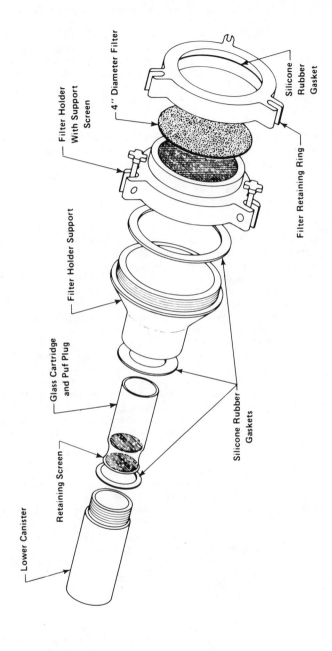

FIGURE 2.   SAMPLING HEAD

been developed (5). This device provides adequate sample for certain applications, especially when suitable high sensitivity analytical techniques can be used. However, a need exists for a high volume (e.g., 8 cubic feet per minute sampling rate) sample that can be operated indoors. A prototype device using an advanced acoustic design has been developed and may become available during the next 6-12 months (8).

ANALYTICAL TECHNIQUES

In general, the analytical techniques employed for the determination of semivolatile and nonvolatile organic compounds in ambient air are the same as are used for other media. Two recent publications review the various approaches which are currently available (9). Typical approaches for some of the most important compound classes are shown in Table 3.

PAHs are determined using either gas chromatography/mass spectrometry (GC/MS) or high performance liquid chromatography (HPLC) with fluorescence or ultraviolet (UV) absorption detectors. In general, the HPLC techniques are somewhat more sensitive than GC/MS, but the operating conditions must be carefully chosen for each compound. GC/MS can detect a wide range of compounds and generally gives adequate sensitivity, especially when multiple ion detection (MID) detection is employed. Consequently, HPLC is often employed when only one or two individual PAHs are of interest, with GC/MS being employed when a wide range of compounds are of interest. PAHs (quinones, phenols) are very polar and can present problems in GC/MS analysis. Consequently, HPLC/fluorescence is used to a greater extent for these compounds. However, nitrosubstituted PAHs are not fluorescent and must be converted to the amine in order to be detected. Consequently, GC/MS is frequently the method of choice for nitro PAH. In this case, negative ion chemical ionization has proven to be extremely sensitive and selective, due to the electron withdrawing properties of the nitrosubstituent.

PCBs and organochlorine pesticides are most frequently determined by gas chromatography with electron capture detection. This approach is much more sensitive than GC/MS, making it especially useful for indoor air studies where a limited volume of air can be collected. For complex samples, a Florisil or silica gel column chromatographic cleanup step is frequently needed to achieve adequate selectivity. Organophosphorous pesticides are generally determined using either flame photometric detection (FPD) or thermionic selective nitrogen/phosporus detection (NPD). The latter detector is much more sensitive, but less selective than the flame photometric detector.

Chlorinated dioxins and furans are of special interest and the analytical approach frequently must be capable of detecting pg/m^3 levels. Consequently, ultrahigh selectivity and sensitivity are required. The usual approach involves complex, multi-step column

TABLE 3.   ANALYSIS TECHNIQUES FOR SELECTED CLASSES OF
SEMIVOLATILE AND NONVOLATILE POLLUTANTS

PAHs

      GC/MS (Gas Chromatography/Mass Spectrometry)

      HPLC/Fluorescence (High Performance Liquid
           Chromatography/Fluorescence Detection)

NO_2 PAH

      GC/MS   (Negative Ion Chemical Ionization)

      HPLC/Fluorescence (After Reduction to the Amine)

PCBs/Pesticides

      GC/MS

      GC/ECD   (Electron Capture Detection)

      GC/FPD   (Flame Photometric Detection)

      GC/NPDC   (Thermionic Nitrogen/Phosphorus Selective
                Detection)

Dioxins

      High Resolution Gas Chromatography/High Resolution
      Mass Spectrometry

chromatographic cleanup. Individual isomers are determined by high
resolution gas chromatography/high resolution mass spectrometry
(HRGC/HRMS) techniques (6).

RESEARCH NEEDS

Although much progress has been made over the past 10-15 years,
many problems persist. Some of the most urgent research needs are
summarized below:

- Low-noise, high volume sampling system suitable for
  indoor use

- Additional study of the interaction of organics with
  reactive gases (e.g., O_3, NO_2)

- Reliable personal monitoring devices to determine human exposure

- Additional study of PUF versus other adsorbents, especially with regard to artifact formation and component stability.

REFERENCES

(1) Coutant, R. W. and R. M. Riggin. "Assessment of Sample
     Integrity and Distribution of Gaseous Particulate-Sorbed
     Organics in Ambient Air". Draft Final Report on Contract
     No. 68-02-3487. U.S. Environmental Protection Agency,
     Research Triangle Park, North Carolina, February, 1984.

(2) Lewis, R. G. "Considerations for Sampling of Semivolatile
     Organic Chemicals in Air". Unpublished Report. October,
     1984.

(3) Cautreels, W. and K. Van Casenberghe. "Experiments on the
     Distribution of Organic Pollutants Between Airborne
     Particulate Matter and the Corresponding Gas Phase".
     Atmospheric Environment., 12: 1133-1141 (1978).

(4) Lewis, R. G., A. R. Brown, and M. D. Jackson. "Evaluation
     of Polymethane Foam for Sampling of Pesticides,
     Polychlorinated Biphenyls, and Polychlorinated
     Naphthalenes in Ambient Air". Analytical Chemistry, 49:
     1668-1672 (1977).

(5) Lewis, R. G., M. D. Jackson, and K. E. MacLeod. "Protocol
     for Assessment of Human Exposure to Airborne Pesticides".
     EPA-600/2-80-180. U.S. Environmental Protection Agency,
     Research Triangle Park, North Carolina, August, 1980.

(6) DeRoos, F. L., J. E. Tabor, S. E. Miller, S. C. Watson,
     and J. A. Hatchel. "Evaluation of the EPA High-Volume
     Sampler for Collection of Polychlorinated Dibenzodioxins".
     Draft Final Report on Contract No. 68-02-3487, U.S.
     Environmental Protection Agency, Research Triangle Park,
     North Carolina, September, 1984.

(7) J. Lewtas, U.S. EPA, private communication.

(8) J. Howes, Battelle, unpublished data.

(9) Riggin, R. M. "Technical Assistance Document for Sampling
     and Analysis of Toxic Organic Compounds in Ambient Air".
     EPA-600/4-83-027, U.S. Environmental Protection Agency,
     Research Triangle Park, North Carolina, June, 1983.

(10) Chuang, C. C. and B. A. Petersen. "State-of-the-Art
     Review and Study Design of Polynuclear Aromatic Compounds
     in Air from Mobile Sources". Draft Final Report on
     Contract 68-02-3487, U.S. Environmental Protection Agency,
     Research Triangle Park, North Carolina, November, 1983.

28.

Organic Chemicals in Indoor Air: A Review of Human Exposure Studies and Indoor Air Quality Studies

Lance A. Wallace

Environmental Scientist, U.S. EPA (Presently Visiting Scholar, Harvard University School of Public Health, Boston, Massachusetts)

Edo D. Pellizzari

Vice President, Research Triangle Institute, Research Triangle Park, North Carolina

Sydney M. Gordon

Head, Mass Spectrometry Facility, IIT Research Institute, Chicago, Illinois

INTRODUCTION

We have learned much in past years about the concentrations and sources of certain indoor air pollutants and rather little about certain others. In the first class, we may include radon [1], formaldehyde [2], particles [3], nitrogen dioxide (NO_2) [4], and carbon monoxide (CO) [5]. In the second class, we may include nitrosamines, polyaromatic hydrocarbons (PAHs), asbestos, pesticides, and until very recently, volatile organic compounds (VOCs).

Within the last several years, however, several studies have added greatly to our knowledge of what VOCs are present at what concentrations in homes and public buildings.

In this discussion, we shall briefly review the rationale and the preliminary results of these studies.

RATIONALE

Many of our most common and useful chemicals are volatile organics. Unfortunately, some of them cause mutations in bacteria and/or cancer in animals or man. Several Federal agencies have authority to regulate such chemicals. However, before such regulation is undertaken, information on the sources, health effects, and human exposure to each chemical must be collected.

Since there are many thousands of chemicals in use, it is necessary to narrow the possible universe from thousands to scores of chemicals. Such efforts are underway in a number of programs.

RESULTS

Two major studies have been undertaken on exposures to volatile organics indoors. The Total Exposure Assessment Methodology (TEAM) Study [6-17] has determined 12-hour integrated exposures and corresponding breath levels of 20-25 target VOCs in 650 households in six cities. The Halocarbon Study [18-19] has determined 12-hour integrated exposures to 20-30 halogenated hydrocarbons in 150 households in three cities. Seven other studies of volatile organics in 10 or more homes have been reported since 1979. Molhave [20] found elevated levels of benzene and toluene in 39 Danish dwellings. Jarke [21] found more complex chromatograms and increased concentrations of organics in 34 Chicago homes. Lebret [22] found that 35 of 35 organics displayed mean indoor-outdoor ratios >1 in 134 Netherlands homes, with 7 mean indoor-outdoor ratios exceeding 10. Tobacco smoking was correlated with increased levels of 10 organics. Factor analysis identified certain clusters of compounds as petroleum distillate-based. Seifert [23] reported that 15 homes in Berlin displayed increased levels of toluene and xylene from printed material. De Bortoli [24] found that 32 of 32 organics had indoor-outdoor ratios >1 in Northern Italy homes, with 6 indoor-outdoor ratios >10. Gammage [25] detected gasoline vapors in 40 east Tennessee homes, most with attached garages. Monteith [26] found increased levels of 18 VOC's in 44 mobile homes in Texas. Sources included plywood, particle board, and carpets.

These nine studies of more than 1000 homes (Table 1) show remarkable agreement on the following points:

1) Essentially, every one of the 40 or so organics studied has higher indoor levels than outdoor.

2) Sources are numerous, including building materials, furnishings, dry-cleaned clothes, cigarettes, gasoline, cleansers, moth crystals, hot showers, printed material, etc.

3) Ranges of concentrations are great, often 2 or more orders of magnitude.

TABLE 1.   STUDIES OF VOLATILE ORGANICS IN HOMES

Year	Location	No. Homes	Investigator	References
1979	Chicago	34	Jarke	21
1979	Denmark	39	Molhave	20
1980-84	Elizabeth, NJ Bayonne, NJ	350	Pellizzari	13-17
	Greensboro, NC Devils Lake, ND	25 25		
	Los Angeles, CA	125		7
	Antioch, CA Pittsburg, CA	75		7
1981-82	Houston, TX	30	Pellizzari	12
	Baton Rouge, LA	75		18-19
	Greensboro, NC	45		
1981-82	Berlin	15	Seifert	23
1981-82	The Netherlands	134	Lebret	22
1983	East Tennessee	40	Gammage	25
1983	Northern Italy	14	De Bortoli	24
1983	Texas	44	Monteith	26
		1070		

Finally, EPA's FY 1982-83 program [27-28] of monitoring air quality in commercial and public-access buildings is identifying hundreds of VOCs in schools, office buildings, hospitals, and homes for the elderly.

Brief descriptions of the EPA studies and their major findings follow.

TEAM Study

In this study, carried out by EPA's Office of Research and Development, personal exposures to 20-25 VOCs were measured in the air and drinking water of 350 residents of Bayonne and Elizabeth, NJ; 25 residents of Greensboro, NC; 25 residents of

Devils Lake, ND; 175 residents of Los Angeles, CA; and 75 residents of Antioch and Pittsburg, CA. The participants were chosen by a statistical design to represent the entire population of their area. (For example, more than 5000 persons were interviewed in Bayonne and Elizabeth before selecting the final 350 participants to represent the combined population of 128,000.)

Participants carried a specially designed personal air monitor to collect two 12-hour samples of air -- one during the day while they engaged in their normal daily activities and one at night while at home. At the end of the 24-hour monitoring period, each participant gave a sample of exhaled breath, which was later analyzed for the same 20-25 target chemicals. A sub-sample of participants (about 25%) were selected to have their backyards monitored by the identical equipment, to compare outdoor air values with indoors.

The estimated frequency distributions of personal air exposures, outdoor air concentrations and concentrations in exhaled breath of several prevalent target compounds are displayed in Figures 1-3. Each figure is plotted in log-normal probability format. That is, the vertical scale is the common logarithm of the concentration (in ug/m^3). The lower bound in all figures is 1 ug/m^3, the approximate quantifiable limit for most of the compounds. The horizontal scale is spaced so that a truly log-normal distribution (one with its logarithms normally distributed) would be plotted as a straight line. Additional data on 11 of the most prevalent of the target-compounds are contained in Tables 2 and 3.

Several general observations based on these figures and tables may be ventured:

1) For all 11 compounds, personal air exposures were greater than outdoor air concentrations. The ratio of personal/ outdoor concentrations increased at the higher percentiles, often exceeding a factor of 10 at the 99th percentile.

2) Daytime and nighttime personal and outdoor air concentrations seldom differed significantly. The major exception was tetrachloroethylene, for which daytime values were consistently higher.

3) Breath levels were roughly 30-40% of personal air concentrations for nine of the 11 compounds. Benzene (75-80%) and tetrachloroethylene (90%) had higher ratios.

4) Indoor concentrations are greater than outdoor concentrations for all 11 prevalent chemicals in Bayonne and Elizabeth. The ratio by which indoor values exceed the outdoor values increases from 2-4 at median levels to factors exceeding 10-20 at high levels (Tables 2-3).

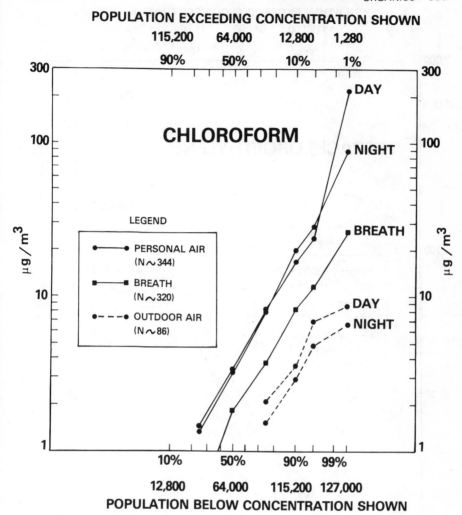

Figure 1. Chloroform: Estimated frequency distributions of personal air exposures, outdoor air concentrations, and exhaled breath values for the combined Elizabeth-Bayonne target population (128,000). All air values are 12-hour integrated samples. The breath value was taken following the daytime air sample (6:00 a.m. - 6:00 p.m.). All outdoor air samples were taken in the vicinity of the participants' homes.

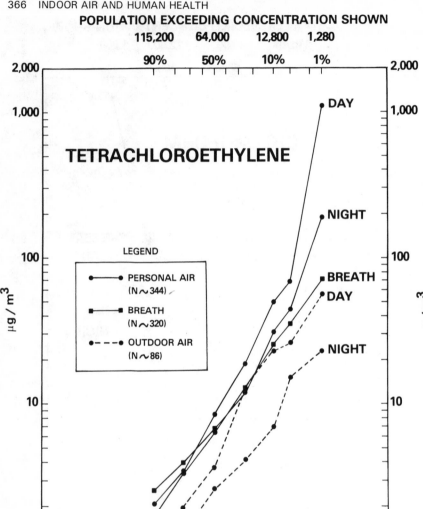

Figure 2. Tetrachloroethylene: Estimated frequency distributions of
personal air exposures, outdoor air concentrations, and exhaled breath
values for the combined Elizabeth-Bayonne target population (128,000).
All air values are 12-hour integrated samples. The breath value was
taken following the daytime air sample (6:00 a.m. - 6:00 p.m.). All
outdoor air samples were taken in the vicinity of the participants'
homes.

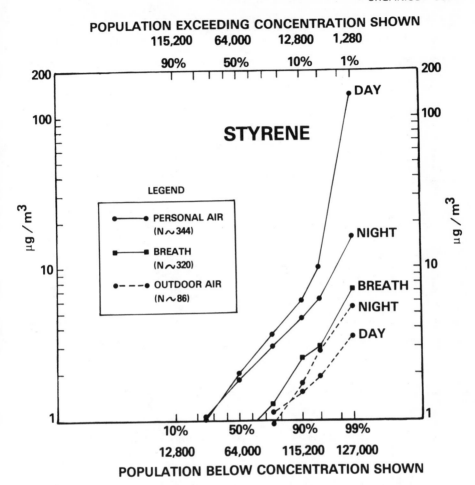

Figure 3. Styrene: Estimated frequency distributions of personal air exposures, outdoor air concentrations, and exhaled breath values for the combined Elizabeth-Bayonne target population (128,000). All air values are 12-hour integrated samples. The breath value was taken following the daytime air sample (6:00 a.m. - 6:00 p.m.). All outdoor air samples were taken in the vicinity of the participants' homes.

This latter finding indicates that multiple sources for these chemicals must be present in the home. In fact, using the questionnaire data, it is possible to associate higher levels in the home with activities of the participant. For example, indoor levels of benzene were higher in homes containing smokers. Indoor levels of 1,1,1-trichloroethane were higher for persons exposed to paint. The largest indoor-outdoor ratio was observed for

TABLE 2.   INDOOR-OUTDOOR RELATIONSHIPS: OVERNIGHT PERSONAL AIR MEANS COMPARED TO OUTDOOR MEANS

Compound	Arithmetic Mean			Geometric Mean		
	Outdoor[a]	Personal[b]	Ratio	Outdoor	Personal	Ratio
Chloroform	1.2[c]	8.7	7	0.55	3.3	6
1,1,1-Trichloroethane	5.4	110	21	3.7	19	5
Benzene	8.6	30	3	4.1	12	3
Carbon Tetrachloride	1.2	14	12	0.8	1.8	2
Trichloroethylene	2.1	7.3	3	1.4	2.6	2
Tetrachloroethylene	3.7	11	3	2.1	6.3	3
Styrene	0.90	2.7	3	0.55	1.5	3
m,p-Dichlorobenzne	1.5	56	37	1.0	5.1	5
Ethylbenzene	3.8	13	3	2.5	6.4	3
o-Xylene	4.0	16	4	2.8	5.3	2
m,p-Xylene	11	55	5	8.3	16	2

a N = 86
b N = 340
c $\mu g/m^3$

TABLE 3. PERSONAL-OUTDOOR AIR COMPARISONS: DAYTIME MEAN VALUES AND RATIOS

Compound	Arithmetic Mean			Geometric Mean		
	Outdoor[a]	Personal[b]	Ratio	Outdoor	Personal	Ratio
Chloroform	1.5[c]	7.4	5	0.58	3.0	5
1,1,1-Trichloroethane	8.6	820	96	3.3	19	6
Benzene	9.5	26	3	3.8	11	3
Carbon Tetrachloride	1.0	4.6	5	0.70	1.3	2
Trichloroethylene	2.4	19	8	1.4	3.0	2
Tetrachloroethylene	8.3	78	9	3.8	9.2	2
Styrene	0.82	15	18	0.61	1.8	3
m,p-Dichlorobenzne	1.9	35	18	0.78	4.7	6
Ethylbenzene	4.3	25	6	2.6	7.3	3
o-Xylene	4.0	17	4	2.6	6.1	2
m,p-Xylene	12	48	4	7.0	18	2

a N = 90
b N = 344
c $\mu g/m^3$

chloroform, indicating a strong indoor source (probably the chlorinated water in the home).

Halocarbon Study

This study, carried out by EPA's Office of Toxic Substances, was nearly identical to the TEAM Study in its use of personal monitors, outdoor monitors, and drinking water and breath samples. Target chemicals were limited to 32 halogenated hydrocarbons, with an overlap of about 15 chemicals in both studies. Three cities -- Baton Rouge, Houston, and Greensboro -- were studied, with 75, 30, and 45 participants, respectively.

The results of the study confirm the conclusion that indoor levels of nearly all of the common target chemicals are higher than outdoor levels.

Air Quality in Commercial and Public Access Buildings

This study aimed at identifying a broad variety of organic chemicals in large buildings where people spend considerable time: schools, homes for the elderly, and office buildings. More than 350 volatile organics were identified in five locations in a Washington, DC home for the elderly.

Of these, 35 chemicals (10%) were present in all five locations. About a dozen of the 35 ubiquitous chemicals were known mutagens or carcinogens (Table 4).

A set of 20 target chemicals was quantified both indoors and outdoors, with results again confirming the finding of the two previous studies -- mean indoor levels were about twice the outdoor levels. Again the sources of the high indoor levels could be partially determined. For example, levels of trichloroethylene in a new office building increased dramatically after the office employees moved in, particularly in the secretaries' bay. A possible source could be the bottles of correction fluid at the desks. On the other hand, levels of 1,1,1-trichloroethane were 100 times outdoor levels before the workers moved in, indicating that this chemical was a part of the building materials or finishing operations.

DISCUSSION

All three EPA studies and most of the other seven American and European studies show that most of the chemicals measured were at higher concentrations -- often much higher -- indoors than outdoors. Therefore, the sources of these chemicals must also be indoors. Although none of the studies had the objective of identifying the sources, we may speculate on the most likely sources, using the TEAM Study as our vehicle.

TABLE 4.   CHEMICALS PRESENT IN ALL INDOOR MONITORING SITES AT
HOME FOR THE ELDERLY IN WASHINGTON, D.C.:   MARCH 1983

Acetone	1-Methylnaphthalene
*Benzene	2-Methylnaphthalene
*Carbon tetrachloride	2-Methylnonane
*Chloroform	2-Methyloctane
Cyclohexane	2-Methylundecane
Decalin	Naphthalene
Decanal	Nonanal
*Decane	Nonane
*m,p-Dichlorobenzene	*Octane
1,4-Diethylbenzene	Pentane
Diethylcyclohexane	Propylbenzene
*1,4-Dioxane	i-Propylbenzene
*Dodecane	Propylcyclohexane
Ethyl acetate	*Styrene
Ethylbenzene	*Tetrachloroethylene
Heptane	Toluene
Hexane	*1,1,1-Trichloroethane
Hexanol	*Trichloroethylene
Methylcyclohexane	Trichlorofluoromethane
Methyl decalin	1,3,5-Trimethylbenzene
4-Methyldecane	1,2,4-Trimethylcyclohexane
5-Methyldecane	1,3,5-Trimethylcyclohexane
*Methylene chloride	2,2,4-Trimethylhexane
2-Methylheptane	2,4,4-Trimethyl-i-pentene
2-Methylhexane	*Undecane
3-Methylhexane	m,p-Xylene
	o-Xylene

*   Mutagenic, carcinogenic, or co-carcinogenic properties.

In the TEAM Study, eleven chemicals were present in more
than half of all samples. For purposes of discussion, we may
divide the 11 chemicals into three groups: Aromatics, chlorinated
solvents, and a miscellaneous group.

Group I: Aromatics

This group of chemicals includes three (ethylbenzene and the
o- and m, p-xylenes) that are ubiquitous components of gasoline,
paints, and similar substances. Benzene, a human carcinogen, is
also contained in gasoline but has been banned from the other products. Styrene is possibly more commonly produced by plastics or
insulation than by gasoline or paints.

Thus, the very high correlations (Tables 5-6) among the
ethyl-benzene, o-xylene and m,p-ylenes give further indication
of their common sources. The somewhat lower correlations with
benzene and styrene are due to the different sources for these

Table—5. Significant (p<.05) Spearman correlations between chemical concentrations in personal air daytime and overnight samples in Bayonne and Elizabeth (all values above limit of detection)

Bayonne (Night) N = 18 - 141   Bayonne (Day) N = 16 - 146
Elizabeth (Night) N = 15 - 197   Elizabeth (Day) N = 15 - 194

Night (6 PM - 6 AM)

Each cell shows Bayonne (top) / Elizabeth (bottom).

| | Group I: Aromatics | | | | | Group II: Chlorinated Solvents | | | Miscellaneous | | |
	A	B	C	D	E	F	G	H	I	J	K
A Benzene		.57* / .44*	.65* / .56*	.64* / .48*	.66* / .54*	.49* / .54*	.44* / .39*	.65* / .50*		.56* / .35*	.28 / .27
B Styrene	.37* / .47*		.67* / .66*	.63* / .59*	.66* / .67*	.48* / .43*	.33 / .33*	.51* / .41*	.31 / .23	.38* / .36*	.25 / .25
C Ethylbenzene	.59* / .51*	.58* / .64*		.94* / .90*	.95* / .93*	.43* / .38*	.43* / .36*	.49* / .47*	.31 / .27	.41* / .37*	.20 / .30*
D o-Xylene	.50* / .47*	.64* / .57*	.93* / .92*		.97* / .91*	.43* / .36*	.40* / .32*	.48* / .46*	.25 / .30*	.35 / .23	.30* / .17
E m+p-Xylene	.55* / .51*	.65* / .63*	.93* / .97*	.95* / .93*		.46* / .40*	.41* / .39*	.49* / .47*	.24 / .30*	.31 / .36*	.18 / .30*
F 1,1,1-Trichloroethane	.47* / .21	.39* / .24	.51* / .37*	.46* / .31*	.49* / .41*		.44* / .36*	.55* / .37*		.44* / .44*	.45* / .34*
G Trichloroethylene	.24 / .40*	.36* / .31*	.46* / .38*	.41* / .31*	.41* / .40*	.57* / .44*		.59* / .45*		.37* / .31*	.41* / .40*
H Tetrachloroethylene	.44* / .36*	.40* / .23	.53* / .40*	.53* / .42*	.55* / .43*	.54* / .44*	.46* / .50*		.18 / .18	.43* / .27	.37* / .30*
I m+p-Dichlorobenzene	.20	.28 / .15	.26	.29 / .16	.27	.20				.25	
J Chloroform	.47* / .33*	.29 / .32*	.23 / .32*	.21	.24 / .25	.23 / .24	.18	.24	.30*		.41* / .46*
K Carbon tetrachloride	.32 / .42*	.32 / .30	.28 / .38*	.25 / .20	.29 / .37*	.29 / .50*	.36 / .42*	.34 / .28	.25	.50* / .52*	

Legend: Bayonne / Elizabeth

* p < .0001

Table 6.   Significant (p<.05) Spearman correlations between chemical concentrations in fixed air daytime and overnight samples in Bayonne and Elizabeth (all values above limit of detection)

Bayonne (Day) N = 12 - 32   Bayonne (Night) N = 14 - 33
Elizabeth (Day) N = 16 - 58   Elizabeth (Night) N = 15 - 57

Night (6 PM - 6 AM)

		Group I: Aromatics					Group II: Chlorinated Solvents			Miscellaneous		
		A	B	C	D	E	F	G	H	I	J	K
A	Benzene		.57	.54 / .72*	.45 / .63*	.42 / .66*	.53 / .43	.65	.42 / .64*		.53	.36
B	Styrene	.56		.66*	.60*	.54	.48 / .56	.40	.55 / .64			.43
C	Ethylbenzene	.62 / .63*	.71*		.94* / .95*	.95* / .91*	.50 / .56	.39	.60 / .77*			.43 / .44
D	o-Xylene	.49 / .64*	.75*	.93* / .96*		.98* / .92*	.50 / .50	.40	.60 / .78*			
E	m+p-Xylene	.52 / .61*	.70*	.94* / .94*	.95* / .96*		.43 / .47	.40	.55 / .76*	.42		
F	1,1,1-Trichloroethane	.41 / .38	.50	.40 / .69*	.56*	.60*		.69* / .35	.62 / .50	.69	.69*	.43
G	Trichloroethylene	.37	.83*	.64* / .73*	.63*	.54* / .73*	.47 / .40		.64 / .62*		.49	.43
H	Tetrachloroethylene	.49 / .48	.67*	.68*	.71* / .68*	.60*	.52 / .37	.48 / .68*			.53	.44
I	m+p-Dichlorobenzene											
J	Chloroform						.57 / .64*		.53			
K	Carbon tetrachloride	.35	.54	.42	.39	.30	.46	.50	.41		.46	

Legend: Bayonne / Elizabeth

* p < .0001

two compounds, and (in the case of benzene) the increased scatter of the measurements due to background contamination of the Tenax.

Group II: Chlorinated Solvents

This group of three chemicals (F, G and H of Tables 5-6) finds use with metal degreasers, dry cleaners, and household maintenance products. The most ubiquitous are 1,1,1-trichloroethane and tetrachloroethylene.

The former has, until recently, been considered the least toxic of the three, but a recent study has indicated possible carcinogenic potential. If so, all three would be animal carcinogens and, therefore, suspected human carcinogens.

Tetrachloroethylene consistently shows the highest breath-air correlations of any chemical in Group I or II. It is also the only Group II chemical that shows strong correlations with the air concentrations of the first three Group I aromatics. Outdoor air concentrations of tetrachloroethylene are also highly correlated with the same aromatics.

Group III: Miscellaneous

Carbon tetrachloride. This chemical shows indications of being more time and site specific than the other chemicals. For example, it is strongly correlated with nine of ten other chemicals in the Elizabeth daytime outdoor samples, but not at all with the Bayonne outdoor samples or the Elizabeth night-time outdoor samples.

m+p-Dichlorobenzene. This chemical is used in moth crystals and deodorants. It shows no strong correlations with any other chemical in outdoor air samples. It is, however, correlated with ethylbenzene and the xylenes in breath and day and night personal air samples in Bayonne, and in the night personal air samples only in Elizabeth.

Chloroform. This chemical is unique among the common volatile organics in being transmitted mainly through drinking water. Thus, its almost complete lack of correlations with the other air chemicals is understandable. Only 1,1,1-trichloroethane is sometimes strongly associated with chloroform levels in outdoor samples.

CONCLUSION

The studies described above have led to major advances in our knowledge of the exposures of the general population to toxic and carcinogenic chemicals. We have identified about a dozen chemicals that appear to be present nearly everywhere, including our homes and even our bodies. We have learned the extent and distribution of exposure of urban-industrial populations to these target chemicals. Finally, we have the somewhat surprising finding that the most important sources of these chemical exposures are likely to be found in our own homes or personal activities.

These findings will have considerable impact on our future research on personal exposures and indoor air quality. The sources of these exposures need to be identified and their emission rates quantified. A wide variety of living areas (suburban and rural) need to be surveyed to better determine the range of exposures nationwide. Ultimately, this information must be combined with estimates of toxic and carcinogenic potency to arrive at estimates of risks to be incorporated in our environmental policies and building codes.

Disclaimer

Opinions expressed are those of the author and do not reflect official positions of the U.S. Environmental Protection Agency. Mention of products or brand names does not imply endorsement by the government.

References

1. Alter, H.W. and R.A. Oswald, "Results of Indoor Radon
    Measurements Using the Trach Etch Method," Health Physics
    45: 425-428, (1983).

2. Anderson, I. "Indoor Air Pollution Due to Chipboard Used as
    Construction Material," Atmos. Environ. 9: 1121-1127, (1975).

3. Spengler, J.D., D.W. Dockery, W.A. Turner, J.M. Wolfson, and
    B.G. Ferris, Jr., "Long-Term Measurements of Respirable
    Sulfates and Particles Inside and Outside Homes," Atmos.
    Environ. 15: 23-30, (1981).

4. Spengler, J.D., B.G. Ferris, Jr., D.W. Dockery, and F.E.
    Speizer, "Sulfur Dioxide and Nitrogen Dioxide Levels Inside
    and Outside Homes and Implications on Health Effects
    Research," Envir. Sci. Tech. 13: 1276-1280, (1979).

5. Hartwell, T.D. et al, Study of Carbon Monoxide Exposure of
    Residents of Washington, D.C., and Denver, Colorado, Final
    Report, EPA Contract #68-02-3679, (1984).

6. Pellizzari, E.D., Erickson, M.D., Giguere, M.T., Hartwell,
    T.D., Williams, S.R., Sparacino, C.M., Zelon, H., and Waddell,
    R.D., "Preliminary Study on Toxic Chemicals in Environmental
    and Human Samples: Work Plan, Vols. I and II, (Phase I),
    U.S. EPA, Wash., D.C. (1980).

7. Pellizzari, E.D., Erickson, M.D., Sparacino, C.M., Hartwell,
    T.D., Zelon, H., Rosenzweig, M., and Leininger, C., "Total
    Exposure Assessment Methodology (TEAM) Study: Phase II:
    Work Plan," U.S. EPA, Wash., D.C. (1981).

8. Wallace, L., R. Zweidinger, M. Erickson, S. Cooper, D.
    Whitaker, and E. Pellizzari, "Monitoring Individual Exposure:
    Measurements of Volatile Organic Compounds in Breathing-Zone
    Air, Drinking Water, and Exhaled Breath," Env. Int. 8,
    269-282, (1982).

9. Pellizzari, E.D., Hartwell, T., Zelon, H., Leininger, C.,
    Erickson, M., Cooper, S., Whittaker, D., and Wallace, L.,
    "Total Exposure Assessment Methodology (TEAM) Prepilot
    Study -- Northern New Jersey," U.S. EPA, Wash. D.C. (1982).

10. Sparacino, C., Leininger, C., Zelon, H., Hartwell, T.,
    Erickson, M., and Pellizzari, E., "Sampling and Analysis
    for the Total Exposure Assessment Methodology (TEAM) Prepilot
    Study," Research Triangle Park, U.S. Environmental Protection
    Agency, Wash. D.C. (1982).

11. Sparacino, C., Pellizzari, E., and Erickson, M., "Quality Assurance for the Total Exposure Assessment Methodology (TEAM) Prepilot Study," U.S. Environmental Protection Agency, Wash. D.C. (1982).

12. Pellizzari, E.D., Hartwell, T.D., Leininger, C., Zelon, H., Williams, S., Breen, J., and Wallace, L., "Human Exposure to Vapor-Phase Halogenated Hydrocarbons: Fixed-Site vs. Personal Exposure," "Proceedings from Symposium on Ambient, Source, and Exposure Monitoring of Non-Criteria Pollutants," May, 1982, Sponsored by Environmental Monitoring Systems Lab., Research Triangle Park, NC (1983) EPA 600/9-83-007.

13. Wallace, L., Pellizzari, E., Hartwell, T., Rosenzweig, M., Erickson, M., Sparacino, C., and Zelon, H., "Personal Exposure to Volatile Organic Compounds: I. Direct Measurement in Breathing-Zone Air, Drinking Water, Food, and Exhaled Breath," Env. Res. 35: 193-211 (1984).

14. Pellizzari, E., Hartwell, T., Sparacino, C., Sheldon, C., Whitmore, R., Leininger, C., and Zelon, II, "Total Eposure Assessment Methodology (TEAM) Study: First Season, Northern New Jersey -- Interim Report," Contract No. 68-02-3679, U.S. EPA, Wash. D.C. (1984).

15. Hartwell, T.D., Perritt, R.L., Zelon, H.S., Whitmore, R.W., Pellizzari, E.D., and Wallace, L., "Comparison of Indoor and Outdoor Levels for Air Volatiles in New Jersey" in Indoor Air, v. 4, pp. 81-86, Swedish Council for Building Research, Stockholm, Sweden, (1984).

16. Pellizzari, E., Sparacino, C., Sheldon, L., Leininger, C., Zelon, H., Hartwell, T., and Wallace, L., "Sampling and Analysis for Volatile Organics in Indoor and Outdoor Air in New Jersey," in Indoor Air, v. 4, Chemical Characterization and Personal Exposure, pp. 221-226. Swedish Council for Building Research, Stockholm (1984).

17. Wallace, L., Pellizzari, E., Hartwell, T., Zelon, H., Sparacino, C., and Whitmore, R., "Analysis of Exhaled Breath of 355 Urban Residents for Volatile Organic Compounds," in Indoor Air, V. 4, pp. 15-20, Swedish Council for Building Research, Stockholm, Sweden (1984).

18. Rosenzweig, M. and T.D. Hartwell, Statistical Analysis and Evaluation of the Halocarbon Survey, Research Triangle Institute, Final Report, EPA Contract #68-01-5848, (1983).

19. Hartwell, T.D., Zelon, H.S., Leininger, C.C., Clayton, C.A., Crowder, J.H., and Pellizzari, E.D., "Comparative Statistical Analysis for Volatile Halocarbons in Indoor and Outdoor

Air," in Indoor Air, v. 4, pp. 57-62, Swedish Council for
Building Research, Stockholm, Sweden (1984).

20. Molhave, L., and Moller, J., "The Atmospheric Environment in
     Modern Danish Dwellings: Measurements in 39 Flats," in
     Indoor Climate, pp. 171-186, Danish Building Research
     Institute, Copenhagen (1979).

21. Jarke, F.H., ASHRAE Rp 183, IITRI, Chicago (1979).

22. Lebret, E., Van de Wiel, H.J., Bos, H.P., Noij, D., and
     Boleij, J.S.M., "Volatile Hydrocarbons in Dutch Homes," in
     Indoor Air, v. 4, pp. 169-174, Swedish Council for Building
     Research, Stockholm, Sweden (1984).

23. Seifert, B., and Abraham, H.J., "Indoor Air Concentrations
     of Benzene and Some Other Aromatic Hydrocarbons," Ecotoxicol.
     Environ. Safety, 6: pp. 190-192 (1982).

24. De Bortoli, M., Knoppel, H., Pecchio, E., Peil, A., Rogora,
     L., Schauenberg, H., Schlitt, H., and Vissers, H., "Integra-
     ting 'Real Life' Measurements of Organic Pollution in Indoor
     and Outdoor Air of Homes in Northern Italy," in Indoor Air,
     v. 4, pp. 21-26, Swedish Council for Building Research,
     Stockholm, Sweden (1984).

25. Gammage, R.B., White, D.A., and Gupta, K.C., "Residential
     Measurements of High Volatility Organics and Their Sources,"
     in Indoor Air, v. 4, pp. 157-162, Swedish Council for Build-
     ing Research, Stockholm, Sweden (1984).

26. Monteith, D.K., Stock, T.H., and Seifert, W.E., Jr., "Sources
     and Characterization of Organic Air Contaminants Inside
     Manufactured Housing," in Indoor Air, v. 4, pp. 285-290,
     Swedish Council for Building Research, Stockholm, Sweden
     (1984).

27. Wallace, L., Bromberg, S., Pellizzari, E., Hartwell, T.,
     Zelon, H. and Sheldon, L., "Plan and Preliminary Results
     of the U.S. Environmental Protection Agency's Indoor Air
     Monitoring Program (1982), in Indoor Air, v. 1, Recent
     Advances in the Health Sciences and Technology," pp. 173-178.
     Swedish Council for Building Research, Stockholm, Sweden (1984).

28. Pellizzari, E., Sheldon, L., Sparacino, C., Bursey, J.,
     Wallace, L., and Bromberg, S., "Volatile Organic Levels in
     Indoor Air," in Indoor Air, v. 4, Chemical Characterization
     and Personal Exposure," pp. 303-308. Swedish Council for
     Building Research, Stockholm, Sweden (1984).

29.

Does Formaldehyde Cause Allergic Respiratory Disease?

Charles E. Reed and Evangelo Frigas

Mayo Graduate School of Medicine, Rochester, Minnesota

INTRODUCTION

The reasons that formaldehyde has been suspected of causing allergic respiratory diseases are: first, formaldehyde irritates the mucus membrane of the eyes, nose and throat; second, it is a highly reactive molecule combining covalently with proteins; and third, it can cause allergic contact dermatitis. The most widely reported symptoms from exposure to high levels of formaldehyde are irritation of the upper respiratory tract, irritation of the eyes, and headaches, all of which are said to be precipitated at a concentration of 2 to 5 parts per million (ppm).

The difficulty in interpreting these symptoms is that most of them are the common everyday symptoms that everyone has at one time or another for a variety of reasons. Furthermore, these symptoms do not lend themselves to objective measurement. Controversy about them persists because interpretation depends more on opinion than on scientifically determined fact. Asthma is the most serious of the diseases that have been attributed to formaldehyde; fortunately it can be studied objectively.

Reports of occupational asthma from exposure to formaldehyde have appeared since 1939 (1). Suspected cases have been workers in industries using formaldehyde (2-4), embalmers (5), and medical and paramedical personnel (6,7). The levels of inhaled formaldehyde gas that might produce asthma are unknown. Inhalation of formaldehyde at concentrations of more than 11 ppm has been reported to cause chemical pneumonitis, pulmonary edema, and death (8). For the past five years, there has been growing speculation in the medical (9-15) and the nonmedical press that exposure to low levels (less than 3 ppm) of formaldehyde gas in homes may produce asthma.

Our interest in the subject was kindled by the case of a middle-aged woman who developed asthma after the walls of her 50-year-old farm home had been insulated with urea-formaldehyde foam (16). This insulation was improperly installed and shortly

379

crumbled to a fine dust that trickled out into the house around windows and electrical outlets. Her symptoms improved after she moved out of the house and became worse when she returned. Laboratory Exposure to this fine buoyant dust provoked cough, wheeze and drop in FEV_1 to 51% of baseline values. She did not respond to similar exposure to an aluminium oxide dust that had provoked asthma in a different patient or to formaldehyde gas up to concentrations of 3.5 ppm. Nor did other asthmatic patients develop airway obstruction from this urea-formaldehyde dust (17). We could not detect formaldehyde in aqueous extracts of the dust. We concluded that the urea-formaldehyde dust was responsible for her positive bronchial challenge but we do not know the mechanism of the reaction. It was not due to formaldehyde (18).

We then went on to perform bronchial challenge with formaldehyde gas in 13 patients who were referred for evaluation of asthma and who suspected that exposure to formaldehyde gas was responsible for their asthmatic symptoms.

MATERIAL AND METHODS

Subjects

The 11 women and 2 men ranged in age from 15 to 70 years. Before our evaluation, all of them had been chronically exposed to formaldehyde gas for variable periods, either at work or at home, in concentrations that ranged from 0.1 to 1.2 ppm. The patients reported symptoms of chest tightness, coughing, or wheezing, which they attributed to exposure to formaldehyde gas. In addition, most of the patients had noted that these symptoms would abate or disappear when they were away from the site of exposure to formaldehyde. Also, several of them had noted nasal congestion, irritation of the eyes, and headache. All except two (cases 5 and 7) had been healthy before exposure. These two patients had had a history of mild asthma that necessitated occasional bronchodilator treatment. After they were exposed to formaldehyde, their symptoms worsened, occurred more frequently, and needed more treatment. Of the patients who had been exposed to formaldehyde in their homes, one patient (case 2) also reported that her husband had had similar symptoms although less severe. Of the five smokers, three had stopped smoking soon after they became symptomatic. At the time of evaluation, five patients (cases 5, 7, 9, 12 and 13) were being treated with bronchodilators, and three of them had also received systemic glucocorticoids. The rest of the patients had not required treatment for their symptoms, although a few were using nonprescription antihistamines. Of our 13 patients, 11 were still living or working in the same environment suspected of causing their symptoms, and 6 had initiated or were contemplating a lawsuit.

None had a history of allergy. Five patients (cases 3, 4, 5, 6 and 13) had negative skin tests to a battery of 40 common

inhalant allergens. The other patients did not undergo skin
testing. No abnormal auscultatory findings were noted on chest
examination, except for wheezing in case 7. All patients had normal chest roentgenograms, and their total blood eosinophil counts
were within normal limits except for case 5 (1,100 cells/mm^3), a
patient who had chronic asthma, nasal polyposis, and hyperplastic
pansinuitis. Two patients (cases 4 and 5) had nasal polyps. The
remaining findings on the physical examinations were unremarkable,
and the patients had no additional symptoms or illnesses except for
one (case 2) who had diabetes.

Bronchial Challenge

Flow-volume loops were determined with maximal forced
exhalations in triplicate the first and second days in order to
establish the baseline for each subject and to exclude effort-
induced bronchoconstriction. We delayed the bronchial challenge
until patients were asymptomatic and able to reproduce baseline
flow-volume loops with a maximal variability of 10%. Pulmonary
obstruction was found in three patients (cases 5, 7 and 12), but
only one patient (case 7) was wheezing. Orally administered or
inhaled bronchodilators were discontinued 24 hours or more before
the beginning of the evaluation. At the time of testing, no one
was taking systemic glucocorticoids or cromolyn sodium.

Beginning on the morning of the second day, each patient
underwent bronchial challenge with room air as a placebo or formaldehyde gas. We tested only one formaldehyde concentration or
the placebo each day. Depending on the time constraints for each
patient, we tested with one or several concentrations of formaldehyde: 0.1, 1, and 3 ppm. Formaldehyde gas or placebo was delivered
from a Dynacalibrator (model 340-23-X, Metronics, Santa Clara, CA)
at a flow rate of 10 to 14 liters/min through Teflon tubing to a
loosely fitting face mask for 20 minutes. Most of the gas over-
flowed around the mask and was exhausted from the chamber, which
had dimensions of 1.5 by 2 by 3 m. Spirometry was performed before
the challenge, immediately after, and 1/4, 1/2, 1, 3, 6 and 24
hours after the challenge. In three of these patients, the
sequence of challenge was double-blind and randomized except that a
larger concentration of formaldehyde was always preceded by a
smaller or by placebo. In the others, it was single-blind. A
bronchial challenge was considered positive when it was followed by
a 20% or more decrease of the baseline FEV_1 in a dose-related
fashion with the various formaldehyde concentrations tested, and
the placebo challenges did not produce the same effect.

RESULTS

Table I shows the greatest decreases in FEV_1 after the
formaldehyde and placebo challenges. In only one patient (case 7)
did the FEV_1 decline by 20% or more after exposure to formaldehyde,
and in this case the placebo challenge induced almost the same

decrease in FEV_1 as did formaldehyde. For the rest of the patients, the FEV_1 did not diminish significantly after challenge with formaldehyde gas.

In addition to the FEV_1, we measured the vital capacity and airflows at 50% and 25 to 75% of the expiratory volume. All of these variables showed changes similar to the FEV_1 values reported in Table I.

The health of all of our patients throughout the period of evaluation and testing remained stable, and they did not require any treatment. Several patients noted subjective symptoms such as irritation of the eyes, nose, and throat and tightness of the chest, but these occurred as frequently with the placebo as with the formaldehyde challenges and were without any distinguishing features. Consequently, we have interpreted all of these challenges as negative.

TABLE I.   Pulmonary Function Before and After Bronchial Challenge

FEV_1 (liters)

Case	Before Formal- dehyde	After Formal- dehyde	% Change	Before Placebo	After Placebo	% Change
1	3.50	3.39	-3.2	3.49	3.28	-6.0
2	3.04	2.84	-6.6	3.24	3.06	-5.6
3	3.21	2.88	-10.3	3.32	3.23	-2.7
4	3.45	3.22	-6.7	3.23	3.34	+3.4
5	1.80	1.55	-13.9	1.85	1.55	-16.2
6	2.70	2.65	-1.9	2.50	2.55	+2.0
7	1.75	1.30	-25.7	1.85	1.40	-24.3
8	2.35	2.00	-14.9	ND	ND	---
9	2.35	2.00	-14.9	2.30	1.95	-18.8
10	2.45	2.15	-12.2	2.45	2.15	-12.2
11	3.10	2.75	-11.3	3.10	2.95	- 4.8
12	1.25	1.05	-16.0	ND	ND	---
13	3.05	2.60	-14.8	3.40	3.40	0

FEV_1=forced expiratory volume in 1 second; ND=not done; % change= the change in the FEV_1 measured after bronchial challenge as a percentage of the baseline FEV_1 before the bronchial challenge. This table was derived from a more detailed tabulation of meas- urements, which is available upon request. (Reprinted with per- mission from Frigas E et al, Mayo Clin. Proc. 59:295, 1984).

Of our 13 patients, 3 (cases 1, 3 and 13) also underwent bronchial challenge with methacholine. Two of these patients (cases 1 and 3) had negative challenges, whereas the third patient

had a positive challenge. Three patients (cases 5, 7 and 12) were
not tested with a methacholine challenge because they had unequivocal histories of asthma and pulmonary obstruction when pulmonary
function tests were done. Also, five patients (cases 4, 9, 10 and
11) had convincing histories of asthma, thus, testing them with a
methacholine bronchial challenge was avoided. The remaining two
patients (cases 2 and 8) were not tested because of time
constraints.

The final diagnoses were as follows: habitual cough (case
1); possible asthma, now inactive and unrelated to formaldehyde
exposure (cases 4, 6, 9 and 10); history of mild exercise-induced
asthma, unrelated to formaldehyde exposure (case 11); mild to
moderate asthma of the nonallergic variety, unrelated to formaldehyde exposure (cases 5, 7, 12 and 13); and no respiratory diagnosis
(cases 2, 3 and 8). In addition, some of the patients had a secondary diagnosis of vasomotor rhinitis and nasal polyps (cases 4, 5,
6, 7, 8 and 13) plus hyperplastic sinusitis (cases 4 and 5).

DISCUSSION

These data deal with the issue of whether formaldehyde-induced
asthma can develop in persons exposed to low levels (less than 3
ppm) of the gas. Besides published case reports (1,6,12) that have
raised the question of formaldehyde-induced asthma, the previous
studies (3,6,7) that have published accounts of "positive" formaldehyde challenges in patients with asthma lacked adequate controls,
and the subjects tested were exposed to unknown but possibly high
concentrations of formaldehyde gas. Accurate delivery of a specific concentration of formaldehyde gas is difficult because of its
extreme solubility in water, its reactivity with rubber or
plastics, and its self-polymerization. With the Dynacalibrator,
we are confident that we delivered the concentrations intended.
The delivered concentration was confirmed by measuring formaldehyde
gas at the end of our delivery system.

Ideally, the bronchoprovocation should be a double-blind test
with placebo challenges interspersed randomly in a sequence of
increasing concentrations of formaldehyde. In this type of testing, placebo controls are essential, both because of psychologic
effects and the fear that surrounds this particular agent (many of
the subjects are involved in litigation) and also because of the
spontaneous variability of indices of pulmonary function in
patients with asthma. A bronchial challenge can be interpreted as
positive when a patient who is asymptomatic, with normal or near-
normal results of pulmonary function tests without medication and
who is able to reproduce a steady baseline FEV_1, shows a sustained
decrease in FEV_1 of 20% or more and in a dose-related fashion after
exposure to the material in question, while placebo produces a
negligible change or no change.

Interpretation of these tests in patients with chronic airway

obstruction is more difficult. Because medications may alter the
response, their use must be discontinued. The test must be
repeated often enough to show consistently that the response after
exposure to the suspected agent is significantly different from the
response after the placebo challenge. When airway caliber is frequently changing because of the disease, any change after a single
stimulus may represent only a chance variation. An apparently
positive test must be confirmed by repeating it.

Another issue is the duration of exposure. It is conceivable
that exposure to 1 ppm of formaldehyde gas for 20 minutes may not
induce asthma, whereas exposure to 1 ppm for several hours may do
so. No precedence exists for this hypothesis from experience with
challenges with other agents. As brief an exposure as five inhalations has become the standard bronchial provocation protocol for
allergens such as ragweed (16). Also, a 20 minute exposure to
toluene diisocyanate, like the procedure described herein for formaldehyde, regularly produced a positive response in workers with
occupational asthma due to diisocyanates (17). In addition, a
rapidly metabolized substance such as formaldehyde is unlikely to
have a cumulative effect.

In conclusion, testing with a formaldehyde bronchial challenge
did not provoke asthma in 13 selected patients with symptoms of
asthma and a history of exposure to formaldehyde gas (18). If formaldehyde causes asthma, it does so rarely.

Returning to our original patient with asthma from the urea-
formaldehyde dust, what future testing could be considered? First
of all, she may be unique; no similar cases have come to our attention. Second, the dust from the walls was abundant and unusually
buoyant; it might have provoked the reaction by purely mechanical
stimulus of rapidly-adopting irritant receptors. Third, the dust
might have contained unknown insect or microbial antigens. Unless
similar cases present themselves for study of these issues, no
final answer will be possible.

REFERENCES

1. Vaughan, W.T. Practice of Allergy (St. Louis, MO: C.V. Mosby
    Co., 1939), p. 677.

2. Harris, D.K. "Health problems in the manufacture and use of
    plastics," Br. J. Ind. Med. 10:255-268 (1953).

3. Popa, V., Teculescu, D., Stanescu, D., Gavrilescu, N.
    "Bronchial asthma and asthmatic bronchitis determined by
    simple chemicals," Dis. Chest 56:395-404 (1969).

4. Schoenberg, J.B., Mitchell, C.A. "Airway disease caused by
    phenolic (phenol formaldehyde) resin exposure," Arch.
    Environ. Health 30:574-577 (1975).

5. Kerfoot, E.J., Mooney, T.F., Jr. "Formaldehyde and paraformaldehyde study in funeral homes," Am. Ind. Hyg. Assoc. J. 36:533-537 (1975).

6. Hendrick, D.J., Lane, D.J. "Formalin asthma in hospital staff," Br. Med. J. 1:607-608 (1975).

7. Hendrick, D.J., Lane, D.J. "Occupational formalin asthma," Br. J. Ind. Med. 34:11-18 (1977).

8. "Occupational Exposure to Formaldehyde," DHEW Publication No. NIOSH 77-126 (Washington, DC: Department of Health, Education and Welfare, Government Printing Office, 1976).

9. Baumann, H. "Formaldehyde in UF-schuim," Plastica 30:72-74 (1977).

10. Bardana, E.J., Jr. "Formaldehyde: hypersensitivity and irritant reactions at work and in the home," Immunol. Allergy Pract. 2:11-23 (1980).

11. Godish, T. "Formaldehyde and building-related illness," J. Environ. Health 44:116-121 (1981).

12. Bernardini, P., Carelli, G., Valentino, R. "Formaldehyde in insulated housing (letter to the editor)," Lancet 2:375 (1981).

13. Breysse, P.A. "The health cost of 'tight' homes (editorial)," JAMA 245:267-268 (1981).

14. "The health hazards of formaldehyde," Lancet 1:926-927 (1981).

15. Hendrick, D.J., Rando, R.J., Lane, D.J., Morris, M.J. "Formaldehyde asthma: challenge exposure levels and fate after five years," J. Occup. Med. 24:893-897 (1982).

16. Frigas, E., Filley, W.V., Reed, C.E. "Asthma Induced by Dust From Urea-Formaldehyde Foam Insulating Material," Chest 79:706-707 (1981).

17. Frigas, E., Filley W.V., Reed, C.E. "UFFI Dust - Nonspecific Irritant Only?" Chest 82:511-512 (1982).

18. Frigas, E., Filley, W.V., Reed, C.E. "Bronchial challenge with formaldehyde gas: lack of bronchoconstriction in 13 patients suspected of having formaldehyde-induced asthma," Mayo Clin. Proc. 59:295-299 (1984).

19. Chai, H., Farr, R.S., Froehlich, L.A., Mathison, D.A., McLean, J.A., Rosenthal, R.R., Sheffer, A.L., II, Spector, S.L.,

Townley, R.G. "Standardization of bronchial inhalation challenge procedures," J. Allergy Clin. Immunol. 56:323-327 (1975).

20. Butcher, B.T., Salvaggio, J.E., Weill, H., Ziskind, M.M. "Toluene diisocyanate (TDI) pulmonary disease: immunologic and inhalation challenge studies," J. Allergy Clin. Immunol. 58: 89-100 (1976).

30.

Volatile Organic Compounds in Indoor Air: An Overview of Sources, Concentrations, and Health Effects

David A. Sterling

IIT Research Institute, Chicago, Illinois

It is now evident that indoor contaminant levels are often higher than outdoors, and at times may exceed ambient and even occupational standards [1,2]. Energy conservation measures which serve to: (1) tighten building structures, resulting in reduced air exchange rates; or (2) decrease the intake of fresh make-up air into ventilation systems, have intensified indoor-air-quality problems.

Contaminants originating outdoors infiltrate through cracks and openings, enter directly with make-up air, or are carried inside on hair, skin, and clothing of occupants. Contaminants are also generated indoors from building materials and interior furnishings (particularly from synthetic materials) and from daily activities such as cooking and cleaning, from the use of various consumer products, and from normal biological processes of people and pets. Types and levels of contaminants found indoors then consist of outdoor pollutants that have infiltrated as well as those generated indoors.

From a health perspective, indoor air quality (IAQ) is more important than quality of air outdoors, because most people spend the majority of their time in enclosed structures, such as office buildings, schools, shopping malls, and homes. The exposed population is diverse, including many especially sensitive individuals. Exposure standards set for working populations are not strictly applicable for all individuals exposed indoors. Employed individuals in general are healthier than the general population, who are composed not only of those "healthy" individuals fit for employment but also of children, the elderly, and those with pre-existing conditions who are potentially more susceptible to the possible adverse effects from exposure to pollutants [2,3,4].

SOURCES AND CONCENTRATIONS

Formaldehyde

Formaldehyde has been extensively investigated as an outdoor air pollutant. There are numerous sources of formaldehyde emissions indoors. (The most common sources are shown in Table 1.) The major reported, as well as publicized, sources of formaldehyde are from urea-formaldehyde foam insulation (UFFI) [5,6], and particle board or pressed-wood products [5,7,8,9]. Formaldehyde is used in consumer paper products treated with UF resins, including grocery bags, waxed papers, facial tissues, and paper towels. Many common household cleaning agents contain formaldehyde. Lachapelle et al. [10] investigated 80 common household cleaning agents and reported 17 to be positive when tested for formaldehyde. As a stiffener, wrinkle resister, and water repellent, UF resins are present in floor coverings, carpet backings, adhesive binders, fire retardants, and permanent-press clothes [11,12]. Other sources are combustion devices used for heating and cooking fuels, such as natural gas and kerosenes [7,13,14,15].

TABLE 1.   Examples of Formaldehyde (Aldehyde) Uses
and Potential Indoor Sources

Source Categories	Examples
Paper products	Grocery bags, waxed paper, facial tissues, paper towels, disposable sanitary products
Stiffeners, wrinkle resisters, and water repellents	Floor coverings (rugs, linoleum, varnishes, plastics), carpet backings, adhesive binders, fire retardants, permanent-press clothes
Insulation	Urea-formaldehyde foam insulation (UFFI)
Combustion devices	Natural gas, kerosene, tobacco
Pressed-wood products	Plywood, particle board, decorative paneling
Other	Cosmetics, deodorants, shampoos, fabric dyes, inks, disinfectants

Findings from various studies show how the numerous sources of formaldehyde indoors produce various indoor concentrations. Results from five selected studies are presented in Table 2 as examples. Twenty-eight residences, which had UFFI from six months to nine years old and which did not have particle board flooring, were monitored by Godish [16]. A mean formaldehyde concentration of 0.07 ppm, with a range of 0.02 to 0.13 ppm, was found. Control homes had a mean of 0.05 ppm and a range of 0.03 to 0.07 ppm. In

TABLE 2.   Selected Examples of Observed
Formaldehyde Concentrations

Sampling Site	Concentration, ppm		
	Range	Mean	Reference
28 Residences (1981)			[16]
UFFI	0.02–0.13	0.07	
Control	0.03–0.07		
78 Structures (1983)			[18]
Apartments		0.08	
UFFI and non–UFFI	0.03–0.20	0.05	
Public buildings		0.04	
3 Residences			[19]
UFFI	0.11–0.16	--	
Non–UFFI	0.06–0.08	--	
Energy–efficient non–UFFI	0.13–0.17	--	
164 Mobile homes (1984)	<0.02–0.78	0.15	[22]
65 Mobile homes (complaint) (1979)	<0.01–3.68	0.47*	[23]
65 Mobile homes (1979)	<0.10–0.80	0.16*	[11]

* Median

another study, Georghiou et al. [17] measured formaldehyde concentrations in 44 UFFI and six control homes.   Up to four measurements in
each residence were obtained over a five–month period.   The homes
with UFFI had statistically significant higher formaldehyde concentrations than the controls.   The age of the insulation was an important long-term factor.   Of the homes where UFFI had been installed
within two years of the measurement, 84% had values greater than
0.06 ppm and 58% had greater than 0.1 ppm.   In homes with UFFI three
years old, 44% of the homes had formaldehyde concentrations greater
than 0.06 ppm and only one had greater than 0.1 ppm.   In homes with
UFFI more than three years old, only 22% had formaldehyde concentrations greater than 0.06 ppm and none had greater than 0.1 ppm.

   TerKonda et al. [18] monitored formaldehyde and total aldehydes
in 78 buildings over a one-year period which included residences,
apartments, mobile homes, offices, and public buildings.   The mean
concentrations for each type of building ranged from a high of 0.2
ppm in mobile homes to a low of 0.03 ppm in office buildings.   The
mean formaldehyde concentration in apartments was reported as 0.08
ppm.   In both UFFI and non–UFFI homes the mean formaldehyde concentrations were 0.05 ppm, and in public buildings they were 0.04 ppm.
Indoor total aldehyde concentrations were 8 to 60 times higher than
outdoors.   The outdoor mean was reported as 0.003 ppm.   Formaldehyde
was typically 40% to 80% of the total aldehydes measured.

Formaldehyde and ventilation rates were monitored daily for a one-year period in three types of residences by Gammage et al. [19]: a ten-year-old home with no particular source of formalde- hyde; a three-year-old UFFI home; and an energy-efficient home with substantial quantities of pressed wood but no UFFI. Daily diurnal variations of formaldehyde levels showed a range of 0.06 ppm to 0.08 ppm in the ten-year-old home, 0.11 ppm to 0.16 ppm in the UFFI home, and 0.13 to 0.17 ppm in the energy-efficient home. Seasonal fluctu- ations of formaldehyde in the UFFI home varied by an order of magni- tude, typically brought about by abrupt changes in environmental conditions both indoors and outdoors. The measured levels were typically lower during the spring and fall periods, when windows were open. Levels were higher during hot and humid weather, when windows were typically closed, and also when the heating system was first used.

Hawthorne et al. [20] monitored formaldehyde levels as well as other indoor contaminants in residences for a ten-month period. Formaldehyde concentrations were associated with age of the home and season. Seasonal effects were noted especially during October, when windows were often open. The range of formaldehyde levels found was 0.025 ppm to 0.4 ppm. Hollowell et al. [21] also found that formal- dehyde levels drop substantially when windows are open.

In a recent study of 164 mobile homes of less than five years of age in four geographic locations in Texas, an overall mean of 0.15 ppm and a range of less than 0.02 ppm to 0.78 ppm were found [22]. Another study of mobile homes, where complaints had been reported, showed a median of 0.47 ppm and a range of less than 0.01 to 3.68 ppm [23]. In yet another study of 65 mobile homes, Hanrahan et al. [11] reported a median of 0.16 ppm and a range of from below detection to 0.80 ppm.

The range of formaldehyde measured in complaint homes, mobile homes, and homes containing large quantities of particle board or UFFI tends to be from 0.02 ppm to 0.80 ppm, with elevated levels as high as 4 ppm in some instances. Older conventional homes typically have formaldehyde values less than 0.05 ppm. At the same time, out- door concentrations of formaldehyde range from 0.002 to 0.006 ppm in remote, unpopulated regions, to 0.010 to 0.020 ppm and sometimes 0.050 ppm in highly populated areas and industrial urban air.

There are various factors which affect the indoor concentra- tions of formaldehyde even in similar type homes, as illustrated by the ranges of values shown in Table 2.

A residential environment, as any other environment, is not static. Temperature, humidity, and air exchange rates vary over the course of a day, from day to day, and seasonally. These environmen- tal changes have subsequent effects on indoor concentrations. Other factors are the types and amounts of material used in construction and furnishings, volume to surface area ratio of those materials, type and method of air control, the "tightness" of the structures, indoor activities, and so on.

Volatile Organic Compounds

Great emphasis is now being placed on the spectrum of volatile organic compounds (VOCs) which may be found indoors and on the potential health effects of exposure to these compounds. Like formaldehyde, other VOCs have numerous indoor sources. Indoor/outdoor comparisons have shown that many more organic vapors are found indoors than outdoors. More than 350 different organic compounds have been identified in concentrations of over 0.001 ppm in indoor air. This large and ever-growing number makes it difficult to associate health and comfort problems with specific compounds.

Organic compounds are part of almost all materials and products in use, such as, for example, in construction materials, furnishings, combustion fuels, consumer products, and pesticides. A large variety of organic compounds are produced from combustion of cooking and heating fuels, tobacco, and human metabolism. Examples of organic compounds and their sources are shown in Table 3.

TABLE 3. Examples of Organic Compound Types and Potential Indoor Sources

Pollutant Type	Example	Indoor Sources
Aliphatic hydrocarbons	Propane, butane, hexane, limonene	Cooking and heating fuels, aerosol propellants, cleaning compounds, refrigerants, lubricants, flavoring agents, perfume base
Halogenated hydrocarbons	Methyl chloroform, methylene chloride, PCBs	Aerosol propellants, fumigants pesticides, refrigerants, and degreasing, dewaxing, and dry cleaning solvents
Aromatic hydrocarbons	Benzene, toluene, xylenes	Paints, varnishes, glues, enamels, lacquers, cleaners
Alcohols	Ethanol, methanol	Window cleaners, paints, thinners, cosmetics, adhesives, human breath
Ketones	Acetone	Lacquers, varnishes, polish removers, adhesives
Aldehydes	Formaldehyde, nonanal	Fungicides, germicides, disinfectants, artificial and permanent-press textiles, paper, particle boards, cosmetics, flavoring agents, etc.

Aliphatic hydrocarbons contain the alkanes, alkenes, and alkynes. Principal sources indoors are: cooking and heating fuels; aerosol propellants such as propane and butane; and cleaning compounds, glues, and thinner solvents such as hexane. Halogenated hydrocarbons are widely used in degreasing, dewaxing, and dry cleaning solvents such as methyl chloroform and methylene chloride in home products. Fluorescent light ballasts, many transformers, and other home appliance dielectrics contain PCBs. Aromatic hydrocarbons such as benzene and toluene come from paints, varnishes, glues, enamels, lacquers, and household cleaners. Alcohols, such as ethanol and methanol, come from window cleaners, paints, paint thinners, cosmetics, and adhesives. Ketones are present in lacquers, varnishes, polish removers, and adhesives.

Many pesticides are commonly used in almost all places where people gather, work, and live. Examples shown in Table 4 include Chlordane, Heptachlor, Malathion, Diazinon, Dursban, Ronnel, and Dichlorvos. There have been many reports of illness of residents after the use of pesticides around the home [24].

TABLE 4.   Measurements of Pesticides
in Residences

Pesticide	Range, $\mu g/m^3$	
MacLeod [25]		
Dichlorvos	0.05 − 28.0	
Malathion	0.10 − 1.0	
Ronnel	0.01 − 10.0	
Diazinon	0.01 − 2.0	
Stevens* [26]		
Dichlorvos	7.50	
Ronnel	0.24	
Malathion	0.17	
Jensen [27]		
Chlordane	0.1 − 10.0	
Ronnel	0.2 − 2.0	
Dursban	0.2 − 2.0	
Dichlorvos (DDVP)	0.5 − 10.0	
Malathion	0.2 − 2.0	
Diazinon	0.2 − 2.0	

*A single residence

Many types of organic compounds from indoor sources have been reported in the literature, but how may these reports be used to target significant compounds? How often are these compounds reported? Which compounds are reported, and at what concentrations? As an example, Table 5 indicates the number of reports where VOCs

TABLE 5. Number of Measurements of Selected Pollutants in 143
Reports of Indoor Air Quality Studies of Working Buildings*

Pollutant	Example	Number of Reports
Aromatic hydrocarbons	Benzene, toluene	58
Alcohols	Ethanol	11
Aldehydes	Acetaldehyde	8
Amines	Analine	10
Chlorinated hydrocarbons	Trichloroethylene	17
Formaldehyde	--	44
Hydrocarbons (aliphatic)	Hexane	77

*Derived from the Building Performance Database of TDS Ltd. [28]

were cited as being measured, out of a total of 143 reports from
office building investigations. This table does not indicate the
percentage of each compound typically found indoors; not all reports
are used, and not all types of compounds were monitored in each
study reported. However, the table does indicate, in general, the
types of compounds which are being found more often than others and
the extent to which VOCs are found indoors. This is an approach
that needs to be considered to isolate important target compounds.
Monitoring for selected compounds and the researchers' use of different collection and analytical techniques makes comparison of the
studies difficult, but not impossible.

Table 6 lists those specific compounds which were identified in
four out of six reports in which identification of the spectrum of
indoor VOCs was accomplished with GC-MS. This list may be considered as an example of an attempt to target important compounds
found indoors. The reports listed totals of 45 to 359 compounds as
being identified. In Table 6, the compounds reported most often are
categorized as alkanes, terpenes, benzene types, chlorinated, and
other hydrocarbons. Where applicable, the OSHA permissible limits
(PELs) are listed.

Note that the OSHA standard concentrations are in mg/m^3, while
concentration ranges of the target compounds are in $\mu g/m^3$. In no
case are the levels within a factor of 100 of the OSHA PELs except
for benzene, which has been established as a human carcinogen.
Although the concentrations reported are low in comparison to OSHA
standards, the standards are set for working, "healthy" populations,
not the diverse population found in the nonoccupational environment.

TABLE 6.   Common Organic Compounds Measured Indoors

Compounds	Ranges of Means, $\mu g/m^3$	OSHA PELs, mg/m^3
Alkanes (C$_6$ to C$_{12}$)	5 - 430	--
Terpenes		
α-Pinene	14 - 122	560*
Limonene	13 - 126	--
Benzenes		
Benzene	10 - 52	4.3
Toluene	10 - 610	820
1,3 and 1,4 Xylene	9 - 92	⌈435
1,2 Xylene	3 - 33	
Ethyl benzene	5 - 40	⌊435
Chlorinated		
Tetrachloroethylene	4 - 198	670
1,1,1-Trichloroethane	15 - 290	1900
Trichloroethylene	2 - 68	535
1,4 Dichlorobenzene	7 - 99	450
Other		
Naphthalene	1 - 26	50
Total	293 - 4937	--
Number of reported identified compounds [29,34]	45 - 359	

*As turpentine

HEALTH EFFECTS

Of the compounds from the six studies used in Table 6, the compounds most frequently identified were toluene, the xylenes, and α-pinene [29-34]. Other studies where specific sampling and analysis were performed have also reported similar results. Among the compounds found indoors are some known and suspected human or animal carcinogens: benzene, tetrachloroethylene, trichloroethane, trichloroethylene, and formaldehyde. Furthermore, limonene has been selected for testing by the National Toxicology Program for carcinogens, and the carcinogenicity data for dichlorobenzene are inconclusive. Limonene, α-pinene, and toluene are also on the EPA's priority list for carcinogenicity testing.

Because of the large number of compounds found indoors and the low concentrations measured, there are few data for many of the compounds. Table 7 lists the number of reported negative and/or

TABLE 7. Mutagenicity and Carcinogenicity of
19 Chemicals Found in Indoor Air*

Chemical	Dose Range,* μg/plate	S9	Reported† Mutagenicity	Reported† Carcinogenicity
Formaldehyde	20 - 500	−	−(5)	+(2)
		+	+(5)	
Benzene	10 - 250	−	−(5)	−(1)
		+	−(5)	+(2)
p-Xylene	10 - 1000	−		(0)
		+	−(4)	
			−(1)	
m-Xylene	10 - 1000	−		
		+		
n-Butanol	50 - 2000	−	−(2)	−(1)
		+	−(1)	
Camphene	10 - 1000	−	−(2)	(0)
		+	−(1)	
Toluene	50 - 2000	−	−(5)	−(1)
		+	−(1)	
Nonanal	50 - 2000	−	−(4)	(0)
		+	−(1)	
α-Pinene	10 - 500	−	−(4)	(0)
		+	−(1)	
Limonene	10 - 500	−	−(3)	(0)
		+	−(1)	
o-Xylene	10 - 1000	−	−(3)	(0)
		+	−(1)	
Naphthalene	100 - 2000	−	−(4)	−(1)
		+	−(1)	
p-Dichlorobenzene	10 - 1000	−	−(3)	(0)
		+	−(1)	
Tetrachloroethylene	50 - 2000	−	−(3)	+(1)
		+	−(1)	
Styrene	50 - 2000	−	−(5)	+(2)
		+	+(1)	
Tetradecane	50 - 2000	−	−(1)	(0)
		+	−(1)	

TABLE 7. Mutagenicity and Carcinogenicity of
19 Chemicals Found in Indoor Air* (continued)

Chemical	Dose Range,* µg/plate	S9	Reported† Mutagenicity	Reported† Carcinogenicity
n–Undecane	50 – 2000	–	–(1)	(0)
		+	–(1)	
m–Dichlorobenzene	50 – 2000	–	–(2)	(0)
		+	–(1)	
o–Dichlorobenzene	50 – 2000	–	–(2)	? (1)
		+	–(1)	

*Derived from Connor et al. [35].

†() Number of References reporting the indicated result.
   References are reported in Connor et al. [35].

positive findings for mutagenicity and carcinogenicity from bacterial strain tests for selected compounds. In most instances there
have been few studies, and for most of the compounds no studies
related to carcinogenicity have been reported. Health effects
observed from exposure of humans to formaldehyde and other organic
compounds come primarily from occupational studies. In these
studies the compound may typically be singled out, and concentration
and exposure estimates may be determined to establish dose response
curves. Health effects observed in these situations are from acute
or chronic exposures at levels much higher than would typically be
found indoors. This is particularly true for organic compounds
other than formaldehyde, which may be found in high concentrations
indoors but at much lower concentrations than in occupational settings.

Various compounds found indoors have been associated with a
number of symptoms. Typical tight-building-syndrome symptoms are:

- Headache
- Irritation of the eyes and of mucous membranes
- Irritation of the respiratory system
- Drowsiness
- Fatigue
- General malaise.

Aldehydes, and many of the other organic compounds such as alcohols
and aromatic hydrocarbons, are eye and upper respiratory irritants.
The range and severity of symptoms are variable and appear to depend
greatly on the sensitivity of an individual at low concentrations.

A number of health effect studies of low-level exposure to formaldehyde have reported:

- Estimates that 10 to 20% of the general population may be sensitive;

- Odor threshold as low as 0.01 ppm;

- Concentrations of 0.1 to 3 ppm, and sometimes as low as 0.03 ppm in sensitive individuals, induce eye, nose, and throat irritation, and aggravate chronic bronchial problems;

- At higher concentrations the intensity of symptoms increases;

- Delayed hypersensitivity in some individuals upon repeated challenge;

- Ocular discomfort, respiratory irritation, head-aches, and skin maladies in residents of homes where complaints have been reported.

Studies using animals have shown strong evidence that formaldehyde may be a potential carcinogen in humans. Human epidemiologic studies of workers in industry and professionals exposed to formald-ehyde show a higher standardized mortality rate for all types of cancers [36]. Also, formaldehyde has been shown to alter DNA struc-ture and produce chromosomal aberrations. Sufficient evidence has not been provided to conclude that formaldehyde is teratogenic or carcinogenic in man, but formaldehyde should be considered at least a potential carcinogen until shown otherwise.

More objective psychological and neurophysical studies have suggested central nervous system effects at approximately 1 ppm. The effects seen were subtle changes in short-term memory, increased anxiety, and slight changes in dark adaptation abilities [37].

Health effects from exposure to organic compounds at the low levels found indoors are difficult to determine. The exposures are numerous, complaint symptoms are common to many compounds, and the population is diverse. In general, VOCs are lipid soluble and easily absorbed through the lungs. Their ability to readily cross the blood-brain barrier may induce CNS depression, a possible causal factor in drowsiness, fatigue, and general malaise. Alcohols, arom-atic hydrocarbons, and aldehydes can irritate mucous membranes. Further, a number of organic compounds have been implicated as proven or suspected human carcinogens.

Another impact on health effects not generally considered is from odor. Organic compounds which are commonly considered primary contributors to indoor odors are:

- Limonene
- α-Pinene
- n-Hexanol
- 1,3-Xylene
- Other terpenes.

Odors, particularly those that are unpleasant or those one is not used to, can act as stressors as well as serving to initiate complaints. Odors may also serve as an indication of unwanted organic compounds indoors.

When attempting to assess health effects from exposure to organic vapors indoors, as well as to most other indoor pollutants, there are a number of considerations:

- No one compound, or no one exposure, may be the cause of adverse health effects or discomfort. There may be additive or synergistic effects;

- The concentration of organic compounds in general is very low. The health effects of chronic exposures versus acute exposures at these levels are not known;

- The exposed population is diverse, and may vary in age, sex, susceptibility, and life style;

- Annoyance from odors may lead to complaints as well as serving as stressors, thus increasing susceptibility to other agents. Odors may also be used as an indication of the presence of organic compounds;

- Exposure to organic compounds may increase susceptibility to effects from exposure to other agents, and vice versa, and/or may aggravate pre-existing conditions; and

- Sensitization and allergic reactions to various compounds may occur over time.

CONCLUSION

There are only a limited number of studies which have attempted broad spectrum analysis of organic compounds indoors. Among these studies, different collection and analytical techniques are used. Because of the large range of organic types, each method is only a window for looking at a specific type. For an analogy, if one looks out a window facing west, there may be something of importance in another direction that is being missed. More work is needed to further identify and isolate the types of compounds consistently found indoors. Because of the difficulty in assessing health effects to indoor nonoccupational exposures at low concentrations, more objective determinations of potential health effects from exposures to these compounds need to be developed.

With the increasing number of studies on VOCs indoors, it may be possible to begin to isolate those compounds consistently found indoors, identify major sources, determine potential health effects, and apply controls.

REFERENCES

(1) Sterling, T. D., and Kobayashi, D., "Exposure to Pollutants in Enclosed 'Living Spaces'," Environ. Res. 13:1–35 (1977).

(2) Yocom, J., "Indoor-Outdoor Air Quality Relationships," J. Air Poll. Control Assoc. 32(5):500–520 (1982).

(3) Holcombe, J., and Kalika, P., "The Effects of Air Conditioning Components on Pollution in Intake Air," ASHRAE Trans. 77:33 (1971).

(4) Halpern, M., "Indoor/Outdoor Air Pollution Exposure Continuity Relationship," J. Air Poll. Control Assoc. 28:689–691 (1978).

(5) Meyer, B., Urea Formaldehyde Resins (Addison-Wesley Pub. Co., Reading, MA, 1979).

(6) NRC, Formaldehyde and Other Aldehydes (National Research Council, National Academy Press, Washington, DC, 1981).

(7) NRC, Indoor Pollutants (National Research Council, National Academy Press, Washington, DC, 1981).

(8) Deppe, H., "Emission of Organic Substances from Wood Raw Materials," Holz Zentralblatt 108:123–126 (1982).

(9) Pickrell, J., Mokler, B., Griffin, L., and Hobbs, H., "Form-aldehyde Release Coefficients From Selected Consumer Products," Environ. Science Technol. 17(12):753–757 (1983).

(10) Lachapelle, J., and Tennsted, D., "Formaldehyde in Household Cleaning Agents," Contact Dermatitis 7(3):166–167 (1978).

(11) Hanrahan, L., Dally, K., Anderson, H., Kanarek, M., and Rankin, J., "Formaldehyde Vapor in Mobile Homes: A Cross Sectional Survey of Concentrations and Irritant Effects," Amer. J. Pub. Health 74(9):1026–1027 (1984).

(12) Bourneard, H., and Seferian, S., "Formaldehyde in Wrinkle
      Proof Apparel Producers: Tears for Milady," Ind. Med. and
      Surg. 28:232-233 (1959).

(13) Gordon, S., Eisenberg, W., and Relwani, S., "Emission Factors
      of Volatile Organic Compounds and Other Air Constituents From
      Unvented Gas Appliances," Proceedings of the Air Pollution
      Control Association Specialty Conference: Measurements and
      Monitoring of Non-Criteria (TOXIC) Contaminants in Air,
      Chicago, IL, March 22-24, 1983.

(14) Girman, J., Geisling, K., and Hodgson, A., "Sources and
      Concentrations of Formaldehyde in Indoor Environments,"
      Presented at the 75th Annual Air Pollution Control Association
      Meeting, New Orleans, LA, June 20-25, 1983.

(15) Traynor, G., Allen, J., Apte, M., Girman, J., and
      Hollowell, C., "Pollutant Emissions from Portable Kerosene-
      Fired Space Heaters," Environ. Science and Technol. 17(6):368-
      371 (1983).

(16) Godish, T., "Formaldehyde and Building Related Illness," J.
      Environ. Health 44 (3):116-121 (1981).

(17) Georghiou, P., Snow, D., and Williams, D., "Formaldehyde
      Monitoring in Urea-Formaldehyde Foam-Insulated Houses in St.
      Johns, Newfoundland, Canada: Correlative Field Evaluation of
      a Real-Time Infrared Spectrophotometric Method," Environ. Int.
      9:279-287 (1983).

(18) TerKonda, P., and Liaw, S., "Monitoring of Indoor Aldehydes,"
      Proceedings of the Air Pollution Control Association Specialty
      Conference: Measurement and Monitoring of Non-Criteria
      (TOXIC) Contaminants in Air, Chicago, IL, March 22-24, 1983.

(19) Gammage, R., Hingerty, B., Matthews, T., Hawthorne, A.,
      Womack, D., and Gupta, K., "Temporal Fluctuations of Formalde-
      hyde Levels Inside Residences," Proceedings of the Air Pollu-
      tion Control Association Specialty Conference: Measurement
      and Monitoring of Non-Criteria (TOXIC) Contaminants in Air,
      Chicago, IL, March 22-24, 1983.

(20) Hawthorne, A., Gammage, R., Dudney, L., Womack, D.,
      Morris, S., Westly, R., and Gupta, K., "Preliminary Results of
      a Forty-Home Indoor Pollution Monitoring Study," Proceedings
      of the Air Pollution Control Association Specialty Conference:
      Measurement and Monitoring of Non-Criteria (TOXIC) Contami-
      nants in Air, Chicago, IL, March 22-24, 1983.

(21) Hollowell, C., and Miksch, R., "Sources and Concentrations of
      Organic Compounds in Indoor Environments," Bull. NY Acad. Med.
      57(10):962-977 (1981).

(22) Stock, T. H., Sterling, D. A., and Monsen, R. M., "A Survey of
      Indoor Air Quality in Texas Mobile Homes," Proceedings of the
      3rd International Conference on Indoor Air Quality and
      Climate, Vol. 4, pp. 331-334, Stockholm, Sweden, August 20-24,
      1984.

(23) Dally, K., Hanrahan, L., Woodbury, M., and Kanarek, M.,
      "Formaldehyde Exposure in Nonoccupational Environments," Arch.
      Environ. Health 36(6):277-284 (1981).

(24) Savage, E. P., Keefe, T. J., and Wheeler, H. W., "National
      Household Pesticide Usage Study, 1976-1977," Fort Collins, CO,
      Colorado State University, 1979.

(25) Lewis, R., and MacLeod, K., "Portable Sampler for Pesticides
      and Semivolatile Industrial Organic Chemicals in Air," Anal.
      Chem. 54(2):310-315 (1982).

(26) Stevens, F. D., "Sampling Methodology for Airborne Semivola-
      tile Organic Pollutants Using Polyurethane Foam," Ph.D.
      Dissertation, University of Texas School of Public Health,
      Houston, TX, June 1984.

(27) Jensen, J., "Potential Pesticide Indoor Air Pollutants," Memo,
      Environmental Protection Agency, November 10, 1980.

(28) Personal Communication, Theodore D. Sterling, Ltd., Vancouver,
      Canada, 1984.

(29) Bortoli, M., Knöppel, H., Pecchio, E., Peil, A., Rogora, L.,
      Schauenburg, H., Schlitt, H., and Vissers, H., "Integrated
      'Real Life' Measurements of Organic Pollution in Indoor and
      Outdoor Air of Homes in Northern Italy," Proceedings of the
      3rd International Conference on Indoor Air Quality and Cli-
      mate, Vol. 4, pp. 21-25, Stockholm, Sweden, August 20-24,
      1984.

(30) Wallace, L., Pellizzari, E., and Hartwell, T., "Analysis of
      Exhaled Breath of 355 Urban Residents for VOCs," Proceedings
      of the 3rd International Conference on Indoor Air Quality and
      Climate, Vol. 5, pp. 15-20, Stockholm, Sweden, August 20-24,
      1984.

(31) Lebret, E., Van de Weil, H., Hoÿ, D., and Boleÿ, J., "Volatile
      Hydrocarbons in Dutch Homes," Proceedings of the 3rd Interna-
      tional Conference on Indoor Air Quality and Climate, Vol. 4,
      pp. 169-174, Stockholm, Sweden, August 20-24, 1984.

(32)   Monteith, D. K., Stock, T. H., and Seifert, Jr., W. E.,
       "Sources and Characterization of Organic Air Contaminants
       Inside Manufactured Housing," Proceedings of the 3rd Interna-
       tional Conference on Indoor Air Quality and Climate, Vol. 4,
       pp. 285-290, Stockholm, Sweden, August 20-24, 1984.

(33)   Pellizzari, E. D., Sheldon, L., Sparacino, C., Bursey, J.,
       Wallace, L., and Bromberg, S., "Volatile Organic Levels in
       Indoor Air," Proceedings of the 3rd International Conference
       on Indoor Air Quality and Climate, Vol. 4, pp. 303-308,
       Stockholm, Sweden, August 20-24, 1984.

(34)   Molhave, L., Anderson, I. B., Lunqvist, G., and Nielson, O.,
       "Gas Emissions from Building Materials: Occurrence and
       Hygienic Assessment," Report by the Danish Building Research
       Institute (SBI), Report #137, 1982.

(35)   Connor, T. H., Theiss, J. D., Hanna, H. A., Monteith, D. K.,
       and Matney, T. S., "Genotoxicity of Organic Chemicals
       Frequently Found in the Air of Mobile Homes," In Press,
       Toxicology Letters, 1985.

(36)   Sterling, T. D., and Arundel, A., "Possible Carcinogenic
       Components of Indoor Air: Combustion Products, Formaldehyde,
       Mineral Fibers, Radiation, and Tobacco Smoke," J. Environ.
       Science and Health: Environmental Carcinogenic Review,
       January 1985.

(37)   Shenker, M., Weiss, S., and Marawski, B., "Health Effects of
       Residents in Homes with Urea Formaldehyde Foam Insulation: A
       Pilot Study," Environ. Int. 8:359-363 (1982).

31.

Volatile Organic Compounds as Indoor Air Pollutants

Lars Mølhave

The Institute of Hygiene, Universitetsparken, DK-8000
Aarhus C, Denmark

INTRODUCTION

The number of complaints about indoor climate has
increased during the last decade, and the complaints are
now wide-spread throughout the non-industrial indoor environment. In consequence the World Health Organization
(1) has defined the Sick Building Syndrome including the
symptoms shown in table 1. The symptoms are related to
irritation of mucous membranes in eyes, nose, and throat,
unspecific neurological and hypersensitivity reactions.
Several atmospheric factors are supposed to be active in relation to these symptoms, and their combined
effect may be multifactorial.

TABLE 1:   THE SICK BUILDING SYNDROME (1)

- Irritation of eye, nose, and throat
- Dry mucous membranes and skin
- Erythema
- Mental fatique, headache
- Airway infections, cough
- Hoarseness, wheezing
- Unspecific hypersensitivity reactions
- Nausea, Dizziness

Among the symptoms, mucous membrane irritation   (irritation of airways, nose, and eye) is very frequent.
These complaints are similar to those reported by persons
exposed to small concentrations of irritating gases and
vapours. Irritating volatile organic compounds may therefore be one of several causes for the irritating symptoms included in the Sick Building Syndrome. This relation between volatile organic compounds and irritation
will be discussed below.

A dose-response model

A simple multifactorial model for irritation in
eye, nose, and throat caused by irritating volatile organic compounds is suggested in figure 1. The building
and atmospheric factors, other than gases and vapours
to be considered, are e.g. temperature, humidity, air
movements, sound, work load etc. These atmospheric and
building factors act together and may have synergistic
or antagonistic interactions on the individual person
who may respond with a number of symptoms according to
his own disposition for developping each of the symptoms included in the sick building syndrome. Further the
intensity of the symptoms may depend on confounders like
age, sex, stresses etc. of the person. Psychosocial feed
back may increase attention to certain symptoms and factors or suppress attention to others.

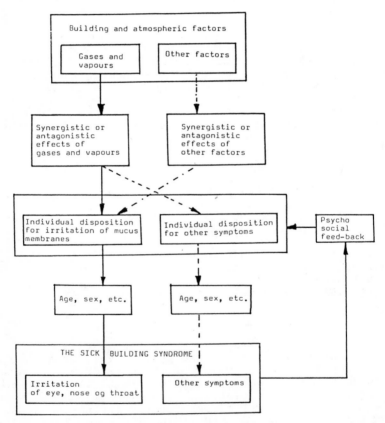

Fig. 1: A general toxicological model for mucous irritation caused by organic gases and vapours.

From the figure a number of characteristics about the Sick Building Syndrome appear. Several causes may be active for each symptom observed in the exposed group. Each person may react differently to the same exposure, and if one factor is eliminated the total number of symptoms in an exposed group of persons may be reduced only weakly.

Depending on their concentration organic gases and vapours of the solvent type is one of several factors from the indoor climate which may cause irritation. From the figure it appears that at least three different types of problems should be considered in relation to irritation caused by gases and vapours. These are the irritation symptoms caused by organic gases and vapours themselves, other symptoms caused by organic gases and vapours, and finally irritation caused by other factors in the environment. The following will, however, only discuss the relation between the total concentration of volatile organic compounds in the non-industrial environment and the irritation symptoms experienced in the eye, nose, and throat.

Human reactions to solvent gases and vapours

Measurements of emission of solvent gases and vapours from 42 building materials (2, 3) showed about 80% of the compounds identified in the air around the materials to be known or suspected mucous membrane irritants.

Irritation in eye, nose, and throat, therefore, may occur in indoor air containing these compounds in sufficient high concentrations. To set an upper limit for acceptable total concentration a pilot exposure experiment with humans was performed at our institute (4, 5). The hypothesis was that typical indoor climate complaints about irritation of eye, nose, and upper airways may result from exposure to organic gases and vapours in small concentrations similar to those found in new buildings. 62 healthy persons participated in the experiment.

The exposure was arranged as a balanced climate chamber experiment with control measurements for each person under neutral conditions. The only varying climate chamber factor was the total concentrations of organic gases and vapours. The concentrations zero, 5 or 25 mg/m^3 were used. These concentrations correspond to clean air, average of normal new houses, and the maximum found in new Danish houses (6). The same mixture of 22 different compounds in fixed concentration ratios were used for the exposure. Compounds were all normal indoor air pollutants.

The experiment focused on the number of different

reactions. Here only subjective voting related to indoor air quality will be discussed. A questionnaire was used to establish the reaction of the subjects to the exposure. Their votings on air quality and odour intensity differed significantly from clean air votings both at 5 and 25 mg/m^3 exposures. Further, a continuous voting showed that the occurrence of dry mucous membranes irritation was significantly increased during exposure to both 5 and 25 mg/m^3. The change of the average voting during exposure to the two concentrations showed that the reaction occurs within the first hour. It was concluded that both the 5 and 25 mg/m^3 organic gases and vapours used in the experiment caused irritation in the nose and upper airways among occupants of non-industrial environments.

In the United States, an outdoor air concentration higher than 0.16 mg/m^3 of unreactive hydrocarbons is considered unacceptable (7). This outdoor air quality standard was established to prevent irritation due to photooxidants resulting from photochemical reactions of the hydrocarbons in the atmosphere. It is not known to what extent these reactions occur in the indoor environment. The limit 0.16 mg/m^3 therefore may be considered an upper limit for acceptable indoor concentrations of organic pollutants. Below this limit no effect is expected due to volatile organic compounds. As no test was performed in our human exposure experiment at total concentrations lower than 5 mg/m^3, this concentration at the present may be considered a limit above which complaints may be expected from some of the persons in an exposed group. The number of symptoms occurring after exposure to concentrations between 0.16 and 5 mg/m^3, may according to the toxicological model in Figure 1, depend on the level of simultaneous exposures to other building or atmospheric factors, the actual chemical composition of the exposure, on the exposed group and the individual sensitivity of the persons in the group.

Volatile organic compounds in the non-industrial indoor environments

Few measurements of the concentration of gases and vapours in the indoor non-industrial environment have been published. Table 2, 3 shows some published results including 12 investigations mentioned in a review by Johansson (8). The range of concentrations in new buildings is from about 0.5 to 19 mg/m^3, which is generally about 10 times larger than the range found in old buildings (range 0.01 - 1,7 mg/m^3). In Ref. 8 it is concluded that the most frequent type of volatile organic compounds in the indoor air is alkanes, most often represented by un-

Table 2: Published Measurements of Total Concentrations of Volatile Organic Compounds in the Non-industrial Indoor Environment.

Investigator	Ref.	Type of room	Number of rooms	Number of occupants	Total concentration mg/m^3		Notes
					Range	Average	
Wang	9	Classroom	1	225-389	0,13-0,18	0,15	b
Wang	9	"	1	No	0,01	0,01	b
Johansson	10	"	2	Occup.	-	0,14	a b
Johansson	10	"	2	No	-	0,05	a b
Johansson	11,12	Kindergarden	4	Occup.	0,29-0,50	-	a b
Mølhave	13,6	New houses	7	No	0,48-18,7	6,2	
Mølhave	13,6	Old houses	39	Occup.	0,02-1,7	0,38	
Frederiksson	14	Problem home	1	"	-	12,9	
Mølhave	15	Experim. house	1	No	-	0,35	
Berglund	16	Home	1	Occup.	-	0,25	a

Table 2: Published Measurements of Total Concentrations of Volatile Organic Compounds in the Non-industrial Indoor Environment. (Continued)

Investigator	Ref.	Type of room	Number of rooms	Number of occupants	Total concentration mg/m³		Notes
					Range	Average	
Berglund	16	Home	1	Occup.	0,63—0,83	–	a
Berglund	17	Kindergar-den	1	"	0,86	–	a
Berglund	17	"	1	"	0,22—0,31	–	a
Mølhave	2,18	Home	1	"	–	1,05	a
Mølhave	2,18	Offices	13	"	1,51—0,09	0,90	
Miksch	19	"	4	"	(0,4—1,6)	–	c

Notes: b: sum of selected compounds
       a: see Johansson, Ref. 8
       c: estimated concentration
       -: no information

Table 3: The Concentration of Organic Gases and Vapours in Non-industrial Indoor Climate (mg/m³).

Occupied	Age	Homes mg/m³	Ref.	Offices etc. mg/m³	Ref.	Schools etc. mg/m³	Ref.
YES	New	12,9	14	–		0,86	17 a
	Old	0,02–1,7 0,25 1,05	13,6 16 2,18	0,09–1,51 0,4–1,6	2,18 19 b	0,13–0,18 0,14 0,29–0,50 0,22–0,31	9 a 10 a 11,12 17
NO	New	0,48–18,7	6,13	–			
	Old	0,24–0,52	15	–		0,01 0,05	9 a 10 a

a: sum of selected compounds   b: estimated   –: no results decane, decane, and nonane. Equally often compounds from the group of alkylated benzenes are found and they are normally represented by toluene which is the overall most frequent seen compound in the indoor air. The third group is the turpines ($C_{10}H_{16}$), whereas the group of other compounds contains alcohols, ketones, and a number of chlorinated hydrocarbons.

Some of the sources in the indoor air for these organic gases and vapours have been identified. Air pollution indoor may originate from the outdoor environment, which is the main source for inorganic air pollutants. The volatile organic compounds are generally found in much higher concentrations indoor than outdoor (8). In some cases human occupancy may increase the concentration of organic gases and vapours. This is often the case for ethyl alcohol (9, 10). Further human activity and processes in the room may emit pollutants. All these sources, however, are only locally or occasionally occurring, and they can hardly explain the nationwide occurrence of the Sick Building Syndrome, which seems to be rather constant in the buildings where it occurs. Left then is degassing from building materials and furniture etc. To investigate this, 42 different types of normally used building materials were examined (2, 3). The same type of compounds appeared to be emitted from the building materials as were found in normal indoor air. Concentrations of gases and vapours around the building materials were similar to those found in new houses. Using a mathematical model, it appeared that the range of concentrations found in modern buildings could be reproduced from the measured emission rates (20). This indicates that these sources are the main contributors for volatile organic compounds in the non-industrial environment. Calculations based on the mathematical model mentioned above (20) showed the concentration of volatile organic compounds in the indoor air to be much more dependent on ventilation for ventilation rates lower than about one air change per hour than for higher air changes (20). Ventilation standards are normally set below one air change per hour (21). A significant increase in indoor air pollution, therefore, may have resulted from the reduced ventilation through the last decade.

CONCLUSION

Based on published measurements of total concentrations of volatile organic compounds in the non-industrial indoor environment, it is argued that total concentrations below 0.16 mg/m^3 may be expected to cause no mucous membrane irritation while total concentration above 5 mg/m^3 are found to do so in a controlled exposure experiment. In the range 0.16 to 5.0 mg/m^3 irritation may

be promoted by other environmental exposures according
to a multifactorial dose respose relation.

REFERENCES

1.   WHO (1983). Indoor Air Pollutants; Exposure and
      Health Effects Assesment. Euro reports and studies
      working group report No. 78; Nördlingen, WHO, Copen-
      hagen.

2.   Mølhave, L., Andersen, I., Lundqvist, G.R., Nielsen,
      P.A., Nielsen, O. (1982). Afgasning fra byggemateri-
      aler - forekomst og hygiejnisk vurdering. SBI-rap-
      port 137, Statens Byggeforskningsinstitut, Hørsholm.

3.   Mølhave, L. Indoor Air Pollution due to Organic Gas-
      es and Vapours of solvents in Building Materials En-
      vironment. Internat. Vol 8; p 117-127, 1982.

4.   Mølhave, L., Bach, B., Pedersen, O.F. (1984). Human
      Reactions during Controlled Exposures to low Concen-
      trations of Organic Gases and Vapours known as nor-
      mal Indoor Air Pollutants. Proceedings of the 3rd
      International Conference on Indoor Air Quality and
      Climate. Stockholm, Aug. 20-24. Vol 3; p 431-437.
      (Edit, Berglund, Lindvall & Sundell. Indoor Air).

5.   Bach, B., Mølhave, L., Pedersen, O.F. (1984). Human
      Reactions during Controlled Exposures to low Concen-
      trations of Organic Gases and Vapours known as In-
      door Air Pollutants: Performance Tests. Proceedings
      of the 3rd International Conference on Indoor Air
      Quality and Climate. Stockholm, Aug. 20-24. Vol 3;
      p 397-403. (Edit, Berglund, Lindvall & Sundall. In-
      door Air).

6.   Mølhave, L., Møller, J., Andersen, I. Luftens ind-
      hold af gasarter, dampe og støv i nyere boliger.
      (Indoor Air Pollution due to Volatile Organic Com-
      pounds and Dust, in Danish). Ugeskr. for Læger,
      141/14; p 956-962, 1979.

7.   DHEW (1970. Air Quality Criteria for Hydrocarbons.
      Dept. of Health, Education and Welfare. N.A.P.C.A.
      Publ. AP 64, Washington, U.S.A.

8.   Johansson, I. (1982). Kemiska luftföroreningar inom-
      hus. En litteratursammanställning. Rapport no. 6/
      1982. Statens Miljømedicinska laboratorium. Stock-

(8)   holm.

9.   Wang, T.C. A Study of Bioeffluents in a College
      Classroom. Ashrae Trans. 81; p 32-44, no. 2328,1975.

10.   Johansson, I. Determination of Organic Compounds in
      Indoor Air with Potential Reference to Air Quality.
      Atmospheric Environment. Vol 12; p 1371-1377, 1978.

11.   Johansson, I., Pettersson, S., Rehn, T. Luftförore-
      ningar i Barn Stugor (Air Pollution in Kindergar-
      dens, in Swedish). VVS Jr 50; p 6-7, 1979.

12.   Johansson, I., Pettersson, S., Rehn, T. Luftförore-
      ningar inomhus (Air Pollution Indoor, in Swedish).
      Svensk VVS 49; p 51-55, 1978.

13.   Mølhave, L., Møller, J. (1979). The Atmospheric En-
      vironment in Modern Danish Dwellings - Measurements
      in 39 Flats; in Fanger & Valbjørn (eds.). Indoor
      Climate; p 171-186, Danish Building Research Insti-
      tute. Copenhagen.

14.   Frederiksson, K. Gifter i Bostadsluft (Indoor Air
      Pollution, in Swedish). Hälsovårds Kontakt - organ
      för svenska Hälsovårdstjänstemanna förbundet 3; 14-
      16, 1979.

15.   Mølhave, L., Andersen, I. Forureningskomponenter i
      indeluften i "nul-energi"-huset, DEH (Air Pollution
      in an Experimental House, in Danish). VARME 45; 121-
      125, 1980.

16.   Berglund, B., Johansson, I., Lindvall, T. Underlag
      för ventilationsnormer. ETAPP II. Final Report to
      STU and BFR, Dec. 1981. The National Institute of
      Environmental Medicine. Stockholm, Sweden.

17.   Berglund, B., Johansson, I., Lindvall, T. A Longi-
      tudinal Study of Air Contaminants in a Newly Built
      Preschool. Report 3, 1982. The National Institute
      of Environmental Medicine. Stockholm, Sweden.

18.   Mølhave, L. Indoor Air Pollution due to Building
      Materials. P 89-110 in: Edet Fanger PO & Valbjørn
      O "Indoor Climate". Danish Building Research Insti-
      tute. Copenhagen 1979.

19.   Miksch, R.R., Hollowell, C.D., Schmidt, H.E. Trace
      Organic Contaminants in Office Spaces. Environmen-
      tal International 8; 129-137, 1982.

20.   Mølhave, L. (1983). The Relation between Emission-
      rate of Organic Gases etc. from Building Materials
      and their Concentrations in the Indoor Environment.
      6th World Congress on the Quality. Vol 2; 345-352.
      Paris.

21.   NKB (1981). Indoor Climate. NKB-report no. 41. Den
      Nordiske Kommité for Bygningsbestemmelser. Bygge-
      styrelsen. Copenhagen.

Summary and Conclusions

Summary and Conclusions

David V. Bates

University of British Columbia
Vancouver, Canada

This has been a timely symposium; the different concerns about the health aspects of indoor air pollution have generated a body of work now large enough for one to begin to attempt some sort of synthesis.

Our speakers and session chairmen have presented and summarized a considerable amount of factual information. I will draw attention to several very very different problems that raise the same general questions. I will then attempt to assess the significance of the data in the context of human health.

Generally, we are trying to understand how to study the effects of low-level exposures. Most of the questions being asked about passive smoking, radon, formaldehyde, and other indoor pollutants have the common characteristic that they involve exposures of inhabitants to low levels of materials over long periods of time. The question of potential health effects presents a considerable challenge. The experimental approach to such questions often hinges on the difficulty of assessing interfering factors. These may be so dominant, as in the case of cigarette smoking and lung cancer, that the detection of other influences is made very complex. At least, however, we know they must be considered. As another example, it is much more difficult to separate the influence of the socioeconomic status of families from the effects resulting from gas cooking in their homes. It seems to me that the major difficulty of measuring the health of low-level exposures resides in the possibility of unknown variables influencing the data; this applies to each of the pollutant categories considered at the symposium.

A second general comment has to do with the difficulty of organizing studies that will take into account the long-term nature of these influences. This means that much of the research has to be particularly painstaking and (no doubt for many years) unrewarding.

Such an aspect of this kind of inquiry is a real barrier to funding.
In the case of some indoor pollutants, the "end point" is ill-
defined. There are often complaints of symptoms like headaches and
upper respiratory ailments that are very difficult to quantify. The
difficulty applies to questions of ill health contingent on exposures to organics and to products of combustion. It does not apply,
of course, to radon daughters where the end point is lung cancer.
In each case the bottom line is some statement based on an estimate
of risk. I would recommend to you the risk analysis of environmental exposure to asbestos recently published by the National Academy
of Science [1]. This report represents one of the best attempts
anywhere to bring together a considerable amount of scientific data
and to extrapolate it to the general population. It gives not only
an estimate of long-continued low-level exposure to asbestos, but
also expresses this in general terms and provides the limits to such
a estimate (which are inevitably wide). One must struggle for risk
estimates to be expressed in this way so that the comparative importance of the subject being treated can be accurately measured.

For many of the pollutants that have been considered, many
advances have been made in measuring typical exposure levels. The
development of personal monitors has given us a much better perception of the profile of indoor exposure to organic chemicals such as
benzene, or to oxides of nitrogen. The new information we are getting from these methods is very important; however, we still face
the great difficulty of integrating such data with the end point,
which may be many years away or require a very long period of observation. I see this as the second major difficulty of all these studies, since it will prove difficult to relate detailed observations
of personal exposure collected over a few months to a later health
outcome. There will, I suspect, have to be considerable "inferen-
tial" reasoning, stopping somewhere short of strict proof. This
situation is often upsetting and, indeed, is sometimes made to look
as if it is somehow the antithesis of good science. Quoting Professor Gilbert Ryle who wrote 30 years ago [2],

> "Unlike playing cards, problems and solutions of
> problems do not have their suits and their denomina-
> tions printed on their faces. Only late in the game
> can the thinker know even what have been trumps."

In evaluating work of this kind, it is very important for the
reader, and indeed for all of us, to keep a fairly strong sense of
perspective. We are not good at predicting where the next important
data will come from; indeed, predictions of this kind by distinguished authorities usually end up only being quoted later,
because it has become apparent that they were ridiculous. However,
I would like to make a few remarks on what we have heard.

I come away from this meeting with a strong sense that we
should be very concerned with the problem of radon-daughter pollution indoors. Difficulties in measurement techniques and the

complex problem of quantifying the dose of radiation delivered to the lung seem to have combined to delay recognition of the seriousness of exposure to the radon daughter. There seems to be a very narrow gap between lifetime exposures in some geological regions and a dose that appears to significantly increase the risk of lung cancer. This seems to be a situation where the effect of cigarette smoking in promoting increased lung cancer has been so dominant that the study of other contributing factors, such as indoor radon exposures, have been made difficult. We should probably not now know, however, that asbestos exposure alone would not have led to an increase of lung cancer had it not been for cigarette smoking in addition to asbestos exposure, which easily led to detectable increases in the incidence of lung cancer. The same may well be true of radon daughters. There are major regional and geological differences in the natural occurrence of radon daughters. In some instances the regions affected are small, and it is notoriously difficult to prove an increased incidence of a disease like lung cancer in a community of a few hundred people. However, the data we are now getting indicate that many cases of lung cancer in nonsmoking individuals may well be attributable to radon daughters. It will be years before the magnitude of their effect becomes established. One is not surprised to have heard here that engineers are beginning to think about ways of building houses to prevent the ingress of radon daughters. This would certainly be an important precaution in those regions of the world where the concentrations are high.

In the case of exposures to organics, there is great difficulty relating cause to effect, although we have learned how sophisticated personal monitoring has become. I was very impressed by the fact that differences in benzene exposures could be individually related to whether or not an individual had filled up the gas tank of his car during the period of observation. Obviously, we are all exposed to very low levels of this material, but I doubt whether we can attribute the natural incidence of leukemia to such general exposures.

The general problems associated with indoor bacteria and fungi seem generally to be rather less well defined and certainly more difficult to measure. The recognition of the importance of contamination of poorly installed and maintained air conditioners was presented dramatically. It is quite obvious that emphasis on general hygiene would certainly diminish these risks considerably.

The questions raised by indoor pollution and respiratory infections in children are not easy to solve, mainly because such infections are so common, which accounts for the contradictory nature of some of the epidemiological evidence. We learned that 65% of children under the age of 2 have had a significant respiratory episode. This makes the problem difficult to study. Nevertheless, the data at present point up the importance of carefully considering proper ventilation above cooking stoves to prevent the buildup of what may be large concentrations of oxides of nitrogen in a kitchen sealed against winter weather. The magnitude of adverse effects is controversial; however, the measured levels of some exposures would certainly suggest that designers of houses should begin to take this possibility into account.

Perhaps the one indoor pollutant about which we should be least concerned, from the point of view of regulation, is passive cigarette smoke. What is needed here is public education. Such a program will, I am sure, be successful and lead to much cleaner atmospheres in places that used to contain a haze of cigarette smoke like taverns and game rooms. Within homes, the slow process of public education should eventually be enough.

I will not suggest that these general observations will have any permanent validity. This symposium has enabled us to review the present status of a very considerable body of work in different fields; no doubt we all have some general conclusions. I have offered some of mine, but I am sure yours may have been different and, very possibly, wiser.

It remains for me to thank the organizers of the symposium for their initiative and thoroughness in its organization. It will be of great interest to look back to our proceedings a few years from now and compare the general problem of indoor air pollution now and then.

REFERENCES

[1]   Ryle, Gilbert. "Dilemmas" [the Tarner lectures (1953)],
       Cambridge University Press, 1960.

[2]   "Non-Occupational Health Risks of Asbestiform Fibers,"
       Committee of the Board on Toxicology and Environmental Health
       Hazards, Commission on Life Sciences, National Research
       Council, National Academy of Sciences (Washington, DC:
       National Academy Press, 1984).

Keyword Index

INDEX

Acremonium 154
adsorbents for sampling
    activated carbon 336,337,342,
        343
    carbon molecular sieve 336,337
    inorganics 336
    XAD resins 337
air exchange rate 273
air monitoring 209
air quality 406,410
airways function
    monitoring techniques 301
aldeyhdes 388,389,396,397
Alternaria 153,161
alveolar macrophage 279,284,
    285–287,291
amplifier 185
analysis techniques 351,356
animal models 279,280
antigens 141,145
aromatic hydrocarbons
    in woodsmoke 267,268
aromatics 371–374
asbestos mortality 21
aspergillosis 5
    allergic bronchopulmonary
        studies 150,166
Aspergillus 150,157
    flavus 158
    fumigatus 152,155,158,166
    niger 153,158
    species 153,158,161
    in epidemiological studies 306,
        309–312
automated field analysis
    gas chromatographs 345,346

mass spectrometers 346
multiple-sensor analysis 346

biological monitoring 207,208,210
Botryosporium 154
Botrytis 154
biomass 265
Brain, Joseph D. 215
breath 362,364,370,374
bronchial dose 70,71,72
bronchial provocation testing
    (bronchial challenge) 380–384
bronchiolitis 319,323,325
building materials 110,111,113
Burge, Harriet A. 139

cancer 361
    other than lung 229,230
carbon dioxide 10,15
carbon monoxide 6,14,15,242,249,
    250,264–266,272
carcinogens 370,371,374,393,394,
    397
    in woodsmoke 267,268
cardiovascular diseases
    and symptoms 241,242,248–250
    anginal pain after exercise 242,
        250,252
    blood pressure 241,242,253
    coronary heart disease 252
    exercise 242,250,252
    heart rate 241,242,250,253
    ischemic heart disease 241,248,
        252
carriers of microorganisms 171,174
causality of associations 228,236

425